Anaesthesia and Pain Management in Large Animals

Guest Editor

Gabrielle C. Musk

MDPI

Basel • Beijing • Wuhan • Barcelona • Belgrade • Novi Sad • Cluj • Manchester

Guest Editor
Gabrielle C. Musk
School of Veterinary and Life
Sciences
Murdoch University
Perth
Australia

Editorial Office
MDPI AG
Grosspeteranlage 5
4052 Basel, Switzerland

This is a reprint of the Special Issue, published open access by the journal *Animals* (ISSN 2076-2615), freely accessible at: https://www.mdpi.com/journal/animals/special_issues/57795575SQ.

For citation purposes, cite each article independently as indicated on the article page online and as indicated below:

Lastname, A.A.; Lastname, B.B. Article Title. *Journal Name* **Year**, *Volume Number*, Page Range.

ISBN 978-3-7258-4291-9 (Hbk)
ISBN 978-3-7258-4292-6 (PDF)
https://doi.org/10.3390/books978-3-7258-4292-6

Cover image courtesy of Gabrielle Musk

Contents

animals

MDPI

Article

Nociception in Chicken Embryos, Part I: Analysis of Cardiovascular Responses to a Mechanical Noxious Stimulus

Larissa Weiss [1], Anna M. Saller [1], Julia Werner [1], Stephanie C. Süß [1], Judith Reiser [1], Sandra Kollmansperger [2], Malte Anders [2], Heidrun Potschka [3], Thomas Fenzl [2], Benjamin Schusser [4] and Christine Baumgartner [1,5,*]

1 Center for Preclinical Research, TUM School of Medicine, Technical University of Munich, 81675 Munich, Germany; larissa.weiss@tum.de (L.W.); anna.saller@tum.de (A.M.S.); julia.werner@tum.de (J.W.); stephanie.suess@tum.de (S.C.S.); judith.reiser@tum.de (J.R.)

2 Clinic for Anesthesiology and Intensive Care, TUM School of Medicine, Technical University of Munich, 81675 Munich, Germany; s.kollmansperger@outlook.de (S.K.); malteanders@gmail.com (M.A.); thomas.fenzl@tum.de (T.F.)

3 Institute of Pharmacology, Toxicology, and Pharmacy, Ludwig-Maximilians-Universität München, 80539 Munich, Germany; potschka@pharmtox.vetmed.uni-muenchen.de

4 Reproductive Biotechnology, TUM School of Life Sciences, Technical University of Munich, 85354 Freising, Germany; benjamin.schusser@tum.de

5 Veterinary Faculty, Ludwig-Maximilians-Universität München, 80539 Munich, Germany

* Correspondence: christine.baumgartner@tum.de

Simple Summary: Chicken embryos are frequently not protected by animal welfare laws. However, they are used in various research areas, and male embryos are commonly killed in food production as an alternative to culling day-old chicks. Increasing knowledge regarding the onset of nociception and pain perception in chicken embryos is fundamental for animal welfare protection. The aim of this exploratory study was to further narrow down the period when chicken embryos acquire the capacity for nociception. Therefore, changes in blood pressure and heart rate after the introduction of a noxious stimulus were assessed during the embryonic development of chickens. Embryos from 16 days of incubation onward showed cardiovascular changes after a noxious mechanical stimulus was introduced at the base of the beak, indicating a nociceptive response.

Abstract: Although it is assumed that chicken embryos acquire the capacity for nociception while developing in the egg, an exact time point has not yet been specified. The present research was an exploratory study aiming to determine when the capacity of nociception emerges during embryonic development in chickens. Changes in blood pressure and heart rate (HR) in response to a noxious mechanical stimulus at the base of the beak versus a light touch on the beak were examined in chicken embryos between embryonic days (EDs) 7 and 18. Mean arterial pressure (MAP) was the most sensitive parameter for assessing cardiovascular responses. Significant changes in MAP in response to a noxious stimulus were detected in embryos at ED16 to ED18, whereas significant changes in HR were observed at ED17 and ED18. Infiltration anesthesia with the local anesthetic lidocaine significantly reduced the response of MAP on ED18, so the measured cardiovascular changes may be interpreted as nociceptive responses.

Keywords: blood pressure; heart rate; nociception; pain; chicken embryo; development; *Gallus gallus domesticus*; poultry

1. Introduction

In present times, animal welfare has increasingly become the focus of public attention regarding farm and laboratory animals. Consequently, the culling of male day-old chickens for economic reasons is increasingly questioned. A large proportion of the male offspring in the layer industry are killed after hatching, as the fattening of male layer-type chickens is not economically profitable [1]. In the EU, 330 million male chicks are killed annually through maceration or gassing [2], which is currently the subject of a major discussion. Germany and France have already adapted their laws and banned the killing of male day-old chicks for economic reasons, although there is not yet an EU-wide regulation [3]. As an alternative, *in ovo* sex determination with subsequent killing of male embryos is already being practiced [4]. However, it is important for animal welfare reasons and for the public acceptance of *in ovo* sex determination that related culling be conducted at an early stage of development when nociception and the perception of pain are not yet possible [4,5]. According to current knowledge, methods of *in ovo* sex determination are reliably applicable from the 9th day of incubation at the earliest [1]. Methods that can be applied in the first trimester of embryonic development are still in development under laboratory conditions [4].

Furthermore, chicken embryos are of great importance for biomedical research because of the advantages they provide in terms of fast growth and because of their good accessibility in various research areas, such as developmental biology, toxicology, cancer research and drug development [6,7]. Under European regulations, interventions and treatments on chicken embryos are not considered animal experiments and even count as a replacement method in the context of the 3R principles [8]. At this time, there are no regulations regarding anesthesia and analgesia of chicken embryos during painful interventions [6,8]. Greater clarity regarding the period during which chicken embryos are capable of nociception and pain sensation would lead to improved animal welfare in research.

In pain research, a fundamental distinction is made between nociception and the perception of pain [9]. Although nociception is the detection of a potentially tissue-damaging stimulus and its transmission by the nociceptive component of the nervous system [10,11], pain is characterized by a subjective, conscious sensory perception, usually triggered by nociception [12,13]. Nociception and pain are progressive adaptive processes that gradually develop throughout the fetal period [14]. It is considered confirmed that the chicken embryo acquires the capacity for nociception at some point during the 21-day developmental period in the egg [8,15]. However, the question of the exact time point at which nociception or even pain sensation can be presumed is controversial. In several publications, researchers agree that nociception and pain perception are not possible in the first trimester of embryonic development in the chicken [4,15]. A requirement for the ability to perceive pain is the existence of functional pathways that enable the transmission of stimuli to the brain [12,14]. Although the first sensory afferent nerve fibers develop on incubation day 4, the closure of multisynaptic reflex arcs does not occur until day 7 [16–18]. It is described in the literature that the chicken embryo develops a functional brain on day 13 [15,19]. However, it is only confirmed that the brain does not show any electrical activity until 6.5 days of incubation [20]. Pain sensation is therefore considered impossible up to incubation day 7, but beyond that, no specific time point can be defined from which the chicken embryo is capable of nociception and pain sensation [4,15].

Because self-reporting, which is the gold standard in humans to detect pain [21], is not possible as a direct method of pain evaluation in animals, indirect methods such as the alteration of physiological and behavioral parameters must be resorted to [22]. Changes in heart rate (HR) and blood pressure are therefore used as clinical indicators of nociception and pain [23].

This study is part of a comprehensive study in which the nociceptive ability of chicken embryos was investigated using cardiovascular parameters, behavioral observations and EEG. Here, we present the results of the cardiovascular study and, in particular, the implemented cardiovascular measurement methods regarding chicken embryos that were designed for investigation of the time point at which chicken embryos are able to respond to a noxious stimulus with a nociceptive cardiovascular response. The corresponding results of the EEG measurements and behavioral observations and the implemented techniques will be presented in further publications.

2. Materials and Methods

2.1. Animals

Fertilized Lohman Selected Leghorn chicken eggs were obtained from the TUM Animal Research Center (Thalhausen) and stored at 15 °C. Embryonic day (ED) 0 was considered as the day when eggs were transferred to the incubator (Favorit-Olymp 192 Spezial, HEKA-Brutgeräte, Rietberg, Germany). The eggs were incubated for 7 to 18 days at 37.8 °C and 55% humidity and turned six times a day until they were fenestrated.

At ED3 of incubation, the eggshell was fenestrated [24]. For this purpose, the egg was placed horizontally for at least two minutes, and then 5 to 7 mL albumen were withdrawn from the apical pole of the egg using a 5 mL syringe and an 18 G needle. The top of the egg was then covered with tape. A hole was cut in the shell, and the vitality of the embryo was verified. Next, 0.5 mL penicillin-streptomycin (10,000 units penicillin, 10 mg streptomycin/mL, P4333-100 mL Sigma-Aldrich, St. Louis, MI, USA) was added; the egg was then resealed with cling film and was further incubated in a horizontal position. The vitality of the embryos was checked daily until the end of the experiment. Experiments were conducted between 9:00 a.m. and 7:00 p.m. so that the variance in the age of the embryos within an ED was limited to a maximum of 10 h.

2.2. Experimental Design

This study was exploratory and was not preceded by an a priori power analysis. At ED12 to ED18, n = 10 embryos of each ED were measured. Due to higher losses in younger embryos, group sizes of n = 6 (ED9) and n = 3 (ED7) embryos were chosen. Furthermore, to study the effect of local anesthesia, n = 6 ED18 embryos were used.

Experiments were performed under standardized conditions in a specially designed heating chamber equipped with a heating lamp (ARTAS GmbH, Arnstadt, Germany) and an air humidifier (HU4811/10 Series 2000, Philips, Amsterdam, The Netherlands). The eggs were placed on a heating mat (ThermoLux, Witte + Sutor GmbH, Murrhardt, Germany) in a bowl filled with warmed Armor Beads (Lab Armor Beads™, Sheldon Manufacturing, Cornelius, NC, USA). The mean temperature and mean humidity during all experiments were 37.7 °C ± 0.8 and 55.5% ± 4.3, respectively.

A schematic representation of the experimental setup is shown in Figure 1. First, the cling film was removed from the egg, and the shell was carefully opened to the level of the chorioallantoic membrane (CAM). Using a microscope (Stemi SV6, Zeiss, Oberkochen, Germany), the allantoic and amniotic membranes were opened over the head of the embryo, avoiding any large vessels so that the beak could be reached in the further course of the experiment. A side branch of the chorioallantoic artery was prepared, temporarily ligated to avoid blood loss, and incised with microsurgical scissors. A microtip catheter (FISO-LS Fiber Optic Pressure Sensor, FOP-LS-PT9-10, FISO Technologies Inc., Quebec, QC, Canada) was then inserted into the vessel and fixed in place with a ligature. Systolic (SAP), diastolic (DAP) and mean arterial pressure (MAP) as well as HR were recorded continuously every four seconds (PLUGSYS module, EIM-B, EIM-A, HAEMODYN Software v 2.0, Hugo Sachs Elektronik-Harvard Apparatus GmbH, March-Hugstetten, Germany, Evolution Software v 2.1.6.0, FISO Technologies Inc., Quebec, QC, Canada). The beak of the embryo was carefully placed on a Desmarres lid retractor. For younger embryos at ED7 and ED9, the beak was carefully placed on a custom-made wire loop.

After implementation of the catheter, a two-minute waiting period followed. Then, two mechanical stimuli were applied at the base of the beak. In randomized order, a noxious mechanical stimulus was applied with a surgical clamp (*Pinch*), and a light touch (*Touch*) was applied as a negative control. The two stimuli were delivered five minutes apart to allow the parameters to return to the baseline between the stimuli. After the second stimulus, measurements were continued for five more minutes. The measurement time between the two stimuli and after the second stimulus was reduced from five to three minutes for embryos at ED13 and younger due to the increasing sensitivity of the organism.

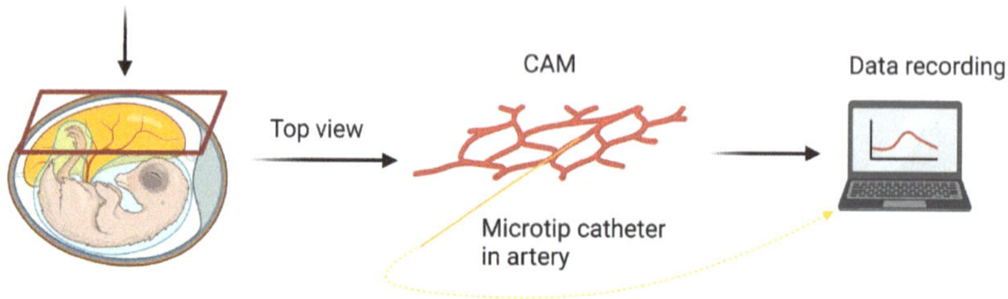

Figure 1. Schematic illustration of the experimental setup. A microtip catheter was inserted into a side branch of the chorioallantoic artery, and the values of blood pressure and heart rate (HR) were recorded every four seconds (created with BioRender.com).

For the *Pinch*, a surgical clamp was applied to the base of the beak and squeezed. For *Touch*, the beak was only lightly touched with the surgical clamp. For both stimuli, a mosquito clamp was used for ED12 to ED18 embryos. For embryos at ED7 and ED9, the surgical clamp was too large, and microsurgical forceps were used instead for both stimuli. To ensure comparability, the stimuli were always applied by the same person. In the further course of the study, an analgesia meter (BIO-RP-M, BioSeb, Vitrolles, France) with customized tips of the mosquito clamp was used to monitor the pressure applied by the mechanical stimuli.

To verify whether the measured cardiovascular responses could be classified as nociceptive responses, a local anesthetic was applied to n = 6 ED18 embryos before stimulation. For this purpose, after the preparation and placement of the microtip catheter, 0.02 mL of lidocaine 2% (Xylocitin® 2%, Mibe GmbH Arzneimittel, Brehna, Germany) were injected into the upper and lower beak using a 30 G needle (*ED18 w/Lido Touch* and *Pinch*). The measurements were carried out following the same experimental protocol as for other ED14 to ED18 embryos with the exception that a waiting period of three minutes was added prior to the measurement. During this time, blood pressure and HR were monitored for the occurrence of side effects of lidocaine, such as bradycardia, arrhythmia or hypotension. As a comparison group without lidocaine, the already measured ED18 embryos were used (*ED18 w/o Lido Touch* and *Pinch*).

Immediately after the end of the experiments, the embryos were euthanized by intravenous injection of pentobarbital sodium (Narcoren®, 16 g/100 mL, Boehringer Ingelheim Vetmedica GmbH, Ingelheim am Rhein, Germany; ED7–ED12: 0.1 mL; ED13–ED19: 0.2 mL) followed by decapitation.

2.3. Analysis

SAP, DAP, MAP and HR were recorded every four seconds. For the evaluation of the reactions to the stimuli, the means of MAP and HR were calculated over one minute before (=baseline) and one minute after the respective stimulus. To avoid any influence of the approach of the clamp, the 15 s immediately before the respective stimuli were introduced were not included as part of the baseline. In embryos showing a hyperacute decrease in HR with a subsequent increase in HR after *Pinch*, the decrease was not included in the calculation and was evaluated separately to avoid negation of opposite reactions. The deviation of the response after the stimulus (*Pinch*/*Touch*) as a percentage of the baseline value was then calculated. Differences in the percent changes to the baseline in MAP and HR after *Pinch* and *Touch* were tested for statistical significance. For normally distributed data, a paired *t*-test (two-tailed) was used. For data that failed the normality test, a Wilcoxon signed-rank test (two-tailed) was performed. For the comparison of multiple groups, either a one-way ANOVA (normally distributed) or a Kruskal—Wallis test (not normally distributed) was used. Additional information on statistical metrics can be found in Table S1.

3. Results

3.1. Increasing Arterial Pressure and Evolution of HR during Embryonic Development of the Chicken

SAP, DAP and MAP in the chorioallantoic artery and HR were recorded over one minute at ED7, ED9 and EDs 12 to 18. SAP, DAP and MAP increased with the age of the embryos (Table 1). ED7 showed the lowest MAP with a value of 2.08 mmHg ± 0.40, and ED18 showed the highest MAP with a value of 17.28 mmHg ± 3.04.

Table 1. Systolic arterial pressure (SAP), diastolic arterial pressure (DAP), mean arterial pressure (MAP) and HR at embryonic day (ED) 7 (n = 3), ED9 (n = 6), and ED12 to ED18 (n = 10). Values are shown as the mean ± standard deviation.

	ED7	ED9	ED12	ED13	ED14	ED15	ED16	ED17	ED18
SAP (mmHg)	3.50 ± 0.65	6.04 ± 1.46	9.19 ± 1.32	9.88 ± 1.52	13.02 ± 1.60	16.54 ± 3.04	21.44 ± 2.78	24.46 ± 5.50	24.65 ± 4.36
DAP (mmHg)	1.07 ± 0.36	1.98 ± 1.10	2.20 ± 1.12	2.96 ± 0.61	3.95 ± 1.14	5.69 ± 1.82	7.80 ± 2.16	10.77 ± 3.53	11.43 ± 2.43
MAP (mmHg)	2.08 ± 0.40	3.44 ± 1.24	4.83 ± 1.05	5.52 ± 0.79	7.32 ± 1.26	10.11 ± 2.45	13.73 ± 2.38	16.79 ± 4.21	17.28 ± 3.04
HR (bpm)	128.97 ± 15.40	147.57 ± 9.03	159.08 ± 26.64	146.61 ± 19.99	179.10 ± 35.06	154.33 ± 33.12	151.35 ± 36.44	179.08 ± 29.18	176.07 ± 35.75

3.2. Increase in MAP in Response to a Noxious Stimulus

The response of MAP to a noxious mechanical stimulus at the base of the beak (*Pinch*) was compared to the response to a light touch at the base of the beak as a negative control (*Touch*) in embryos between EDs 7 and 18. As shown in Figure 2, a significant increase in MAP was observed as a reaction to *Pinch* in embryos on ED16 (p = 0.0008, r = 0.857), ED17 (p = 0.0020, r = 0.627) and ED18 (p = 0.0048, r = 0.778). ED18 embryos showed the strongest response in MAP, with an increase of 15.52% ± 12.36 from the baseline. In comparison, a deviation from the baseline of only 1.30% ± 0.94 was detected in response to *Touch* on ED18. In embryos at ED7, ED9 and EDs from 12 to 15, no significant differences between the MAP responses to *Pinch* and *Touch* were detected, which can be seen in Figures S1 and S2.

3.3. Changes in HR in Response to a Noxious Stimulus

Regarding HR, two reaction patterns were observed, particularly in ED17 and ED18 embryos. In some embryos, HR immediately increased after *Pinch*. In other embryos, a hyperacute decrease in HR followed by an increase was observed in response to *Pinch*, as shown in Figure 3d–f. A change in HR of at least −15% with a subsequent increase of at least 5% from the baseline mean value after *Pinch* was observed in 80% of ED18 embryos and in 30% of ED17 embryos and was not detected after *Touch*. In embryos at ED18, HR decreased by up to −48.54% ± 19.71 over 9.50 s ± 6.02 on average after *Pinch*. At ED17, these embryos showed a decrease in HR by up to −41.87% ± 8.32 over 16.00 s ± 6.93 on average after *Pinch*. Simultaneous with the hyperacute decrease in HR, a slight decrease in MAP was also observed, particularly when the decrease in HR was large. In embryos at ED15 and ED16, the observations were inconsistent and could not be clearly distinguished from physiological variations in HR. In younger embryos, no hyperacute decrease in HR was observed in response to *Pinch*.

Significant increases in HR in response to *Pinch* compared to *Touch* were detected in embryos at ED17 (p = 0.0148, r = 0.708) and ED18 (p = 0.0154, r = 0.705) (Figure 3a–c). Embryos at ED18 showed the largest increase in HR after *Pinch*, with a deviation of 5.14% ± 3.60 from the baseline, compared to a deviation of only 2.07% ± 1.20 from the baseline after *Touch*. At ED7, ED9 and EDs 12 to 16, no significant changes in HR were observed, as shown in Figures S3 and S4.

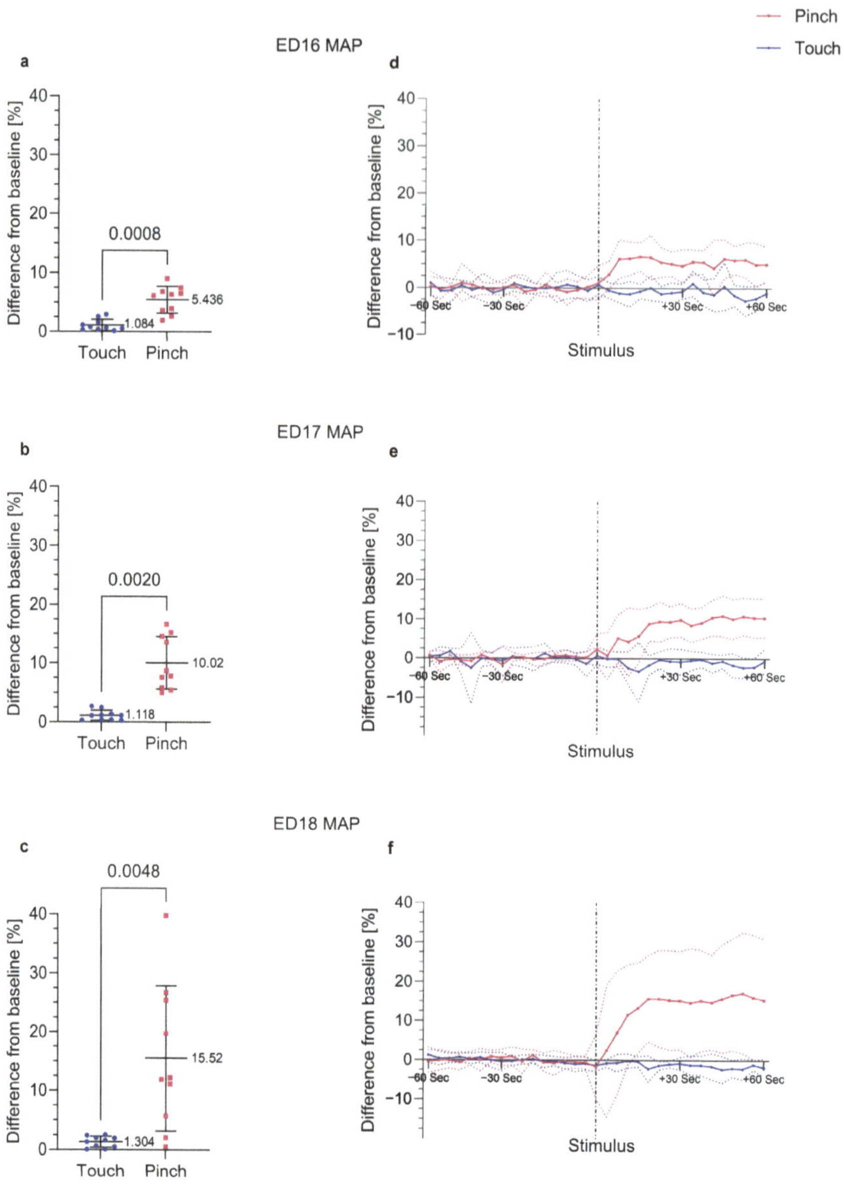

Figure 2. Percent change in MAP post *Touch* and *Pinch*. Embryos at EDs 16 to 18 (n = 10) received a noxious mechanical stimulus (*Pinch*) and a light touch as a negative control (*Touch*) at the base of the beak in randomized order. (**a–c**) Percent change from baseline MAP after *Pinch* compared to *Touch*. Displayed as the mean ± standard deviation. Paired *t*-test (normally distributed: (**a**,**c**)) or Wilcoxon signed-rank test (not normally distributed: (**b**)). Mean and *p* values shown; (**a**): $p = 0.0008$, (**b**): $p = 0.0020$, (**c**): $p = 0.0048$. (**d–f**) Percent change from the baseline mean value of MAP over time; values recorded every four seconds for one minute before and one minute after stimulation (*Pinch* and *Touch*); values shown as the mean ± standard deviation (shaded).

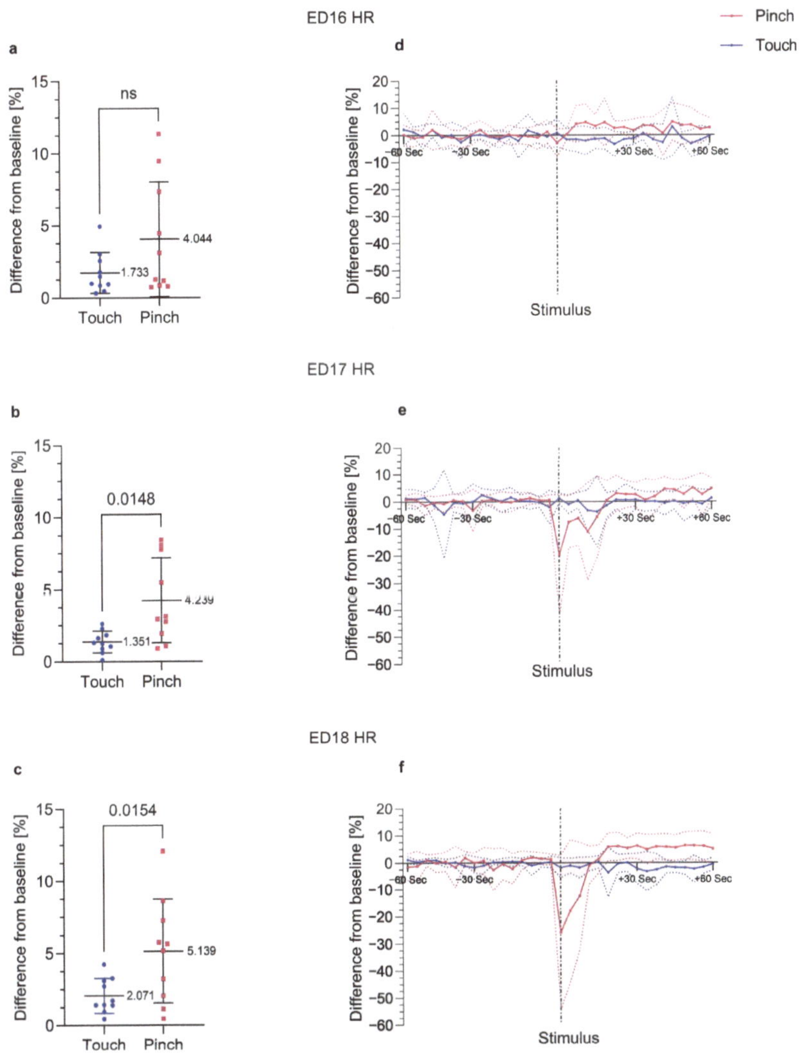

Figure 3. Percent change in HR post *Touch* and *Pinch*. Embryos at EDs 16 to 18 (n = 10) received a noxious mechanical stimulus (*Pinch*) and a light touch as a negative control (*Touch*) at the base of the beak in randomized order. (**a**–**c**) Percent change from baseline HR after *Pinch* compared to *Touch*. Displayed as the mean ± standard deviation. Paired *t*-test (normally distributed: (**b**,**c**)) or Wilcoxon signed-rank test (not normally distributed: (**a**)). Mean and *p* values shown; (**b**): $p = 0.0148$, (**c**): $p = 0.0154$; ns = no significant difference between the groups (**a**). (**d**–**f**) Percent change from the baseline mean value in HR over time; values recorded every four seconds for one minute before and one minute after stimulation (*Pinch* and *Touch*); values shown as the mean ± standard deviation (shaded).

3.4. Reduction of Cardiovascular Response by Local Anesthesia

The application of the local anesthetic lidocaine (*Lido*) at the base of the beak prior to stimulation significantly reduced the MAP increase in response to *Pinch* in embryos at ED18. Compared to the group without local anesthesia (*ED18 w/o Lido*), which showed an increase of 15.52% ± 12.36 post *Pinch*, the increase in MAP was reduced to 5.00% ± 3.42 in the group that received lidocaine (*ED18 w/Lido*). As represented in Figure 4a,c, the *ED18 w/o Lido Pinch* group showed the largest increase in MAP in response to *Pinch*, exceeding those of the *ED18 w/o Lido Touch* (p = 0.0007), *ED18 w/Lido Touch* (p = 0.0031) and *ED18 w/Lido Pinch* (p = 0.0397) groups, with an effect size of η^2 = 0.467.

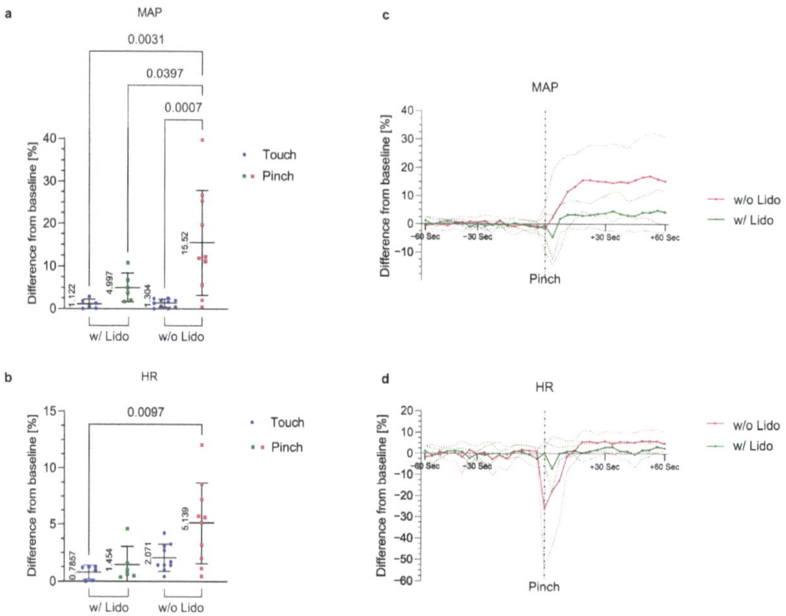

Figure 4. Local anesthesia control group. Percent change in MAP and HR post *Touch* and *Pinch*. ED18 embryos either received a lidocaine injection (*ED18 w/Lido*; n = 6) or no lidocaine injection (*ED18 w/o Lido*; n = 10) at the base of the beak prior to stimulation (*Touch* and *Pinch*). (**a**) Percent change from baseline MAP after *Pinch* in the group without lidocaine (*ED18 w/o Lido Pinch*) compared to *ED18 w/o Lido Touch, ED18 w/Lido Touch* and *ED18 w/Lido Pinch*. Displayed as the mean ± standard deviation. One-way ANOVA (normally distributed); mean and p values shown. (**b**) Percent change from baseline HR after *Pinch* in the group without lidocaine (*ED18 w/o Lido Pinch*) compared to *ED18 w/o Lido Touch, ED18 w/Lido Touch* and *ED18 w/Lido Pinch*. Displayed as the mean ± standard deviation. Kruskal—Wallis test (not normally distributed); mean and p values shown. (**c**,**d**) Percent change from the baseline mean value in MAP and HR after *Pinch* over time; values recorded every four seconds for one minute before and one minute after stimulation (*ED18 w/o* or *w/Lido Pinch*); values shown as the mean ± standard deviation (shaded).

The changes in HR in response to *Pinch* were slightly reduced by the application of lidocaine. However, a significant difference in HR was observed only between *ED18 w/o Lido Pinch* and *ED18 w/Lido Touch* (p = 0.0097), as displayed in Figure 4b. In the group treated with lidocaine (*ED18 w/Lido*), no embryo showed a hyperacute change in HR of −15% with a subsequent increase of 5% from the baseline mean value after the stimulus, but this reaction pattern was observed in 80% of the embryos in the *ED18 w/o Lido Pinch* group. A slight decrease in HR after *Pinch* was also observed in the local anesthetic group (*ED18 w/Lido Pinch*), but this decrease could not be distinguished from physiological variations in HR (Figure 4d).

4. Discussion

This study successfully developed methods to record blood pressure and HR in chicken embryos between EDs 7 and 18. Cardiovascular changes in response to a noxious mechanical stimulus at the base of the beak were investigated with the aim of identifying the onset of nociception during embryonic development in chickens.

Although there are many well-established noninvasive methods for determining HR in chicken embryos [25–27], direct intra-arterial measurement is the gold standard for recording blood pressure [28]. In the past, blood pressure in chicken embryos was measured using glass capillaries or needle catheters inserted into an embryonic artery [29–31]. Corresponding to prior descriptions in the literature [29–31], an increase in arterial blood pressure with increasing age of the embryos was observed in the present study, but there were no major differences in HR between the EDs. Thus, the optical measurement of arterial blood pressure and HR with a microtip catheter represents a reliable method for invasive measurement of blood pressure and HR in chicken embryos. However, insertion of the catheter was particularly challenging at ED7 and ED18 due to the small size of the chorioallantoic vessels at ED7 and the beginning regression of the chorioallantoic vessels at ED18.

Because self-reporting is not possible in animals, it is difficult to evaluate their pain perception [22]. On the other hand, nociceptive reactions to a noxious stimulus can be measured [13]. The recording of cardiovascular parameters is well suited to the clinical evaluation of nociception in animals, including birds [32,33]. In the present study, the acquisition of cardiovascular parameters could be established for chicken embryos between EDs 7 and 18. Blood pressure and HR are mainly influenced by the autonomic nervous system [34]. Transmission of a noxious stimulus to the central nervous system results in activation of the sympathetic nervous system, which usually leads to an increase in blood pressure and HR [34]. Therefore, recording cardiovascular variables is considered the gold standard for the detection of nociception under anesthesia [35].

As a means of assessing the cardiovascular response of the chicken embryo to a noxious mechanical stimulus at the base of the beak, MAP was found to be the most sensitive parameter in the present study. Significant differences in MAP between *Pinch* and *Touch* were detected earliest on ED16 (Figure 2), whereas significant changes in HR were only observed in ED17 and ED18 embryos (Figure 3). Effect sizes were high, indicating the clinical relevance of the findings. Although there was a distinct increase in MAP in response to *Pinch* that reached over 10% deviation from the baseline in ED17 and ED18 embryos, the changes in HR were variable, and there were not necessarily any associations between changes in MAP and HR. Similar observations have been reported in adult chickens [36]. MAP has also been described in other studies concerning nociceptive responses in mammals as the most sensitive indicator of nociception [34,37].

A prerequisite for cardiovascular response to external stimuli is functional regulation of the cardiovascular system. Blood pressure in the chicken embryo is mainly regulated by the sympathetic nervous system [38]. The adrenergic tone in the cardiovascular system is considered to be present from a point in time that is halfway through the incubation period [39,40]. Therefore, the sympathetic influence on blood pressure is expected to be functional from approximately ED10 [39]. In the heart, adrenergic and cholinergic receptors are already functional on ED4 [41]. Changes in HR due to alterations in environmental conditions such as oxygen levels and temperature have already been observed on ED3 [42]. In the present study, significant changes in HR after a noxious stimulus was introduced were not observed until ED17 (Figure 3).

Another prerequisite for the assessment of a nociceptive response is functioning stimulus transmission. Despite some differences in the nervous system, the processing of noxious stimuli in birds is comparable to that in mammals [13]. C-fibers and A-delta fibers have been found in chickens, innervating the beak, nasal and buccal mucosa as well as the legs [11,43]. High-threshold mechanothermal nociceptors are polymodal and respond to mechanical lesions, elevated temperatures and chemical insult [13]. It is believed that injuries to the beak can be highly painful for the bird [43], because the beak tip is an intensely innervated area [44], and both the upper and the lower portions of the beak contain nociceptors [45]. Reflective reactions such as movements of the head to mechanical and thermal stimuli and to needle punctures appear for the first time in the skin area of the beak on ED7 [46]. Therefore, in the present study, the application of a noxious stimulus to the base of the beak was chosen to evoke the highest possibility for a nociceptive response.

Regarding HR, irregularities appeared spontaneously over the whole measurement period, even at the baseline. Mainly short decelerations in HR were observed, whereas MAP was not affected. It has already been reported in several publications that HR irregularities physiologically occur at the end of the second week of incubation [47–50]. Nevertheless, the HR irregularities did not have a great influence on the calculation of the mean. Minor changes in DAP corresponded to the HR irregularities, but the analysis showed that MAP was not affected. In contrast to physiological variations in HR, a hyperacute decrease in HR with a subsequent increase could be clearly identified as a response to *Pinch* in 30% of ED17 and 80% of ED18 embryos. This reaction pattern could be distinguished from physiological variations in HR by the finding that after *Pinch*, HR decreased by at least −15%, followed by a sustained increase in HR by at least 5% from the baseline mean value. The decrease in HR after *Pinch* was also accompanied by a short decrease in MAP followed by an increase. A decrease in HR as a reaction to a noxious stimulus has been reported in adult chickens [36] and in mammals [51,52] and may be due to a vasovagal reflex to a noxious stimulus [53]. However, only a few individual embryos showed a hyperacute decrease in HR after *Pinch*, which shows that the response in HR to a noxious stimulus varies among individuals. Variable responses in HR after a noxious stimulus have also been described in adult chickens [36]. Considering these different observations regarding HR, it is difficult to draw conclusions about the presence of nociception. Thus, HR should not be used as a single parameter for evaluating a nociceptive response in chicken embryos; however, MAP was shown to be a more sensitive parameter in the present study.

In addition to a nociceptive response, it must also be considered that the measured cardiovascular changes may be induced by other factors that influence the autonomic nervous system [54] or by embryonic movements. Especially in birds, physiological variables can be influenced by many external factors, such as temperature, light, or handling [54]. A correlation between fetal movements and HR irregularities has been described in human fetuses [55]. In the present study, movements of the embryo induced minor variations in HR and DAP, but MAP was not affected. No sustained increase in MAP and HR as observed in response to *Pinch* could be attributed to movements.

Infiltration anesthesia at the base of the beak could be used to verify that the measured changes in MAP and HR may be classified as a nociceptive response and were not caused by embryonic movements or factors that influence the autonomic nervous system. The application of local anesthetics is one of the best methods to prevent the generation and transmission of nociceptive impulses [56]. These anesthetics act by blocking sodium channels in the nerve axon [54]. The application of lidocaine or bupivacaine has been described as an effective method of analgesia in birds [57]. However, the time of onset of action and the duration of action are not defined for birds [54]. In the present study, lidocaine was used because it has a rapid onset of action in mammals [56], as well as a short onset of action for spinal anesthesia in chickens [58]. Given that higher sensitivity to local anesthetics is expected in birds than in mammals [59], embryos were intensively monitored for the occurrence of toxic effects. No signs of side effects such as bradycardia, arrhythmia or hypotension were observed in the tested embryos. Because the increase in MAP was significantly reduced by the injection of lidocaine (Figure 4), the cardiovascular reactions to *Pinch* in the embryos that did not receive local anesthesia might be interpreted as a nociceptive response to the noxious stimulus. A limitation and possible explanation for the incompletely suppressed reaction in some embryos was that injection into the beak of the moving embryo was challenging, and infiltration of the entire beak area could not always be assured. It must be mentioned that the present study was exploratory and the size of the group receiving local anesthesia was rather small. Further investigations would need to be performed to verify the effect of local anesthesia and to ultimately exclude other factors as the cause of the measured cardiovascular changes. However, assuming that it is a nociceptive response, further studies regarding anesthesia and analgesia protocols are necessary to provide improved animal welfare for chicken embryos in research. Cardiovascular variations are commonly used to determine the need for analgesia or sedatives [23], and thus far, there are no EU-wide regulations regarding anesthesia and analgesia for chicken embryos in research.

Although no significant difference between *Pinch* and *Touch* was reached at ED15 in MAP (Figure S1f) and HR (Figure S3f), individual responses could be observed. Occasionally, embryos at ED15 showed reactions in MAP (Figure S2f) and HR (Figure S4f) after *Pinch*. The measurements of these embryos were performed late in the day. The development of the embryos could therefore have been more advanced compared to embryos examined in the morning. In addition, embryonic development can be influenced by various factors, and some embryos might progress faster in development than others [39]. Therefore, it must be assumed that a nociceptive cardiovascular response is possible in individual embryos at ED15.

A limitation of the study was that intra-arterial measurement of blood pressure and HR is an invasive method. The measurements had to be performed on the fenestrated egg, making it necessary to open the egg membranes. Because chicken embryos are highly sensitive to external factors [29,42], special care was taken to maintain standardized environmental conditions and to avoid blood loss during preparation. In some embryos, severe bradycardia and hypotension were observed, or HR frequently decreased to zero. These embryos had to be excluded from the analysis because reliable measurements could not be completed. At ED7, reaching the beak was challenging, and a measurement could only be performed in three embryos; severe arrhythmias affecting MAP were observed. The microtip catheter is designed to measure low pressures, but the measurement accuracy of 2 mmHg, according to the manufacturer, reached its limits with the occurrence of extremely low blood pressure in ED7. The results from ED7 should therefore be interpreted with caution.

5. Conclusions

In conclusion, significant differences and large effect sizes in a cardiovascular response to a mechanical noxious stimulus at the base of the beak compared with a light touch at the base of the beak were detected in chicken embryos on EDs 16 to 18. For individual embryos, cardiovascular changes after the introduction of a noxious mechanical stimulus have already been observed on ED15. MAP was found to be the most sensitive parameter in the present study, whereas variable observations were made regarding HR. Infiltration anesthesia with the local anesthetic lidocaine (2%) significantly reduced the reactions of MAP to a noxious mechanical stimulus at the base of the beak in ED18 embryos, indicating that the measured cardiovascular changes may be interpreted as nociceptive responses. However, it must be mentioned that this was an exploratory study with a correspondingly small group size. To assess response to a noxious stimulus, a multiparametric approach should be adopted and several parameters should be assessed in their entirety [60]. Thus, to properly evaluate a nociceptive response in the chicken embryo, other parameters, such as movement analysis, should be taken into account in addition to hemodynamic parameters.

Supplementary Materials: The following supporting information can be downloaded at: https://www.mdpi.com/article/10.3390/ani13172710/s1, Figure S1: Percent change in MAP post *Touch* and *Pinch*; Figure S2: Percent change from the baseline mean value in MAP over time; Figure S3: Percent change in HR post *Touch* and *Pinch*; Figure S4: Percent change from the baseline mean value in HR over time; Table S1: Statistical metrics.

Author Contributions: Conceptualization, C.B., A.M.S., J.W., J.R., T.F. and B.S.; methodology, C.B., A.M.S., J.W., J.R., S.K., M.A., T.F., B.S., L.W., S.C.S. and H.P.; validation, A.M.S., J.W. and C.B.; formal analysis, L.W., A.M.S. and J.W.; investigation, L.W., S.C.S., A.M.S., J.W. and J.R.; resources, C.B. and B.S.; data curation, L.W. and A.M.S.; writing—original draft preparation, L.W.; writing—review and editing, L.W., A.M.S., J.W., S.C.S., J.R., S.K., M.A., T.F., B.S. and C.B.; visualization, A.M.S., J.W. and

L.W.; supervision, C.B., T.F., B.S. and H.P.; project administration, C.B.; funding acquisition, C.B., T.F. and B.S. All authors have read and agreed to the published version of the manuscript.

Funding: This research was funded by the German Federal Ministry of Food and Agriculture (BMEL, https://www.bmel.de (accessed on 24 August 2023)) based on a decision of the Parliament of the Federal Republic of Germany, granted by the Federal Office for Agriculture and Food (BLE, https://www.ble.de (accessed on 24 August 2023), grant number: 2821HS005).

Institutional Review Board Statement: According to Directive 2010/63/EU of the European Parliament and the German Animal Welfare Law no ethical approval was required for the use of chicken embryos in the experiments. The experiments were performed in an AAALAC-certified animal facility. All experiments and the euthanasia of embryos were conducted in strict accordance with the institutional Guidelines for Care and Use of Laboratory Animals and under general animal welfare principles.

Informed Consent Statement: Not applicable.

Data Availability Statement: Raw data are available from the corresponding author upon reasonable request.

Acknowledgments: The authors thank the scientific advisory board with Michael Erhard, Wolf Erhardt, Harald Luksch, Heidrun Potschka, Hans Straka and Britta Wirrer for their excellent scientific contribution as well as Johannes Fischer, Stefanie Fitzner and Hicham Sid for their technical support.

Conflicts of Interest: The authors declare no conflict of interest. The funders had no role in the design of the study; in the collection, analyses, or interpretation of data; in the writing of the manuscript; or in the decision to publish the results.

References

1. Weissmann, A.; Reitemeier, S.; Hahn, A.; Gottschalk, J.; Einspanier, A. Sexing domestic chicken before hatch: A new method for in ovo gender identification. *Theriogenology* **2013**, *80*, 199–205. [CrossRef]
2. Vinci, C. *Male Chick Culling*; European Parliamentary Research Service: Brussels, Belgium, 2022.
3. Gautron, J.; Réhault-Godbert, S.; Van de Braak, T.G.H.; Dunn, I.C. Review: What are the challenges facing the table egg industry in the next decades and what can be done to address them? *Animal* **2021**, *15* (Suppl. S1), 100282. [CrossRef]
4. Krautwald-Junghanns, M.E.; Cramer, K.; Fischer, B.; Förster, A.; Galli, R.; Kremer, F.; Mapesa, E.U.; Meissner, S.; Preisinger, R.; Preusse, G.; et al. Current approaches to avoid the culling of day-old male chicks in the layer industry, with special reference to spectroscopic methods. *Poult. Sci.* **2018**, *97*, 749–757. [CrossRef]
5. Reithmayer, C.; Danne, M.; Mußhoff, O. Look at that!-The effect pictures have on consumer preferences for in ovo gender determination as an alternative to culling male chicks. *Poult. Sci.* **2021**, *100*, 643–653. [CrossRef]
6. Seabra, R.; Bhogal, N. In vivo research using early life stage models. *In Vivo* **2010**, *24*, 457–462.
7. Rashidi, H.; Sottile, V. The chick embryo: Hatching a model for contemporary biomedical research. *Bioessays* **2009**, *31*, 459–465. [CrossRef]
8. Aleksandrowicz, E.; Herr, I. Ethical euthanasia and short-term anesthesia of the chick embryo. *Altex* **2015**, *32*, 143–147. [CrossRef]
9. International Association for the Study of Pain (IASP). Pain Terms and Definitions. Available online: https://www.iasp-pain.org/resources/terminology/ (accessed on 11 April 2023).
10. Julius, D.; Basbaum, A.I. Molecular mechanisms of nociception. *Nature* **2001**, *413*, 203–210. [CrossRef]
11. Sneddon, L.U. Comparative Physiology of Nociception and Pain. *Physiology* **2018**, *33*, 63–73. [CrossRef]
12. Bell, A. The neurobiology of acute pain. *Vet. J.* **2018**, *237*, 55–62. [CrossRef]
13. Douglas, J.M.; Sanchez-Migallon Guzman, D.; Paul-Murphy, J.R. Pain in Birds: The Anatomical and Physiological Basis. *Vet. Clin. N. Am. Exot. Anim. Pract.* **2018**, *21*, 17–31. [CrossRef] [PubMed]
14. Bellieni, C.V. New insights into fetal pain. *Semin. Fetal Neonatal Med.* **2019**, *24*, 101001. [CrossRef] [PubMed]
15. Bjørnstad, S.; Austdal, L.P.; Roald, B.; Glover, J.C.; Paulsen, R.E. Cracking the Egg: Potential of the Developing Chicken as a Model System for Nonclinical Safety Studies of Pharmaceuticals. *J. Pharmacol. Exp. Ther.* **2015**, *355*, 386–396. [CrossRef] [PubMed]
16. Rosenbruch, M. The sensitivity of chicken embryos in incubated eggs. *Altex* **1997**, *14*, 111–113. [PubMed]
17. Eide, A.L.; Glover, J.C. Developmental dynamics of functionally specific primary sensory afferent projections in the chicken embryo. *Anat. Embryol.* **1997**, *195*, 237–250. [CrossRef]
18. Eide, A.L.; Glover, J.C. Development of the longitudinal projection patterns of lumbar primary sensory afferents in the chicken embryo. *J. Comp. Neurol.* **1995**, *353*, 247–259. [CrossRef]
19. Mellor, D.J.; Diesch, T.J. Birth and hatching: Key events in the onset of awareness in the lamb and chick. *N. Z. Vet. J.* **2007**, *55*, 51–60. [CrossRef]

20. Peters, J.J.; Vonderahe, A.R.; Powers, T.H. The functional chronology in developing chick nervous system. *J. Exp. Zool.* **1956**, *133*, 505–518. [CrossRef]
21. Herr, K. Pain assessment strategies in older patients. *J. Pain* **2011**, *12*, 3–13. [CrossRef]
22. Prunier, A.; Mounier, L.; Le Neindre, P.; Leterrier, C.; Mormède, P.; Paulmier, V.; Prunet, P.; Terlouw, C.; Guatteo, R. Identifying and monitoring pain in farm animals: A review. *Animal* **2013**, *7*, 998–1010. [CrossRef]
23. Bellieni, C.V. Pain assessment in human fetus and infants. *AAPS J.* **2012**, *14*, 456–461. [CrossRef] [PubMed]
24. Spurlin, J., 3rd; Lwigale, P. A technique to increase accessibility to late-stage chick embryos for in ovo manipulations. *Dev. Dyn.* **2013**, *242*, 148–154. [CrossRef] [PubMed]
25. Phuphanin, A.; Sampanporn, L.; Sutapun, B. Smartphone-Based Device for Non-Invasive Heart-Rate Measurement of Chicken Embryos. *Sensors* **2019**, *19*, 4843. [CrossRef]
26. Aleksandrovich, A.; Nikolaevich, S.; Aleksandrovich, A.; Yakovlevich, S. Non-invasive monitoring of avian embryo heart rate. *J. Anim. Behav. Biometeorol.* **2019**, *7*, 119–122. [CrossRef]
27. Youssef, A.; Berckmans, D.; Norton, T. Non-Invasive PPG-Based System for Continuous Heart Rate Monitoring of Incubated Avian Embryo. *Sensors* **2020**, *20*, 4560. [CrossRef]
28. Acierno, M.J.; da Cunha, A.; Smith, J.; Tully, T.N.; Guzman, D.S.; Serra, V.; Mitchell, M.A. Agreement between direct and indirect blood pressure measurements obtained from anesthetized Hispaniolan Amazon parrots. *J. Am. Vet. Med. Assoc.* **2008**, *233*, 1587–1590. [CrossRef]
29. Girard, H. Arterial pressure in the chick embryo. *Am. J. Physiol.* **1973**, *224*, 454–460. [CrossRef] [PubMed]
30. Van Mierop, L.H.; Bertuch, C.J., Jr. Development of arterial blood pressure in the chick embryo. *Am. J. Physiol.* **1967**, *212*, 43–48. [CrossRef]
31. Tazawa, H. Measurement of blood pressure of chick embryo with an implanted needle catheter. *J. Appl. Physiol. Respir. Environ. Exerc. Physiol.* **1981**, *51*, 1023–1026. [CrossRef]
32. Gentle, M.J. Pain in Birds. *Anim. Welf.* **1992**, *1*, 235–247. [CrossRef]
33. Woolley, S.C.; Gentle, M.J. Physiological and behavioural responses in the hen (*Gallus domesticus*) to nociceptive stimulation. *Comp. Biochem. Physiol. A Comp. Physiol.* **1987**, *88*, 27–31. [CrossRef]
34. Haga, H.A.; Tevik, A.; Moerch, H. Electroencephalographic and cardiovascular indicators of nociception during isoflurane anaesthesia in pigs. *Vet. Anaesth. Analg.* **2001**, *28*, 126–131. [CrossRef] [PubMed]
35. Ruíz-López, P.; Domínguez, J.M.; Granados, M.D.M. Intraoperative nociception-antinociception monitors: A review from the veterinary perspective. *Vet. Anaesth. Analg.* **2020**, *47*, 152–159. [CrossRef] [PubMed]
36. Gentle, M.J.; Hunter, L.N. Physiological and behavioural responses associated with feather removal in Gallus gallus var domesticus. *Res. Vet. Sci.* **1991**, *50*, 95–101. [CrossRef] [PubMed]
37. Haga, H.A.; Dolvik, N.I. Electroencephalographic and cardiovascular variables as nociceptive indicators in isoflurane-anaesthetized horses. *Vet. Anaesth. Analg.* **2005**, *32*, 128–135. [CrossRef] [PubMed]
38. Mueller, C.; Burggren, W.; Tazawa, H. *The Physiology of the Avian Embryo*; Springer: Denton, TX, USA, 2015; pp. 739–766.
39. Burggren, W.; Crossley, D.A., 2nd. Comparative cardiovascular development: Improving the conceptual framework. *Comp. Biochem. Physiol. A Mol. Integr. Physiol.* **2002**, *132*, 661–674. [CrossRef] [PubMed]
40. Crossley, D., 2nd; Altimiras, J. Ontogeny of cholinergic and adrenergic cardiovascular regulation in the domestic chicken (*Gallus gallus*). *Am. J. Physiol. Regul. Integr. Comp. Physiol.* **2000**, *279*, R1091–R1098. [CrossRef]
41. McCarty, L.P.; Shideman, F.E.; Lee, W.C. Measurement of the inotropic effects of drugs on the innervated and noninnervated embryonic chicken heart. *J. Pharmacol. Exp. Ther.* **1960**, *129*, 315–321.
42. Andrewartha, S.J.; Tazawa, H.; Burggren, W.W. Embryonic control of heart rate: Examining developmental patterns and temperature and oxygenation influences using embryonic avian models. *Respir. Physiol. Neurobiol.* **2011**, *178*, 84–96. [CrossRef]
43. Gentle, M.J. Pain issues in poultry. *Appl. Anim. Behav. Sci.* **2011**, *135*, 252–258. [CrossRef]
44. Dubbeldam, J.L. The sensory trigeminal system in birds: Input, organization and effects of peripheral damage. A review. *Arch. Physiol. Biochem.* **1998**, *106*, 338–345. [CrossRef]
45. Kuenzel, W.J. Neurobiological basis of sensory perception: Welfare implications of beak trimming. *Poult. Sci.* **2007**, *86*, 1273–1282. [CrossRef]
46. Chumak, V.I. Dinamika Reflektomykh Reaktsii i Vklyuchenie Retseptornykh Apparatov u Embriona Kuritsy (Dynamics of Reflex Reactions and Initiation of Receptor Systems in the Chick Embryo). In *Voprosy Fiziologii i Patologii Tsentral'noi Nervnoi Sistemy Cheloveka i Zhivotnykli v Ontogeneze*; Sbornik: Moskva, UdSSR, 1961; pp. 63–68.
47. Chiba, Y.; Fukuoka, S.; Niiya, A.; Akiyama, R.; Tazawa, H. Development of cholinergic chronotropic control in chick (Gallus gallus domesticus) embryos. *Comp. Biochem. Physiol. A Mol. Integr. Physiol.* **2004**, *137*, 65–73. [CrossRef] [PubMed]
48. Akiyama, R.; Matsuhisa, A.; Pearson, J.T.; Tazawa, H. Long-term measurement of heart rate in chicken eggs. *Comp. Biochem. Physiol. A Mol. Integr. Physiol.* **1999**, *124*, 483–490. [CrossRef] [PubMed]
49. Moriya, K.; Höchel, J.; Pearson, J.T.; Tazawa, H. Cardiac rhythms in developing chicks. *Comp. Biochem. Physiol. A Mol. Integr. Physiol.* **1999**, *124*, 461–468. [CrossRef]
50. Höchel, J.; Akiyama, R.; Masuko, T.; Pearson, J.T.; Nichelmann, M.; Tazawa, H. Development of heart rate irregularities in chick embryos. *Am. J. Physiol.* **1998**, *275*, H527–H533. [CrossRef] [PubMed]

51. Lehmann, H.S.; Musk, G.C.; Laurence, M.; Hyndman, T.H.; Tuke, J.; Collins, T.; Gleerup, K.B.; Johnson, C.B. Mitigation of electroencephalographic and cardiovascular responses to castration in Bos indicus bulls following the administration of either lidocaine or meloxicam. *Vet. Anaesth. Analg.* **2017**, *44*, 1341–1352. [CrossRef] [PubMed]
52. Saller, A.M.; Werner, J.; Reiser, J.; Senf, S.; Deffner, P.; Abendschön, N.; Weiß, C.; Fischer, J.; Schörwerth, A.; Miller, R.; et al. Local anesthesia in piglets undergoing castration-A comparative study to investigate the analgesic effects of four local anesthetics on the basis of acute physiological responses and limb movements. *PLoS ONE* **2020**, *15*, e0236742. [CrossRef]
53. van Lieshout, J.J.; Wieling, W.; Karemaker, J.M.; Eckberg, D.L. The vasovagal response. *Clin. Sci.* **1991**, *81*, 575–586. [CrossRef]
54. Paul-Murphy, J.; Ludders, J.W. Avian analgesia. *Vet. Clin. N. Am. Exot. Anim. Pract.* **2001**, *4*, 35–45. [CrossRef]
55. Aladjem, S.; Rest, J.; Stojanovic, J. Fetal heart rate responses to fetal movements. *Br. J. Obstet. Gynaecol.* **1977**, *84*, 487–491. [CrossRef] [PubMed]
56. Lascelles, B.D.X.; Kirkby Shaw, K. An extended release local anaesthetic: Potential for future use in veterinary surgical patients? *Vet. Med. Sci.* **2016**, *2*, 229–238. [CrossRef] [PubMed]
57. Machin, K. Controlling Avian Pain. *Compend. Contin. Educ. Pract. Vet. N. Am. Ed.* **2005**, *27*, 299–309.
58. Khamisabadi, A.; Kazemi-Darabadi, S.; Akbari, G. Comparison of Anesthetic Efficacy of Lidocaine and Bupivacaine in Spinal Anesthesia in Chickens. *J. Avian Med. Surg.* **2021**, *35*, 60–67. [CrossRef]
59. Machin, K. Avian Analgesia. *Semin. Avian Exot. Pet. Med.* **2005**, *14*, 236–242. [CrossRef]
60. Sneddon, L.U.; Elwood, R.W.; Adamo, S.A.; Leach, M.C. Defining and assessing animal pain. *Anim. Behav.* **2014**, *97*, 201–212. [CrossRef]

animals

MDPI

Article

Nociception in Chicken Embryos, Part II: Embryonal Development of Electroencephalic Neuronal Activity *In Ovo* as a Prerequisite for Nociception

Sandra Kollmansperger [1,†], Malte Anders [1,2,†], Julia Werner [3], Anna M. Saller [3], Larissa Weiss [3], Stephanie C. Süß [3], Judith Reiser [3], Gerhard Schneider [1], Benjamin Schusser [4], Christine Baumgartner [3] and Thomas Fenzl [1,*]

1 Department of Anaesthesiology and Intensive Care, School of Medicine, Technical University Munich, 81675 Munich, Germany; s.kollmansperger@outlook.de (S.K.); malteanders@gmail.com (M.A.); gerhard.schneider@tum.de (G.S.)

2 Clinical Development and Human Pain Models, Fraunhofer Institute for Translational Medicine and Pharmacology ITMP, 60596 Frankfurt, Germany

3 Center for Preclinical Research, Technical University of Munich, 81675 Munich, Germany; julia.werner@tum.de (J.W.); anna.saller@tum.de (A.M.S.); larissa.weiss@tum.de (L.W.); stephanie.suess@tum.de (S.C.S.); judith.reiser@tum.de (J.R.); christine.baumgartner@tum.de (C.B.)

4 Department of Molecular Life Sciences, Reproductive Biotechnology, School of Life Sciences Weihenstephan, Technical University Munich, 85354 Freising, Germany; benjamin.schusser@tum.de

* Correspondence: thomas.fenzl@tum.de; Tel.: +49-89-4140-9043

† These authors contributed equally to this work.

Simple Summary: Even today, we do not know from which point the chicken embryo is able to process and feel pain. This is of special interest as worldwide, millions of male embryos are killed before hatching. This work aimed to examine when during the development of the embryo the brain shows normal activity, based on EEG recordings. The data strongly suggest developmental day 13 as the earliest embryonal stage being able to process pain. These results may support legislative processes establishing updated laws on animal welfare.

Abstract: Chicken culling has been forbidden in Germany since 2022; male/female selection and male elimination must be brought to an embryonic status prior to the onset of nociception. The present study evaluated the ontogenetic point at which noxious stimuli could potentially be perceived/processed in the brain *in ovo*. EEG recordings from randomized hyperpallial brain sites were recorded *in ovo* and noxious stimuli were applied. Temporal and spectral analyses of the EEG were performed. The onset of physiological neuronal signals could be determined at developmental day 13. ERP/ERSP/ITC analysis did not reveal phase-locked nociceptive responses. Although no central nociceptive responses were documented, adequate EEG responses to noxious stimuli from other brain areas cannot be excluded. The extreme stress impact on the embryo during the recording may overwrite the perception of noniceptive stimuli. The results suggest developmental day 13 as the earliest embryonal stage being able to receive and process nociceptive stimuli.

Keywords: EEG; nociception; pain; embryo; development; *Gallus gallus domesticus*

1. Introduction

Chicken culling has been forbidden in Germany since January 2022. Until this date, around 45 million male birds were killed every year directly after hatching, as raising male layer-type chickens is not profitable for the industry [1]. In recent years, our society gained a clear understanding that sex selection must be brought forward to the embryo status, leaving hatched birds untouched. During *in ovo* sex determination, the sex can be identified early before hatching for example using endocrinological or spectroscopic procedures, so that the incubation of eggs containing male chickens can be prevented [2] as early as possible. Animal welfare and our ethical conscience admonish our society to make sure that sex selection and reasonable killing at this developmental stage must exclude nociceptive perception.

Several studies have indicated that birds can perceive pain in the same way as mammals [3]. This is not limited to behavioural and physiological responses to various nociceptive stimuli, which elicit similar responses as observed in mammals [4,5]. More importantly, it also includes the fact that in birds, cutaneous mechanical, thermal, chemical and polymodal nociceptors have been identified [4,6,7]. As pain includes a subjective component, it is virtually impossible to quantify pain perception deriving from the neuronal activity of nociceptors, especially without communicating through speech. To overcome this limitation, pain research distinguishes between automatic, unconscious recognition of stimuli and thereby induced transmissions of neuronal signals (nociception) and conscious perception of pain. While nociception is the peripheral recognition of potentially tissue-damaging (noxious) stimuli by nociceptors and their transmission through the nociceptive nervous system towards the central nervous system [8], pain is characterized by a subjective, conscious and central sensation, usually triggered by nociception. Pain only arises through the subjective, conscious perception of nociception, which requires a functional brain and its centralized interpretation of nociception [9,10]. In animal research, the most reliable and accepted method to record nociceptive stimuli together with their adequate neuronal answer is the electroencephalogram [6,11–13].

The development of the chicken embryo and its nervous system is a gradual process, e.g., from the fifth day of incubation, spontaneous movements of the embryo are possible [14]. However, as the nervous system of the chicken embryo is still less developed at this time of embryogenesis, nociception is highly unlikely [15,16]. Previous studies investigating the onset of the first spontaneous EEG activity were inconsistent in their results; thus, developmental day 11 [17], day 12 [18] and day 13 [19] have been identified as the EEG onset.

Summarized, due to a lack of consistent data on the development of the neuronal system of chicken embryos including peripheral receptors together with sensory/motor pathways and central processing, the currently very limited knowledge does not allow a precise statement on the physiological onset of nociception. This is even more true for the potential capability of pain perception at the central level. The main focus of the present study was to determine the onset of the EEG signal in chicken embryos as a physiological prerequisite for nociception. The second goal was to evaluate whether standardized painful stimuli may trigger subtle changes in epidural EEG signals.

2. Materials and Methods

2.1. Animals

A total of 361 Lohman Selected Leghorn chicken embryos (TUM Animal ResearchCenter, Versuchsstation Thalhausen, Technical University of Munich) between developmental days 7–19 (ED7–ED19) were used in the experiments (see Table 1 for details). Sexing was performed macroscopically in 280 embryos (♂ = 144 (51.43%), ♀ = 136 (48.58%)) between ED12–ED19.

The fertilized eggs were disinfected (Röhnfried Desinfektion Pro), labeled and stored at 15 °C (embryogenesis put on hold) for further treatment. Within a week, the experimental animals were moved to an incubator (Favorit Olymp 192 Spezial, HEKA—Brutgeräte, Rietberg, Germany, temperature 37.8 °C, air humidity 55%) and assigned to the developmental day ED0. After three days the eggs were windowed, treated with 0.5 mL Penicillin-Streptomycin (10,000 units penicillin, 10 mg streptomycin/mL, P4333—100 mL Sigma-Aldrich, Darmstadt, Germany) and kept in the incubator until starting the first EEG recordings at day ED07. After termination of the experiments, a lethal anaesthesia was applied via intravenous injection of Pentobarbital-Sodium (Narcoren: ED07–ED19: 16 g/100 mL in 0.1–0.2 mL), followed by decapitation. Developmentally critical stages

(ED12, ED13, ED19, see results for details) were additionally defined by a more detailed staging method [20].

Table 1. Summarized number of animals used for the different experimental approaches.

Stage (ED)	ED7	ED8	ED9	ED10	ED11	ED12	ED13	ED14	ED15	ED16	ED17	ED18	ED19	TOTAL
Animals total	36	9	8	9	18	56	57	14	13	17	18	40	66	361
Discarded	5	-	-	-	5	12	9	1	2	3	1	16	22	76
Evaluated	10	9	8	9	13	21	32	13	9	14	17	21	29	205 [1]
Random EEG	-	-	-	-	-	-	-	-	-	-	-	2	13	15
Onset EEG	-	-	-	-	-	-	12	14	-	-	-	-	-	26
Electrical stimulation	8	6	4	5	9	6	15	6	4	7	11	12	10	103
Thermal stimulation	2	3	4	4	4	3	3	7	5	7	6	7	6	61
Histology	21	-	-	-	-	23	16	-	2	-	-	3	15	80

Discarded: From 361 embryos initially prepared, 76 were used for the establishment of the recording routines or discarded due to a low online signal/noise ratio. Evaluated: A total of 229 recording sessions deriving from ED07–ED19 were pre-analyzed for further analysis. Pre-analysis resulted in 15 animals for Random EEG (random hyperpallial EEG placements), 26 animals for Onset EEG (additional recording to evaluate ED12 and ED13), 103 animals for Electrical Stimulation, 61 animals for Thermal stimulation ([1] 24 recordings from the initially evaluated animals were rejected after re-evaluation of the EEG quality). For histology, 80 embryos were used, partly originating from animals used for EEG recordings, partly especially prepared for histology.

2.2. EEG Hardware and EEG Recordings

During experiments the chicken embryos were transferred from the incubator to the experimental setup and kept at a mean temperature of 37.5 °C (±2 °C) and a room humidity of 42%. The embryos were given 5 min before preparation to adapt to the environmental changes in light and humidity. The head of the embryo was gently grabbed and brought to the edge of the egg and then fixed for EEG recordings in a way that kept the head and body on the same horizontal level to minimize additional strain on the vascular system [21,22].

After fixation of the head, the EEG electrodes were placed epidurally at the cerebellum (electrode EEG1), rostrally at multiple hyperpallial sites (electrode EEG2) and above the optic lobe (reference electrode REF). Additional recordings were performed with EEG1 on the optic lobe [18]. Custom-made gold electrodes were applied, for details refer to [23–28]. Each EEG recording consisted of 2 min of basal EEG recordings, followed by 8 min stimulation and another 2 min of basal EEG (Figure 1). The raw EEG data were processed through a pre-amplifier (custom-made, amplification: 1×, npi electronics, Tamm, Germany), amplified (DPA-2FL, npi electronics, Tamm, Germany) with an amplification rate of 1000× (hardware bandpass filter: 0.1 Hz–100 Hz, notch filter @50 Hz) and digitalized @500 Hz (Power1401, CED, Cambridge Electronic Design Limited, Milton/Cambridge, UK) for offline analysis. The recording software (Spike2, CED, Cambridge Electronic Design Limited, Milton/Cambridge, UK) was TTL-synchronized with the stimulation hardware.

Room temperature and room humidity, egg temperature and moisture of the embryonal brain surface was monitored very closely, as dehydration of the brain surface and electrodes may lead to artifacts and changes in electrical activity [19,29].

2.3. Standardized Thermal and Electrical Stimulation

For thermal stimulation, 40 subsequent thermal stimuli were applied with an interstimulus interval of 10 s, using a Peltier-element-based thermal stimulation device (TCS, QST.Lab, Strasbourg, France). The stimulation device had 0.25 cm^2 at the tip of the probe, a heating speed of 41 °C/s (see Figure 1) and a final temperature of 51 °C. Peak temperature during contact heat stimulation was chosen from the literature for nociceptive thermal thresholds [6,30,31]. For electrical stimulation, an electrically isolated constant current stimulator (ISO-STIM 01B, npi electronics, Tamm, Germany), with a constant stimulation current of 1 mA [32,33] (pulse duration: 150 µs, inter-pulse interval: 5 ms, pulse train duration: 40 ms, inter-stimulus interval: 5 s) was used. The composition of the pulse train derived from previous studies applying electrical microstimulations [34–36].

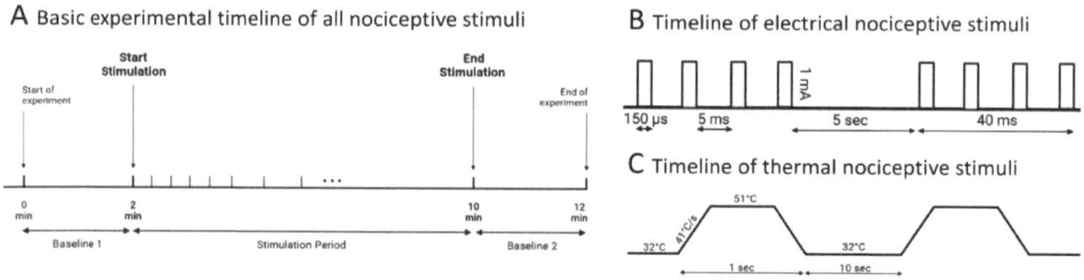

Figure 1. Experimental timeline for EEG recordings in chicken embryos. (**A**) The total duration of a single recording was 12 min, starting and ending with 2 min of baseline EEG recordings. The stimulation duration was 8 min. (**B**) Electrical stimulation was administered at 1 mA, pulse duration: 150 µs, inter-pulse interval: 5 ms, pulse train: 40 ms at 5 s, 90 repetitions/stimulus. (**C**) Thermal stimuli were given at 51 °C with a heating rate of 41 °C/s for 1 s and repeated every 10 s for 40 times. Basal temperature was kept at 32 °C.

2.4. EEG Data Processing

Raw EEG recordings below 25 µV during a minimum of 90% individual recording time and EEG recordings with amplitudes exceeding 500 µV were rejected.

Selected EEG text files were transferred into a vector file (MatLab, MathWorks, Natick, MA, USA), and subsequently imported to the MATLAB toolbox EEGLAB [37] for analysis. As most automated artifact rejection routines such as artifact subspace reconstruction (ASR) are only validated for human data [38], datasets that exceeded ±500 µV in amplitude for more than 10% of the recording time were manually rejected for analysis. From all manually selected datasets, the EEG signal from −1 s to +2 s around the onset of each stimulus was epoched. Event-related spectral perturbation (ERSP) and inter-trial coherence (ITC) were calculated [39,40] using EEGLAB's *newtimef*-function. A divisive baseline from −1 s to 0 s, a resolution in time of 400 points from −1 s to +2 s and a frequency resolution of 200 points between the frequencies of 3 Hz and 100 Hz [41,42] were chosen. EEG signals from EEG1 were analyzed with a wavelet transform portion of the *newtimef*-function with 3 cycles at the lowest frequency of 3 Hz and 20 cycles at the highest frequency of 100 Hz. In the ERSP results, any deactivation or activation below a threshold of −2 dB or above a threshold of +2 dB was considered as a response to the stimulus [39,40,43]. For the analysis of all baseline EEG recordings and stimuli-locked EEGs, 6 randomly selected datasets from each development stage (ED07–ED19) were selected for further analysis. Spectral EEG parameter were analyzed as power spectral density (PSD) with the *pwelch* function from the MATLAB Signal Processing Toolbox and plotted as density spectral arrays (DSA) in a logarithmic (log10) average across all embryos of a particular developmental day.

2.5. Physiological Anticipations

The median ERSP and ITC data are only shown for d19 embryos, as any EEG response on external stimuli was anticipated at the latest development stage, shortly before hatching. The phase-locked response in the EEG after electrical stimulation was expected rather immediately after the onset of the stimulus (below 100 ms). The EEG response after thermal stimulation was expected well after the onset of the stimuli due to $\Delta T_{heating}$ of the Peltier-element (between 100 ms and 800 ms) [39,40,43].

2.6. Statistics

Only the minimum/maximum ERSP and ITC values and their respective 25% and 75% quartiles, as well as the time and frequency at which they occurred, are presented, as this does not depend on the chosen window size when extracting ERSP data. For the evaluation of differences in the spectral power features, calculation of the area under the curve (AUC) of the receiver-operating characteristic (ROC) for each bin with a frequency resolution [(125/512) Hz] of PSD and 10 k-fold bootstrapped 95% confidence intervals (CI) were performed using the MATLAB-based MES toolbox. A difference between the two distributions was considered significant if the 95% CI did not contain levels above 0.5. The significance level was set to $p < 0.05$. For the statistical comparison of the PSD averages from the small sample size (n = 6), the non-parametrical Mann–Whitney *U* test [44] with its suitability for the analysis of EEG data [45] was applied. Some relevant data may have been missed but the influence of the testing procedure and its statistical results did not have any influence on the general EEG findings and spectral analyses.

2.7. Histological Procedures

Following termination of the EEG recordings, brains from ED7, ED12, ED13 and ED19 were removed from the skull and transferred to paraformaldehyde (4% PFA @1x PBS, Sigma-Aldrich) for at least 24 h. After transferring the brains to sucrose solution (30%), the brains were kept at 4 °C. Before slicing (cryotome @100 µm), the brains were mounted in gelatine (60 g gelatine, 50 g sucrose, 0.25 mL Triton X100, 500 mL mQ H_20) and transferred again to the PFA and sucrose bath. For anatomical analysis, a standard Nissl-staining protocol (cresyl violet staining) was applied to the anatomical slices.

3. Results

3.1. Basal EEG Activity

In Figure 2, representative 15 s sections of raw EEG data from three randomly selected datasets (3 × 13 embryos) are shown for D7–D19. The onset of prominent EEG activity can be clearly attributed to ED13.

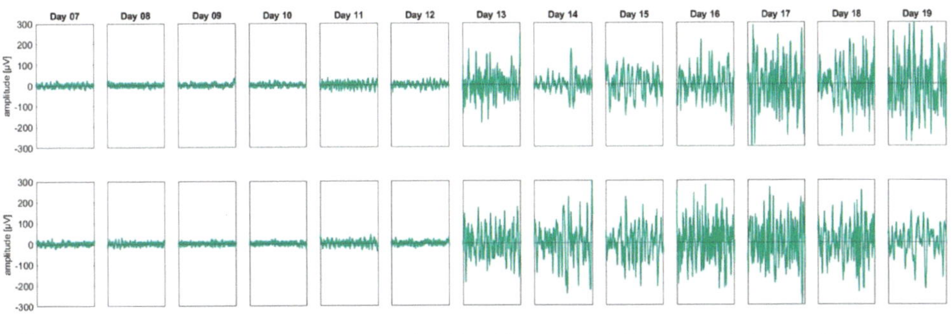

Figure 2. Raw EEG data: An overview of 15 s of raw EEG from three randomly chosen embryo datasets at development stages ED07-ED19. An onset of physiological EEG signatures is prominently visible from ED13 and onwards. The raw EEG amplitudes from ED07 until ED12 partly exceeded ±50 µV, but never exceeded ±100 µV, randomly fluctuating around baseline (0 µV). A strong increase in the EEG signal can be seen from ED13-ED19, with an amplitude regularly exceeding ±200 µV. The plots do not represent longitudinal recordings from ED07-ED19 within one embryo. For each day an individual embryo was recorded and added to a longitudinal graphical presentation representing the global findings.

Table 2 lists the median and the percentiles (25%; 75%) of the delta band (1–4 Hz) for each development stage from all 6 embryos included in Figure 3. Between ED12 and ED13, the average power of the delta band increased by more than 20 dB in absolute terms (p = 0.0022). An additional significant increase of approximately 6 dB (p = 0.0043) in the delta band was found between ED10 and ED11.

Table 2. Medians and percentiles [25%, 75%] of the PSD delta band (1–4 Hz) from the 6 randomly chosen embryos at each developmental stage, as shown in Figure 3.

STAGE (ED)	ED07	ED08	ED09	ED10	ED11	ED12	ED13	ED14	ED15	ED16	ED17	ED18	ED19
MEDIAN DELTA POWER	4.970	1.194	1.532	1.322	7.427	10.555	30.032	28.602	28.988	28.846	32.167	30.848	28.796
[25%] PERCENTILE	3.231	0.338	0.181	0.788	4.582	6.274	28.906	26.767	26.640	28.003	28.551	30.564	27.526
[75%] PERCENTILE	5.972	1.968	2.141	3.377	10.873	12.567	30.656	28.842	30.442	30.497	32.530	31.861	33.304

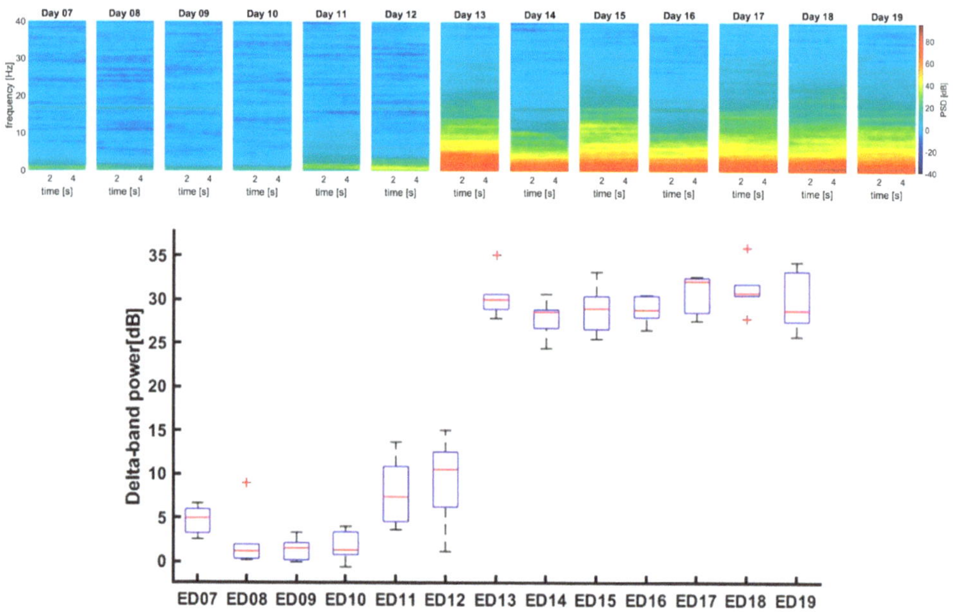

Figure 3. Top. Spectral power density: ED07-ED12 revealed no prominent power in all relevant frequency bands, apart from some minor but consistent oscillations in the low delta regions around 1–2 Hz and an isolated signal at 16.33 Hz. The onset of slow delta oscillations is visible from ED13 onwards. For each developmental day, data from 6 representative EEGs were processed. **Bottom**: Boxplots illustrating the median (red line), the 25% and 75% percentiles (lower and upper box end) and the minimum/maximum values (lower and upper whisker) for the delta band power. Red crosses indicate outliers. AUCED10/ED11: 0.97 [0.83, 1], AUCED12/ED13: 1 [1, 1], only significant AUCs reported (refer to Table 2 for other data).

3.2. Electrical and Thermal Stimulation

Figure 4 represents the median event-related spectral perturbation and inter-trial coherence for thermal and electrical stimulation. After applying a threshold of ±2 dB to unmask stimulus-related EEG activities, a single local maximum of 2.79 dB [−2.15 dB/5.16 dB] at 6.41 Hz and 1052 ms was measured after thermal stimulation. After electrical stimulation, a local maximum of 2.23 dB [0.68 dB/2.23 dB] at 13.24 Hz and 543 ms was detected. No further ERSP responses were found.

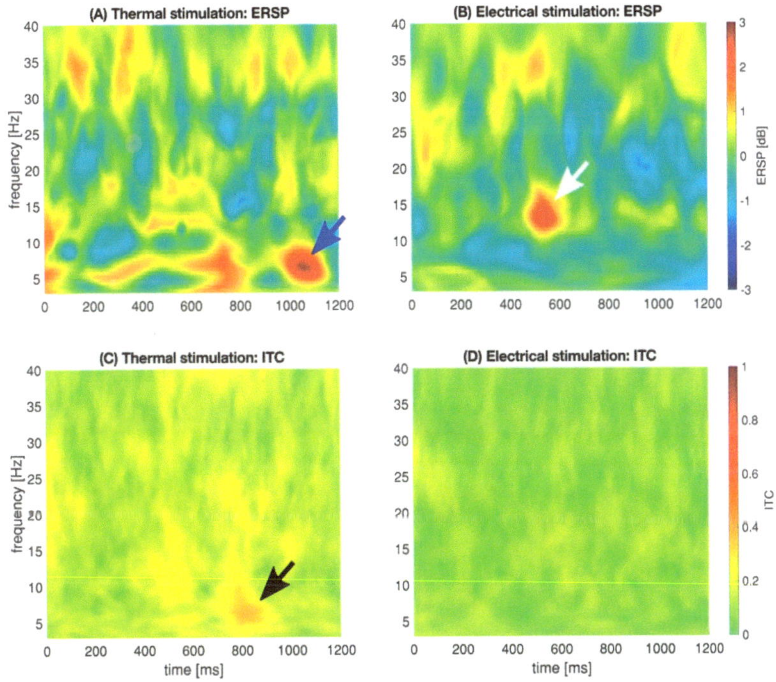

Figure 4. Median oscillatory responses as event-related spectral perturbation (ERSP, (**A**,**B**)), indicating the phase response, i.e., the oscillatory changes at a given time and frequency as a response to the stimulus. Inter-trial coherence (ITC, (**C**,**D**)), indicating the degree of phase-locking, i.e., the phase distribution of the stimulus across all trials. Blue arrow: local spectral maximum of 2.79 dB [−2.15 dB/5.16 dB] at 6.41 Hz and 1052 ms. White arrow: local spectral maximum of 2.23 dB [0.68 dB/2.23 dB] at 13.24 Hz and 543 ms. Black arrow: local ITC maximum of 0.38 [0.27/0.42] occurred at 6.41 Hz and 797 ms.

A local ITC maximum for an thermal stimulation of 0.38 [0.27/0.42] was detected at 6.41 Hz and 797 ms; the local ITC maximum for an electrical stimulation of 0.23 [0.11/0.29] was present at 28.35 Hz and 0.33 Hz. Both local ITC maxima did not correspond to an ERSP response, i.e., a deactivation or activation that exceeds −2 dB or 2 dB, respectively. ITC analysis revealed a low degree of phase locking in the analyzed range of time and frequency. The barely visible local ITC maximum deriving from thermal stimulation indicates that the phase of the oscillation following our stimulus is not completely random.

3.3. Histological Verification

The onset of meaningful EEG activity seems to correspond roughly with the histological data. At ED13, the embryonal development of central neuronal structures is well advanced, anticipating the expression of all neuronal structures within the embryonal brain at ED19 (see Figure 5).

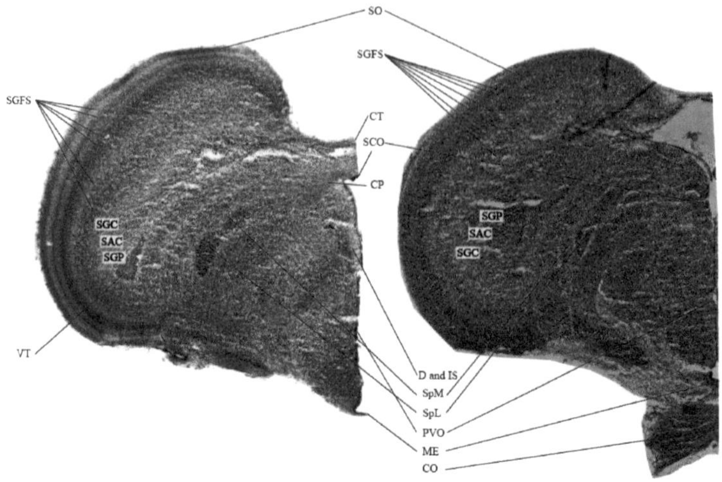

Figure 5. Anatomical differences of the embryonal brain: The two frontal sections (100 μm) from the central brain area represent the neuronal development of the embryonal brain from ED13 (**right**) and ED19 (**left**). Abbreviations: CO: Chiasma opticum, CP: Commissura posterior [caudalis] (Posterior commissure), CT: Commissura tectalis, D: Nucleus of Darkschewitsch; Nucleus paragrisealis centralis mesencephali (ICAAN), IS: Nucleus interstitialis (Cajal), ME: Eminentia mediana (Median eminence), PVO: Organum paraventriculare (Paraventricular organ), SAC: Stratum album centrale, SCE: Stratum cellulare externum, SCO: Organum subcommissurale (Subcommissural organ), SGC: Stratum granimalsiseum centrale, SGFS: Stratum griseum et fibrosum superficiale, SGP: Stratum griseum periventriculare, SO: Stratum opticum, SpL: Nucleus spiriformis lateralis, SpM: Nucleus spiriformis medialis, VT: Ventriculus tecti mesencephalic. The anatomical nomenclature was referred to anatomical atlases [15,46].

4. Discussion

The present study evaluated the neuronal development of an embryonal chicken brain at the level of the EEG. A relatively clear onset of a physiologically meaningful EEG activity could be attributed to ED13. The manifestation of this neuronal activity was shown in the present study until ED19. Electrical and thermal stimuli did not elicit any notable temporal and spectral changes in the corresponding EEGs.

4.1. Basal EEG

The onset of physiologically relevant brain activity in the present study could be reliably demonstrated in various anatomical areas of the hyperpallium from ED13 onwards. Compared to raw EEG signals recorded 2 days after hatching [47], the EEG amplitudes in the embryo show similar temporal and spectral features. In 2-dayold chickens as well as in embryonal stages ED13–ED19, the EEG reach amplitudes of ± 100 μV to ± 200 μV, although the highest amplitudes were present prior to hatching and not during the first few days after hatching. Interestingly, such decreasing EEG amplitudes were also documented between 2-day-old chickens and 8-week-old chickens with an average amplitude below ± 100 μV [47]. Averaged frequency spectra from 2-day-old chickens show a low power maximum around 5–10 Hz with decreasing power towards 40 Hz [47] at frontal recording sites. The embryonal spectral maxima at ED19 were slightly lower within a range from 0.1 Hz to 6 Hz with a maximum at the delta band. A shift from the embryonal delta band towards a dominant but blurry theta/alpha band immediately after hatching may be due to the potential role of the alpha band to act as an attentional suppression mechanism during the selection or elimination of objects or features during cognitive tasks [48]. Whether the embryonal dominant delta band resembles sleep-likes states such as slow-wave sleep [24,49–51] remains a functional enigma. Early findings from embryonal EEG recordings demonstrated spontaneous neuronal activity between ED13 and ED16 [52]. In contrast, the dominant frequency band at developmental day 15 was around 4 Hz to 7 Hz [52], shifting towards higher frequencies close to hatching, which is in line with our findings and may resemble post-embryonal findings from others [47]. Why we found the earliest spontaneous EEG activity already around ED13, whereas Peters and co-workers did not report any electrical discharges of the cerebral lobes before day 14, is not clear. One reason could be that we applied EEG recordings continuously on every developmental day from ED7 to ED19. Peters and co-workers only reported data from day 6, day 8, day 10, day 13, day 16 and the first post-embryonal day. Important but minor developments may have been missed. In our studies, we used consecutive numbering starting with ED0 at the first embryonal breeding day. Whether this numbering and the corresponding staging was applied, or numbering began at ED1 was not documented in other studies [52]. Another reason for these differences may be due to the breeding lines used in different experiments and its potential subtle temporal aberrancy in their embryogenesis of the brain. For example, it is known from adult mice that temporal and spectral features of the EEG differ significantly between closely related breeding lines (Huber, 2000 #6179). It is conceivable that neuronal embryogenesis may also differ between different chicken breeding lines *in ovo*.

Apart from such differences, the global features of the late embryonal EEG are similar to basal EEGs derived from adult chickens. Interestingly, amplitudes above ±200 μv as recorded from our ED17–ED19 embryos were reported in resting adult chickens (Ookawa T., 1965 #8834). At this behavioral state, the dominant frequencies in adult chickens were 3–4 Hz and 6 Hz–12 Hz, respectively. Similar data were reported from newly hatched chickens [19]. Nevertheless, even at ED19, the individual EEGs were highly variable, which is consistent with earlier publications and observations in adolescent chickens [17,21,53,54].

The neuronal development, as expressed in the global embryonal EEG activity reported in the present study, seems to correlate with the development of various brain structures. From developmental day 8 onwards, a mass migration of neuroblasts takes place, which is completed around day 11 with segregation along the dorsolateral walls of the cerebrum [55]. By developmental day 12, the diencephalon has undergone a complete differentiation of nuclei [52,56], potentially setting the stage for physiological neuronal activities as presented for ED13.

From ED07 to ED12, a consistent signal at 16⅔ Hz was frequently recorded, which was covered by more dominant domains from ED13 onwards. Although the literature is very sparse [57] and partly not-peer-reviewed [58], one external source for this very particular frequency recorded may have been subway tracks run by 15 kV AC at 16⅔ Hz in close vicinity to our laboratory.

4.2. Electrical and Thermal Stimulation

Although the EEG maturates during the last week *in ovo*, variations in the electrical patterns are not always correlated with spontaneous motor activity [59] and motility patterns persisted unchanged in total absence of the cerebral EEG [21], raising the principal question in how far an embryonal EEG also does not mirror peripheral sensory input. The situation is much clearer in the adult bird. Physiological responses including spectral changes in the EEG to nociceptive stimuli have been described for awake birds [4,5,60], which are consistent with those observed in awake mammals [13,39,40,43,60]. The neuroanatomical prerequisites such as cutaneous mechanical, thermal, chemical and polymodal nociceptors are present and respond to external stimulation similar to mammalian nociceptors [3,60,61].

Stimuli-related electrical potentials in the mammalian brain are typically found in the somatosensory, insular, cingulate, frontal and parietal cortical network [62]. The avian hyperpallium, nidopallium and mesopallium have been proposed to be homologous to the mammalian somatosensory cortex [60,63–65].

In mammals and birds, somatosensory information is processed from the deeper layers of the thalamus [63,65] with its various nuclei based on their location, pattern of sensory inputs and its embryological derivation [66–68] towards the pallium (hyperpallium in birds). The hyperpallium apicale (and the caudomedial nidopallium) seem to be promising areas for avian stimulus and nociceptive processing [64,69–71], assumingly representing mammalian sensomotoric functionality.

Even so, the hyperpallial and nidopallial recording sites seem appropriate, no global EEG responses to the noxious stimuli could be recorded from others [30,31] and in the present study. The requirements to record *in ovo* EEG signal may create their own limitations. Assuming that the embryo is capable of processing nociceptive sensations and stimuli from ED13, the preparation of the embryo itself to record a clear EEG may drive the perceptive capacity already to its limits. Experimentally applied nociceptive stimuli may then have a perceptive threshold below the impact of the preparation of the embryo itself. These nociceptive stimuli could as well be above a given threshold, but the EEG may already be enhanced with nociceptive sensations from the preparation. Either way, subtle changes in the EEG according to an experimental stimulus may be uncovered. At present, no stereotaxic atlas of the chicken embryo is available to record along the spinothalamic tract and its thalamic and striatal projections to overcome this dilemma.

4.3. Conscious Pain Perception

Assuming that changes in the cortical activity due to nociceptive stimulation are based on the cognitive perception of pain [60,72], experiments applying a minimal anesthesia protocol were performed in several species [60,73–77]. To our knowledge, only one study used this anesthetic protocol in birds and found no consistent evidence of nociception after thermal, electrical, or mechanical stimulation [60]. These results demonstrate either the absence of nociceptive-driven spectral changes in birds or, more likely, a conscious perception of noxious stimuli [60]. This may raise the question of how far embryos and fetuses possess consciousness. The present literature widely spreads from consciousness being only present immediately after birth [78] across the morality of embryo usage in research [79] towards the general and unsolved question what consciousness really means [80–84]. This question by far goes beyond the scope of the present study, especially when we ask about potential consciousness *in ovo*.

4.4. Selection of the Embryonal Timeframe for EEG Recordings

In birds, C and Aδ axons along the spinothalamic tract terminate at peripheral nociceptive receptors connecting the peripheral nervous system with central regions of the avian brain [85]. These afferent fibers start developing around day 4 in the embryo, including functional multisynaptic reflex arcs and sensomotoric coupling *in ovo* around day 7 [16,78,86], excluding EEG recording before this stage. This neuroanatomical gestation is in line with the selective start of EEG recordings at ED7.

4.5. Histological Verification

Assuming that at developmental day 19 the neuronal prerequisites to detect, transmit and process nociceptive stimuli are fully established, a simple anatomical comparison as shown in Figure 5 indicates a general ability for nociception already around day 13. All major structures of the hyperpallium [15] are clearly visible, suggesting also a physiologically similar EEG at day 13 and day 19. The anatomical part of the study was not designed to focus on the morphological development of the brain, but rather being anatomically supportive for a functional EEG. The anatomical development per se would be very interesting, but this would have been out of the focus of a functional EEG study. Further acute slice recording and neuroanatomical verification is needed in the future to support this assumption.

5. Conclusions

The present work suggests the onset of a meaningful EEG at the developmental ED13 in the chicken embryo. Is this an adequate indicator for the processing of nociceptive stimuli or even the perception of pain? The literature suggests a central processing of nociceptive information to establish the sensation of acute pain. Based on the present data, this seems unlikely to be before ED13. A direct EEG-based documentation of central nociceptive processing or the perception of pain was not possible in the chicken embryo *in ovo*. To overcome this limitation, we suggest establishing in vivo recordings of neuronal activity upon nociceptive stimuli starting at the level of the peripheral receptors, proceed along the ascending projections towards the developing central nervous system. The establishment of a stereotactic embryonal atlas and acute slice electrophysiology along the embryogenesis together with the present findings have the potential to overcome this limitation.

Author Contributions: Conceptualization: T.F.; Data curation: S.K., M.A. and T.F.; Formal analysis: S.K. and M.A.; Funding acquisition: C.B., T.F. and B.S.; Investigation: S.K., M.A., L.W. and S.C.S.; Methodology: T.F. and M.A.; Project administration: C.B.; Resources: C.B., B.S., G.S. and T.F.; Supervision: T.F.; Writing—original draft: S.K. and M.A.; Writing—review: T.F., J.W., A.M.S., J.R., S.K., M.A., G.S., B.S. and C.B.; Editing S.K. and T.F. All authors have read and agreed to the published version of the manuscript.

Funding: This work was financially funded by the German Federal Ministry of Food and Agriculture (BMEL) based on a decision of the Parliament of the Federal Republic of Germany, granted by the Federal Office for Agriculture and Food (BLE; grant number 2821HS005).

Institutional Review Board Statement: According to Directive 2010/63/EU of the European Parliament and the German Animal Welfare Law, no ethical approval was required for the use of chicken embryos in the experiments. The experiments were performed in an AAALAC-certified animal facility. All experiments and the euthanasia of embryos were conducted in strict accordance with the institutional Guidelines for Care and Use of Laboratory Animals and under general animal welfare principles.

Informed Consent Statement: Not applicable.

Data Availability Statement: Raw data are available upon reasonable request to the corresponding author.

Acknowledgments: The authors thank the scientific advisory board with Michael Erhard, Wolf Erhardt, Harald Luksch, Heidrun Potschka, Hans Straka and Britta Wirrer for their excellent scientific contribution as well as Johannes Fischer and Hicham Sid for their technical support. Duy Minh Nguyen for his support during the establishment of the *in ovo* recording setup. The authors also want to thank Matthias Kreuzer for his support in all statistical calculations.

Conflicts of Interest: The authors declare no conflict of interest.

References

1. Reithmayer, C.; Mußhoff, O. Consumer preferences for alternatives to chick culling in Germany. *Poult. Sci.* **2019**, *98*, 4539–4548. [CrossRef] [PubMed]
2. Krautwald-Junghanns, M.E.; Cramer, K.; Fischer, B.; Förster, A.; Galli, R.; Kremer, F.; Mapesa, E.U.; Meissner, S.; Preisinger, R.; Preusse, G.; et al. Current approaches to avoid the culling of day-old male chicks in the layer industry, with special reference to spectroscopic methods. *Poult. Sci.* **2017**, *97*, 749–757. [CrossRef]
3. Sandercock, D. Putative nociceptor responses to mechanical and chemical stimulation in skeletal muscles of the chicken leg. *Brain Res. Rev.* **2004**, *46*, 155–162. [CrossRef]
4. Gentle, M.J.; Hunter, L.N. Physiological and behavioural responses associated with feather removal in *Gallus gallus* var domesticus. *Res. Vet. Sci.* **1990**, *50*, 95–101.
5. Woolley, S.C.; Gentle, M.J. Physiological and behavioural responses in the hen (*Gallus domesticus*) to nociceptive stimulation. *Comp. Biochem. Physiol. A Comp. Physiol.* **1987**, *88*, 27–31. [CrossRef] [PubMed]
6. Gentle, M.J.; Tilston, V.; McKeegan, D.E. Mechanothermal nociceptors in the scaly skin of the chicken leg. *Neuroscience* **2001**, *106*, 643–652. [CrossRef]
7. Necker, R.; Reiner, B. Temperature-Sensitive Mechanoreceptors, Thermoreceptors and Heat Nociceptors in the Feathered Skin of Pigeons. *J. Comp. Physiol.* **1980**, *135*, 201–207.

8. Mischkowski, D.; Palacios-Barrios, E.E.; Banker, L.; Dildine, T.C.; Atlas, L.Y. Pain or nociception? Subjective experience mediates the effects of acute noxious heat on autonomic responses. *Pain* **2018**, *159*, 699–711. [CrossRef]
9. Sneddon, L.U. Comparative physiology of nociception and pain. *Physiology* **2017**, *33*, 63–73.
10. Woo, C.-W.; Schmidt, L.; Krishnan, A.; Jepma, M.; Roy, M.; Lindquist, M.A.; Atlas, L.Y.; Wager, T.D. Quantifying cerebral contributions to pain beyond nociception. *Nat. Commun.* **2017**, *8*, 14211.
11. Livingston, A. Physiological basis for pain perception in animals. *J. Vet. Anaesth.* **1994**, *21*, 73–77.
12. Johnson, C.B. Research Tools for the Measurement of Pain and Nociception. *Animals* **2016**, *6*, 71. [CrossRef] [PubMed]
13. Murrell, C.; Johnson, C.B. Neurophysiological techniques to assess pain in animals. *J. Vet. Pharmacol. Therap.* **2006**, *29*, 325–335.
14. O'Donovan, M.S.E.; Sholomenko, G.; Ho, S.; Antal, M.; Yee, W. Development of Spinal Motor Networks in the Chick Embryo. *J. Exp. Zool.* **1992**, *261*, 261–273. [PubMed]
15. Bellairs, R.; Osmond, M. *The Atlas of Chick Development*, 3rd ed.; Elsevier Ltd.: Amsterdam, The Netherlands; Academic Press: Cambridge, MA, USA, 2014; p. 693.
16. Eide, A.L.; Glover, J.C. Development of the Longitudinal Projection Patterns of Lumbar Primary Sensory Afferents in the Chicken Embryo. *J. Comp. Neurol.* **1995**, *353*, 247–259. [CrossRef]
17. Peters, J.J.; Vonderahe, A.R.; Schmid, D. Onset of cerebral electrical activity associated with behavioral sleep and attention in the developing chick. *J. Exp. Zool.* **1965**, *160*, 255–261. [CrossRef]
18. Katori, M. The development of the spontaneous electrical activity in the brain of a chick embryo and the effects of several drugs on it. *Jpn. J. Pharmacol.* **1962**, *12*, 9–25. [CrossRef]
19. Garcia-Austt, E., Jr. Development of Electrical Activity in Cerebral Hemispheres of the Chick Embryo. *Proc. Soc. Exp. Biol. Med.* **1954**, *86*, 348–352.
20. Hamburger, V.; Hamilton, L. A series of normal stages in the development of the chick embryo. *Dev. Dyn.* **1992**, *195*, 231–272.
21. Corner, M.A.; Bakhius, W.L. Developmental patterns in the central nervous system of birds. V. Cerebral electrical activity, forebrain function and behavior in the chick at the time of hatching. *Brain Res.* **1969**, *13*, 541–555. [CrossRef]
22. Corner, M.A.; Schade, J.P.; Sedlácek, J.; Stoeckart, R.; Bot, A.P. Developmental patterns in the central nervous system of birds. I. Electrical activity in the cerebral hemisphere, optic lobe and cerebellum. *Prog. Brain Res.* **1966**, *26*, 145–192. [CrossRef]
23. Schwitalla, J.C.; Pakusch, J.; Mücher, B.; Brückner, A.; Depke, D.A.; Fenzl, T.; De Zeeuw, C.I.; Kros, L.; Hoebeek, F.E.; Mark, M.D. Controlling absence seizures from the cerebellar nuclei via activation of the Gq signaling pathway. *Cell. Mol. Life Sci.* **2022**, *79*, 197. [CrossRef]
24. Fritz, E.M.; Kreuzer, M.; Altunkaya, A.; Singewald, N.; Fenzl, T. Altered sleep behavior in a genetic mouse model of impaired fear extinction. *Sci. Rep.* **2021**, *11*, 8978. [PubMed]
25. Kreuzer, M.; Polta, S.; Gapp, J.; Schuler, C.; Kochs, E.F.; Fenzl, T. Sleep scoring made easy—Semi-automated sleep analysis software and manual rescoring tools for basic sleep research in mice. *MethodsX* **2015**, *2*, 232–240. [CrossRef]
26. Polta, S.; Fenzl, T.; Jakubcakova, V.; Kimura, M.; Yassouridis, A.; Wotjak, C. Prognostic and Symptomatic Aspects of Rapid Eye Movement Sleep in a Mouse Model of Posttraumatic Stress Disorder. *Front. Behav. Neurosci.* **2013**, *7*, 60. [CrossRef] [PubMed]
27. Fulda, S.; Romanowski, C.P.N.; Becker, A.; Wetter, T.C.; Kimura, M.; Fenzel, T. Rapid eye movements during sleep in mice: High trait-like stability qualifies rapid eye movement density for characterization of phenotypic variation in sleep patterns of rodents. *BMC Neurosci.* **2011**, *12*, 110. [CrossRef]
28. Fenzl, T.; Romanowski, C.P.N.; Flachskamm, C.; Honsberg, K.; Boll, E.; Hoehne, A.; Kimura, M. Fully automated sleep deprivation in mice as a tool in sleep research. *J. Neurosci. Methods* **2007**, *166*, 229–235. [CrossRef]
29. Sharma, K.; Dua, S.; Singh, B.; Anand, B. Electro-ontogenesis of cerebral and cardiac activities in the chick embryo. *Electroencephalogr. Clin. Neurophysiol.* **1964**, *16*, 503–509. [CrossRef]
30. Hothersall, B.; Caplen, G.; Nicol, C.J.; Taylor, P.M.; Waterman-Pearson, A.E.; Weeks, C.A.; Murrell, J.C. Development of mechanical and thermal nociceptive threshold testing devices in unrestrained birds (broiler chickens). *J. Neurosci. Methods* **2011**, *201*, 220–227. [CrossRef]
31. Hothersall, B.; Caplen, G.; Parker, R.M.A.; Nicol, C.J.; Waterman-Pearson, A.E.; Weeks, C.A.; Murrell, J.C. Thermal nociceptive threshold testing detects altered sensory processing in broiler chickens with spontaneous lameness. *PLoS ONE* **2014**, *9*, e97883. [CrossRef]
32. Bogdanov, O.V.; Smetankin, A.A.; Saraev, S.Y.; Mikhailenok, E.L.; Ved, V.V. Electrical activity of the chick embryo brain during development of stable rearrangements of movement. *Neurosci. Behav. Physiol.* **1984**, *14*, 79–87. [CrossRef] [PubMed]
33. Ring, C.; Kavussanu, M.; Willoughby, A.R. Emotional modulation of pain-related evoked potentials. *Biol. Psychol.* **2013**, *93*, 373–376. [CrossRef] [PubMed]
34. Fenzl, T.; Schuller, G. Periaqueductal gray and the region of the paralemniscal area have different functions in the control of vocalization in the neotropical bat, Phyllostomus discolor. *Eur. J. Neurosci.* **2002**, *16*, 1974–1986. [PubMed]
35. Fenzl, T.; Schuller, G. Echolocation calls and communication calls are controlled differentially in the brainstem of the bat Phyllostomus discolor. *BMC Biol.* **2005**, *3*, 17. [CrossRef]
36. Fenzl, T.; Schuller, G. Dissimilarities in the vocal control over communication and echolocation calls in bats. *Behav. Brain Res.* **2007**, *182*, 173–179. [CrossRef]
37. Delorme, A.; Makeig, S. EEGLAB: An open source toolbox for analysis of single-trial EEG dynamics including independent component analysis. *J. Neurosci. Methods* **2004**, *134*, 9–21. [CrossRef]

38. Chang, C.Y.; Hsu, S.H.; Pion-Tonachini, L.; Jung, T.P. Evaluation of Artifact Subspace Reconstruction for Automatic EEG Artifact Removal. *Annu. Int. Conf. IEEE Eng. Med. Biol. Soc.* **2018**, *2018*, 1242–1245. [CrossRef]
39. Anders, M.; Anders, B.; Dreismickenbecker, E.; Hight, D.; Kreuzer, M.; Walter, C.; Zinn, S. EEG responses to standardised noxious stimulation during clinical anaesthesia: A pilot study. *BJA Open* **2023**, *5*, 100118.
40. Anders, M.; Dreismickenbecker, E.; Fleckenstein, J.; Walter, C.; Enax-Krumova, E.K.; Fischer, M.J.; Kreuzer, M.; Zinn, S. EEG-based sensory testing reveals altered nociceptive processing in elite endurance athletes. *Exp. Brain Res.* **2022**, *241*, 341–354.
41. Grandchamp, R.; Delorme, A. Single-trial normalization for event-related spectral decomposition reduces sensitivity to noisy trials. *Front. Psychol.* **2011**, *2*, 236. [CrossRef]
42. Herrmann, C.S.; Rach, S.; Vosskuhl, J.; Strüber, D. Time-frequency analysis of event-related potentials: A brief tutorial. *Brain Topogr.* **2014**, *27*, 438–450. [CrossRef]
43. Anders, M.; Anders, B.; Kreuzer, M.; Zinn, S.; Walter, C. Application of referencing techniques in EEG-based Recordings of Contact Heat Evoked Potentials (CHEPS). *Front. Hum. Neurosci.* **2020**, *14*, 527. [CrossRef]
44. Mishra, A.; Kumar, Y.; Kumar, T.; Singh, R.; Jha, K. Electroencephalographic characterization of a case of infantile spasm with atypical presentation. *Indian J. Child Health* **2019**, *6*, 42–45.
45. Maris, E.; Oostenveld, R. Nonparametric statistical testing of EEG-and MEG-data. *J. Neurosci. Methods* **2007**, *164*, 177–190. [CrossRef]
46. Kuenzel, W.; Masson, M. A Stereotaxic Atlas of the Brain of the Chick (*Gallus domesticus*). *Poult. Sci. Fac. Publ. Present.* **1988**. Available online: https://scholarworks.uark.edu/poscpub/1/ (accessed on 1 July 2023).
47. Hunter, M.; Batillana, M.; Bragg, T. EEG as a Measure of Developmental Changes in the Chicken Brain. *Dev. Psychobiol.* **2000**, *36*, 23–28. [CrossRef]
48. Foxe, J.J.; Snyder, A.C. The Role of Alpha-Band Brain Oscillations as a Sensory Suppression Mechanism during Selective Attention. *Front. Psychol.* **2011**, *2*, 154. [CrossRef]
49. Franken, P.; Dijk, D.-J.; Tobler, I.; Borbély, A.A. Sleep deprivation in rats: Effects on EEG power spectra, vigilance states, and cortical temperature. *Am. J. Physiol.-Regul. Integr. Comp. Physiol.* **1991**, *261*, R198–R208. [CrossRef]
50. Dijk, D.-J. EEG slow waves and sleep spindles: Windows on the sleeping brain. *Behav. Brain Res.* **1995**, *69*, 109–116. [CrossRef]
51. Fenzl, T.; Touma, C.; Romanowski, C.P.; Ruschel, J.; Holsboer, F.; Landgraf, R.; Kimura, M.; Yassouridis, A. Sleep disturbances in highly stress reactive mice: Modeling endophenotypes of major depression. *BMC Neurosci.* **2011**, *12*, 29. [CrossRef]
52. Peters, J.V.R.; Powers, T.H. The functional chronology in the developing chick nervous system. *J. Exp. Zool.* **1956**, *133*, 505–518. [CrossRef]
53. Mellor, D.J.; Diesch, T.J. Birth and hatching: Key events in the onset of awareness in the lamb and chick. *New Zealand Vet. J.* **2007**, *55*, 51–60. [CrossRef] [PubMed]
54. Peters, J.; Vonderahe, A.R.; McDonough, J.J. Electrical changes in brain and eye of the developing chick during hyperthermia. *Amer. J. Physiol.* **1964**, *207*, 260–264. [CrossRef] [PubMed]
55. Peters, J.J.; Vonderahe, A.R.; Huesman, A.A. Chronological development of electrical activity in the optic lobes, cerebellum, and cerebrum of the chick embryo. *Physiol. Zool.* **1960**, *33*, 225–231. [CrossRef]
56. Kuhlenbeck, H. The ontogenetic development of the diencephalic centers in a bird's brain (chick) and comparison with the reptilian and mammalian diencephalon. *J. Comp. Neurol.* **1937**, *66*, 23–75. [CrossRef]
57. Kanz, K.-G.; Kay, M.V.; Biberthaler, P.; Russ, W.; Wessel, S.; Lackner, C.K.; Mutschler, W. Susceptibility of automated external defibrillators to train overhead lines and metro third rails. *Resuscitation* **2004**, *62*, 189–198. [CrossRef]
58. Hadrian, W. Die magnetischen Stoerfelder des elektrischen Bahnbetriebes. *Elektrotechnik Und Informationstechnik Ei* **2006**, *1*, 46–49. [CrossRef]
59. Corner, M.; Schadé, J. Developmental Patterns in the Central Nervous System of Birds: IV. Cellular and Molecular Bases of Functional Activity. *Prog. Brain Res.* **1967**, *26*, 237–250.
60. McIlhone, A.E.; Beausoleil, N.J.; Kells, N.J.; Mellor, D.J.; Johnson, C.B. Effects of noxious stimuli on the electroencephalogram of anaesthetised chickens (*Gallus gallus domesticus*). *PLoS ONE* **2018**, *13*, e0196454. [CrossRef]
61. Gentle, M.J. Pain in birds. *Anim. Welf.* **1992**, *1*, 235–247. [CrossRef]
62. Legrain, V.; Iannetti, G.D.; Plaghki, L.; Mouraux, A. The pain matrix reloaded—A salience detection system for the body. *Prog. Neurobiol.* **2010**, *93*, 111–124. [CrossRef]
63. Butler, A.B.; Cotterill, R.M. Mammalian and Avian neuroanatomy and the question of consciousness in birds. *Biol. Bull.* **2006**, *211*, 106–127. [CrossRef]
64. Kuenzel, W. Neurobiological basis of sensory perception: Welfare implications of beak trimming. *Poult. Sci.* **2007**, *86*, 1273–1282. [CrossRef]
65. Reiner, A.; Yamamoto, K.; Karten, H.J. Organization and Evolution of the avian forebrain. *Anat. Rep. Part A* **2005**, *287A*, 1080–1102. [CrossRef]
66. González, G.; Puelles, L.; Medina, L. Organization of the mouse dorsal thalamus based on topology, calretinin immnunostaining, and gene expression. *Brain Res. Bull.* **2002**, *57*, 439–442. [CrossRef]
67. Butler, A.B. The dorsal thalamus of jawed vertebrates: A comparative viewpoint. *Brain Behav. Evol.* **1995**, *46*, 209–223. [CrossRef]
68. Butler, A.B. The evolution of the dorsal pallium in the telencephalon of amniotes: Cladistic analysis and a new hypothesis. *Brain Res. Rev.* **1994**, *19*, 66–101. [CrossRef]

69. Wild, J.M. The avian somatosensory system: Connections of regions of body representation in the forebrain of the pigeon. *Brain Res.* **1987**, *412*, 205–223. [CrossRef]
70. Delius, J.D.; Bennetto, K. Cutaneous sensory projections to the avian forebrain. *Brain Res.* **1972**, *37*, 205–221. [CrossRef]
71. Lierz, M.; Korbel, R. Anesthesia and analgesia in birds. *J. Exot. Pet Med.* **2012**, *21*, 44–58. [CrossRef]
72. Bromm, B. *Pain Measurement in Man: Neurophysiological Correlates of Pain*; Elsevier Publishing Company: Amsterdam, The Netherlands, 1984.
73. Gibson, T.; Johnson, C.; Stafford, K.; Mitchinson, S.; Mellor, D. Validation of the acute electroencephalographic responses of calves to noxious stimulus with scoop dehorning. *New Zealand Vet. J.* **2007**, *55*, 152–157. [CrossRef]
74. Johnson, C.B.; Sylvester, S.P.; Stafford, K.J.; Mitchinson, S.L.; Ward, R.N.; Mellor, D.J. Effects of age on the electroencephalographic response to castration in lambs anaesthetized with halothane in oxygen from birth to 6 weeks old. *Vet. Anaesth. Analg.* **2009**, *36*, 273–279. [CrossRef]
75. Johnson, C.B.; Wilson, P.R.; Woodbury, M.R.; Caulkett, N.A. Comparison of analgesic techniques for antler removal in halothane-anaesthetized red deer (*Cervus elaphus*): Electroencephalographic responses. *Vet. Anaesth. Analg.* **2005**, *32*, 61–71. [CrossRef]
76. Kells, N.J.; Beausoleil, N.J.; Chambers, J.P.; Sutherland, M.A.; Morrison, R.S.; Johnson, C.B. Electroencephalographic responses of anaesthetized pigs (*Sus scrofa*) to tail docking using clippers or cautery iron performed at 2 or 20 days of age. *Vet. Anaesth. Analg.* **2017**, *44*, 1156–1165. [CrossRef] [PubMed]
77. Murrell, J.C.; Johnson, C.B.; White, K.L.; Taylor, P.M.; Haberham, Z.L.; Waterman–Pearson, A.E. Changes in the EEG during castration in horses and ponies anaesthetized with halothane. *Vet. Anaesth. Analg.* **2003**, *30*, 138–146. [CrossRef] [PubMed]
78. Mellor, D.J.; Diesch, T.J. Onset of sentience: The potential for suffering in fetal and newborn farm animals. *Appl. Anim. Behav. Sci.* **2006**, *100*, 48–57. [CrossRef]
79. McMahan, J. Killing embryos for stem cell research. *Metaphilosophy* **2007**, *38*, 170–189. [CrossRef]
80. Zeman, A. What in the world is consciousness? *Prog. Brain Res.* **2005**, *150*, 1–10.
81. Dehaene, S.; Lau, H.; Kouider, S. What is consciousness, and could machines have it? *Robot. AI Humanit. Sci. Ethics Policy* **2021**, 43–56. [CrossRef]
82. Koch, C. What is consciousness. *Nature* **2018**, *557*, S8–S12. [CrossRef]
83. Wallace, R. *What is Consciousness?* Springer: Berlin/Heidelberg, Germany, 2005.
84. Solms, M. What is consciousness? *J. Am. Psychoanal. Assoc.* **1997**, *45*, 681–703. [CrossRef]
85. Douglas, J.M.; Guzman, D.S.-M.; Paul-Murphy, J.R. Pain in Birds: The anatomical and physiological basis. *Vet. Clin. Exot. Anim. Pract.* **2018**, *21*, 17–31. [CrossRef]
86. Rosenbruch, M. [The sensitivity of chicken embryos in incubated eggs][Article in German]. *ALTEX-Altern. Anim. Exp.* **1997**, *14*, 111–113.

animals

Article

Nociception in Chicken Embryos, Part III: Analysis of Movements before and after Application of a Noxious Stimulus

Stephanie C. Süß [1], Julia Werner [1], Anna M. Saller [1], Larissa Weiss [1], Judith Reiser [1], Janie M. Ondracek [2], Yury Zablotski [3], Sandra Kollmansperger [4], Malte Anders [4], Heidrun Potschka [5], Benjamin Schusser [6], Thomas Fenzl [4] and Christine Baumgartner [1,7,*]

1 Center for Preclinical Research, TUM School of Medicine, Technical University of Munich, 81675 Munich, Bavaria, Germany; stephanie.suess@tum.de (S.C.S.); julia.werner@tum.de (J.W.); anna.saller@tum.de (A.M.S.); larissa.weiss@tum.de (L.W.); judith.reiser@tum.de (J.R.)
2 Chair of Zoology, TUM School of Life Sciences Weihenstephan, Technical University of Munich, 85354 Freising, Bavaria, Germany; janie.ondracek@tum.de
3 Clinic for Swine, Center for Clinical Veterinary Medicine, Ludwig-Maximilians-Universität München, 85764 Oberschleißheim, Bavaria, Germany; y.zablotski@med.vetmed.uni-muenchen.de
4 Clinic for Anesthesiology and Intensive Care, TUM School of Medicine, Technical University of Munich, 81675 Munich, Bavaria, Germany; s.kollmansperger@outlook.de (S.K.); malteanders@gmail.com (M.A.); thomas.fenzl@tum.de (T.F.)
5 Institute of Pharmacology, Toxicology and Pharmacy, Ludwig-Maximilians-Universität München, 80539 Munich, Bavaria, Germany; potschka@pharmtox.vetmed.uni-muenchen.de
6 Reproductive Biotechnology, TUM School of Life Sciences Weihenstephan, Technical University of Munich, 85354 Freising, Bavaria, Germany; benjamin.schusser@tum.de
7 Veterinary Faculty, Ludwig-Maximilians-Universität München, 80539 Munich, Bavaria, Germany
* Correspondence: christine.baumgartner@tum.de

Simple Summary: Chicken embryos are currently being increasingly used in various areas of research but are frequently not covered by animal protection legislation. In the food industry, it is often even common practice to kill male embryos because they are of no economic use. In both cases, there is a lack of knowledge about the sentience of these chicken embryos, especially their ability to perceive pain. The purpose of this exploratory study was to investigate whether—and if so, on which developmental day—a chicken embryo shows a behavioral change or a specific response to a noxious stimulus, both of which would be indicative of functioning nociception. Two complementary approaches were used for the evaluation: human observation and computer-assisted evaluation using a markerless pose estimation software (DeepLabCut). Through a combination of both approaches, it became apparent that developmental day 15 was the earliest stage at which a response to the applied stimulus was detectable. This result thus represents a contribution to the future improvement of animal welfare as it suggests that from developmental day 15 a chicken embryo in the egg has the capacity to show a nocifensive reaction.

Abstract: Many potentially noxious interventions are performed on chicken embryos in research and in the poultry industry. It is therefore essential and in the interest of animal welfare to be able to precisely define the point at which a chicken embryo is capable of nociception in ovo. The present part III of a comprehensive study examined the movements of developing chicken embryos with the aim of identifying behavioral responses to a noxious stimulus. For this purpose, a noxious mechanical stimulus and a control stimulus were applied in a randomized order. The recorded movements of the embryos were evaluated using the markerless pose estimation software DeepLabCut and manual observations. After the application of the mechanical stimulus, a significant increase in beak movement was identified in 15- to 18-day-old embryos. In younger embryos, no behavioral changes related to the noxious stimulus were observed. The presented results indicate that noxious mechanical stimuli at the beak base evoke a nocifensive reaction in chicken embryos starting at embryonic day 15.

Keywords: behavior; movement; nociception; pain; chicken embryo; development; *Gallus gallus domesticus*

1. Introduction

The behavior of birds can profoundly differ from the behavior of mammals, especially in terms of indications of pain [1]. For a long time, birds were not believed to feel pain [1]. At present, it is generally accepted that birds are capable of nociception and can feel pain [1,2]. Several studies have established that birds have mechanothermal, mechanical and thermal nociceptors with high stimulus thresholds [2,3]. Furthermore, peripheral and central processing of a potentially noxious stimulus in birds occurs in a similar manner to that in mammals [4]. Raja et al. defined pain as an aversive experience of an individual that includes both sensory perception and emotional aspects [5]. This experience may be caused by a potential or actual lesion of the tissue [5]. Nociception, on the other hand, is described as the detection of a potentially damaging stimulus by primary sensory neurons and its processing in the nervous system [5,6]. The inability to communicate does not exclude the possibility that pain is felt, for example, by animals or neonates [1,5]. Another definition of pain more suitable for assessing pain in animals includes changes in species-specific behavior as a possible consequence of a painful experience [7]. Because pain is a subjective experience, its assessment is difficult in humans and is even more challenging in animals [1,5]. Detection and quantification of pain in animals involves inference from parameters associated with pain in humans [1].

Birds show only subtle behaviors of discomfort or pain due to the disadvantage of showing weakness in a social group or as a prey species in general as well as the potential predominance of the flight reflex [8]. In addition, bird behavior varies greatly among species and individuals, making it necessary to closely examine the typical behavior of the observed individual. This makes it possible to assess deviations in typical behavior as a sign of pain [9]. Although pain-associated behavior is difficult to identify, its major advantage is that it can be observed immediately and noninvasively [3,9]. This makes behavioral observation an essential part of a comprehensive pain assessment in birds.

Behavioral studies have been conducted in a variety of avian species [10]. Many of these studies used chickens (*Gallus gallus domesticus*) and evaluated nociceptive responses to procedures that are assumed to be painful or elicit discomfort [10,11]. The typical behavior of chicken embryos has long attracted scientific interest [10,11]. In the 1960s, the motility of chicken embryos was intensively studied. Movements and motility patterns, along with other aspects, were observed from days 3.5 to 20 of incubation [12–15]. In contrast, little is known about nociception in the chicken embryo or about nocifensive behavioral responses. According to current understanding, nociception in chicken embryos does not occur before the seventh day of incubation [16–18].

The results presented are part of a comprehensive study investigating the developmental day at which chicken embryos are capable of nociception and pain perception. The aim of the present part III of the study was to evaluate the acute behavioral responses of chicken embryos at different developmental stages to a noxious mechanical stimulus. The markerless pose estimation software DeepLabCut (DLC) and manual observations were used to analyze embryonic behavior [19–21]. In addition, cardiovascular [22] and electrophysiological [23] parameters were investigated in parts I and II of the comprehensive study.

2. Materials and Methods

2.1. Animals and Incubation

Chicken embryos from ED9 to ED18 were analyzed. An overview of the experimental groups is provided in Table 1. Fertilized Lohman Selected Leghorn eggs were obtained from the Technical University of Munich (TUM) Animal Research Centre, Thalhausen. Eggs were disinfected (Röhnfried Desinfektion Pro, Dr. Hesse Tierpharma GmbH & Co., Ltd. KG, Hohenlockstedt, Germany), weighed and stored in a refrigerator at 15 °C until use. The maximum storage time from the day of laying until the start of the incubation was seven days. Before incubation, the eggs were placed at room temperature for 24 h. On the day of incubation, eggs were transferred at 8:30 am into a standard incubator (HEKA Favorit-Olymp 192 Spezial, HEKA-Brutgeräte, Rietberg, Germany) and incubated under the following conditions: 37.8 °C temperature and 55% humidity. The eggs were turned six times a day until fenestration on ED3. The first day of incubation was defined as ED0.

Table 1. Number of chicken embryos. Overview of the number of chicken embryos analyzed on each embryonic day and the sex distribution.

	ED9	ED12	ED13	ED14	ED15	ED16	ED17	ED18	ED18 w/ Lido
Amount of embryos (n)	10	10	10	16	16	16	16	16	5
Sex male/female	5/5	3/7	5/5	9/7	7/9	7/8	7/9	7/9	2/3

On ED3, eggs were placed horizontally for two minutes, and 5–7 mL of albumin was withdrawn through a small hole at the pointed pole using a cannula. A small window was cut in the top of the eggshell, and 0.5 mL of penicillin-streptomycin (10,000 units penicillin, 10 mg streptomycin/mL, P4333–100 mL, Sigma-Aldrich, St. Louis, MO, USA) was added. Eggs were sealed with plastic film and tape. With the eggs in a horizontal position, the incubation proceeded until the desired embryonic day [24].

At the end of the experiments, the embryos were euthanized by an intravenous injection of pentobarbital-sodium (Narcoren, 16 g/100 mL, Boehringer Ingelheim Vetmedica GmbH, Ingelheim am Rhein, Germany; ED9: 0.05 mL, ED12 to ED15: 0.1 mL and ED16 to ED18: 0.2 mL), followed by decapitation. Afterward, the sexes of the ED12 to ED18 embryos were identified macroscopically by the assessment of the gonads. For the ED9 embryos, sexing was performed with PCR of genomic DNA samples isolated from pectoral and wing muscle. Screening was performed according to an established protocol [25] using primers targeting the Z chromosome [5′ AAGCATAGAAACAATGTGGGAC 3′ (forward) and 5′ AACTCTGTCTGGAAGGACTT 3′ (reverse)] and female-specific primers targeting the W chromosome [5′ CTATGCCTACCACMTTCCTATTTGC 3′ (forward) and 5′ AACTCTGTCTGGAAGGACTT 3′ (reverse)]. The expected lengths of the DNA fragments were 250 bp and 375 bp, respectively, for female embryos and 250 bp for male embryos. An overview of the sex ratio in each ED is shown in Table 1.

2.2. Preparation Process

All experiments were performed between 9:00 am and 7:30 pm by the same two persons to standardize the procedure. To keep the environmental conditions as similar as possible to typical brooding conditions, experiments were conducted in a special heated chamber. The chamber was equipped with a heat mat (ThermoLux Wärmeunterlage, Witte + Sutor GmbH, Murrhardt, Germany), a heat lamp (Wärmestrahlgerät, Taschenlampenwerk ARTAS GmbH, Arnstadt, Germany) and an air humidifier (Series 2000 Luftbefeuchter HU4811/10R1, Philips, Amsterdam, The Netherlands). Humidity was kept at a constant level of 55.5% ± 4.5. Additionally, the eggs were embedded in warm (38.0 °C) Armor Beads (Lab Armor Beads™, Sheldon Manufacturing, Cornelius, NC, USA). In this manner, the inner egg temperature was kept at 37.9 °C ± 0.9 during the entire experiment. To observe the entire embryo, the window in the eggshell was enlarged. Next, the chorioallantoic membrane (CAM) was carefully cut open and removed from the field of view. If necessary, blood vessels were ligated to prevent bleeding. However, to the extent possible, ligating or cutting vessels was avoided to prevent disruption of blood circulation. To gain access to the embryo and improve visibility, the amnion was carefully opened. A Desmarres lid retractor (Fuhrmann GmbH, Much, Germany) was carefully placed underneath the beak of the embryo to ensure beak visibility. In the case of ED9 embryos, a small wire loop was used.

2.3. Experimental Setup

All experiments were filmed with a camera (Panasonic LUMIX DC-G110V with a Panasonic Lumix G 30 m lens, Matsushita Electric Industrial Co., Ltd., Osaka, Japan; for ED9 to ED16: HOYA SUPER PRO1 Revo Filter SMC Cir-PL, Kenko Tokina Co., Ltd., Tokyo, Japan) with a frame rate of 50 frames per second.

After preparation, a resting period of three minutes was allotted. Baseline behavior was recorded for two (ED15 to ED18) or three (ED9 to ED14) minutes; subsequently, two stimuli were applied in a randomized order. The stimuli used were a noxious mechanical stimulus (*Pinch*) using a manual instrument and a light touch (*Touch*) as a negative control. Both were applied at the base of the beak. For the ED15 to ED18 embryos, a mosquito clamp (Fine Science Tools, Foster City, CA, USA) was used to administer the stimulus. To better monitor the applied force, a mosquito clamp combined with an analgesia meter (Rodent Pincher Analgesia Meter, Bioseb, Vitrolles, France) was used for experiments conducted with ED12 to ED14 embryos. Stimulus 1 (*Pinch* or *Touch*) was administered, followed by an observation duration of three minutes. After a second baseline period, stimulus 2 (*Touch* or *Pinch*) was administered, followed by another three minutes of observation. Because of their small size, microsurgical anatomical forceps (Fine Science Tools, Foster City, USA) had to be used to administer the stimulus to ED9 embryos. An additional group of ED18 embryos (ED18 w/Lido) was injected with 0.02 mL of lidocaine (Xylocitin® 2%, Mibe GmbH Arzneimittel, Brehna, Germany) in the upper and lower beak region five minutes before the first baseline. Experiments were then performed according to the above protocol.

2.4. Analyses: Hardware, Software and Statistical Analyses

All videos were edited in the same way using the "daVinci Resolve" software (Blackmagic Design Pty. Ltd., Port Melbourne, Australia) before analysis. For each embryo, four single videos were cut referring to the sections of the experimental design: *Baseline Pinch*, *Baseline Touch*, *Post Pinch* and *Post Touch*. An overview of the experimental procedure is shown in Figure 1.

2.5. DeepLabCut

To track the body parts of the embryo, the Python-based markerless pose estimation software DLC (version 2.2.1.1) [19,21] was used on a computer (MSI MAG Infinite 11TC-1222AT, Intel Core i7–11700F, 16 GB RAM, nVidia GeForce RTX3060, Micro-Star Int'l Co., Ltd., New Taipei City, Taiwan). The neural network was trained for each ED individually with video footage according to the protocol provided by the developers [21]. Manual labeling was always performed by the same person. The training was performed with the default settings and using a ResNet-50-based neural network [26,27]. A test error below 8.5 was obtained for every ED. After the model training was completed, the four experimental videos (*Baseline Pinch, Baseline Touch, Post Pinch* and *Post Touch*) were analyzed for each embryo. For each labeled body part, DLC created three outputs for each frame of the video: an x coordinate, a y coordinate and a likelihood value. These values were analyzed with custom-written code using MATLAB (MATLAB Version: 9.12.0.1927505 (R2022a) Update 1, MathWorks). In all cases, a likelihood value cutoff of 0.75 was used.

2.5.1. Visualization of the Data Clusters

In the analysis, the focus was on the following body parts:

- Beak;
- Head;
- Limbs;
- Stationary points on the egg, the Desmarres lid retractor, and the wire loop (for ED9) were used as reference controls.

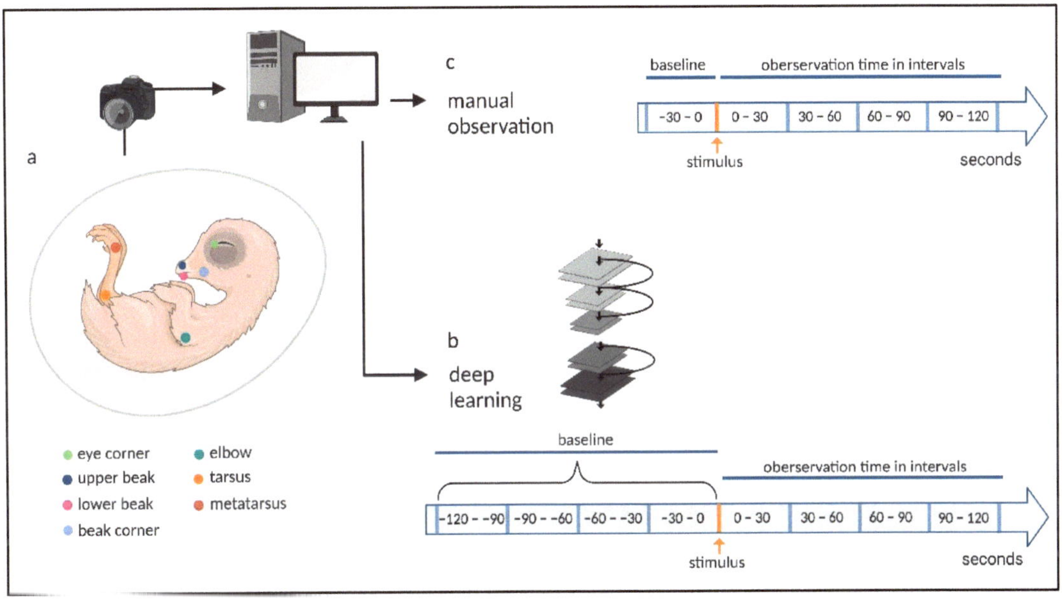

Figure 1. Flowchart of the experimental procedures. (**a**) Recordings of the embryo were collected in ovo, and video data were transferred to a computer for editing. The body parts of chicken embryos tracked by DLC are labeled in the schema. (**b**) The neural network was trained, and the video material was analyzed according to the timeline. (**c**) The video material was manually analyzed according to the timeline. (Created with BioRender.com, accessed on 6 September 2023).

As a first step, the labeled data clusters for each analyzed body part were visualized in the x–y coordinate space. This enabled refinement of the dataset through identification of outliers or mislabeled body parts. The videos were then checked for errors, and if any real outlier was found in a frame, its value was manually excluded.

Distance between the Upper and Lower Beak

The distance between the upper and lower boundaries of the beak was calculated in terms of the Euclidian distance d between two points:

$$d = \sqrt{\left[(x_u - x_l)^2 + (y_u - y_l)^2\right]}$$

where x_u is the x coordinate of the upper beak label, x_l is the x coordinate of the lower beak label, y_u is the y coordinate of the upper beak label and y_l is the y coordinate of the lower beak label. The Euclidian distance was calculated (in pixels) for every frame of the video.

Angle between the Upper and Lower Beak

The angle between the upper and lower beak was computed by calculating the angle α between two lines, P_0 to P_1 and P_0 to P_2, where P_0 is the fulcrum between the beak parts, P_1 is the upper beak point and P_2 is the lower beak point. The angle was then calculated as follows:

$$\alpha = atan2(norm(det([n_2; n_1])), dot(n_2, n_1))$$

where $atan2$ is the four-quadrant inverse tangent, det is the matrix determinant, dot is the dot product, and n_2, n_1 are the Euclidean normalized vectors for P_0 to P_1 and P_0 to P_2, respectively. The angle between the upper and lower beak was calculated for all frames of the video in radians and then converted to degrees.

Movement

The movement of the body parts of interest was calculated in terms of the Euclidean distance *d* between identical labels across consecutive frames:

$$d = \sqrt{\left[\left(x_{f1} - x_{f2}\right)^2 + \left(y_{f1} - y_{f2}\right)^2\right]}$$

where x_{f1} is the x coordinate in frame 1, x_{f2} is the x coordinate in frame 2, y_{f1} is the y coordinate in frame 1 and y_{f2} is the y coordinate in frame 2. The distances were calculated for all consecutive frames. From ED12 to ED18, movements of the medial eye corner, elbow and metatarsus were analyzed. For the body movements on ED9, the tarsus (instead of the metatarsus) was used to assess leg movement, as the tissue of the metatarsus was translucent and prone to errors in tracking.

2.5.2. Analysis

To simplify the analyses, 30 s intervals were evaluated. For each parameter, i.e., *Beak Distance, Beak Angle, Movement Eye Corner, Movement Elbow* and *Movement Metatarsus*, the sum of the 1500 frame values of the interval was calculated. In Post Stimulus, this resulted in four intervals: 0–30, 30–60, 60–90 and 90–120 s. The beginning of the first poststimulus interval was defined as the moment from which the clamp was no longer in contact with the beak. The median of the four 30-s intervals prior to the stimulus was considered the baseline. Missing values, which arose after the exclusion of low likelihood values, were manually imputed. For each missing value series, the median was determined for half of the adjacent data and used in place of the missing value. If more than 5% of the data in an interval were missing, the interval was excluded from the analysis. Due to a lack of visibility, one ED14 embryo and one ED18 embryo were completely excluded from the DLC analysis. A precise overview of the number of datasets ultimately included in the analysis is provided in Table S1.

Due to the presence of repeated measures, generalized linear mixed effects models with the individual embryo as a random effect were chosen for analysis. Due to the violation of numerous model assumptions (normality of residual distribution, heteroscedasticity of residuals, heterogeneity of variances between groups and presence of outliers), only robust linear mixed-effects models were applied for all analyses (R package-robustlmm). All contrasts (differences) between particular groups were assessed after model-fitting by the estimated marginal means (R package–emmeans) with Tukey's *p* value correction for multiple comparisons. The results with a *p* value < 0.05 were considered statistically significant. Data analysis was performed using R 4.2.1 (23 June 2022). Detailed information about the statistical analyses, including the corresponding effect size for each reported *p* value, can be found in Table S2.

2.6. Manual Observation

The same video footage as used in the DLC analyses was used for manual observations. Since preliminary observations and data from the DLC analyses indicated that changes in beak position were frequent after *Pinch*, manual observations focused on beak movements. Four different patterns of beak movements were identified from the video material:

- *Beak Shift*—a small horizontal shift of the upper and lower beaks against each other;
- *Mandibulation*—a small vertical opening of the beak, often executed several times, and reminiscent of a chewing movement;
- *Beak Opening*—single, swift, vertical opening of the beak;
- *Wide Beak Opening*—single, wide, vertical opening of the beak, accompanied by a characteristic tongue movement.

In an analogous approach to the one described above, the baseline and poststimulus observations were divided into intervals of 30 s. For manual observations, the 30 s before the stimulus were used as a baseline. For each interval, the occurrences of the described beak movements were counted.

3. Results

3.1. Beak Movements in Response to a Noxious Stimulus

To analyze the movements of chicken embryos, the markerless pose estimation software DLC was used. The angle (*Beak Angle*) and distance (*Beak Distance*) between the upper and lower beak were calculated to reflect the opening of the beak as a potential response to a noxious mechanical stimulus applied at the base of the beak. The mechanical stimulation of the beak led to a change in the beak position at embryonic day (ED) 9 and ED12; thus, evaluation with DLC was distorted and could not be interpreted. At ED13 and ED14, *Beak Distance* did not differ between any time intervals during the two minutes after the control touch stimulus (hereafter, *Post Touch*) and the time intervals during the two minutes after the noxious pinch stimulus (hereafter, *Post Pinch*) (Figure S1). At ED15, significant increases in *Beak Distance* as a response to *Pinch* were detected (Figure 2). Additionally, in ED15 embryos, beak movements *Post Pinch* increased significantly over the first 120 s compared to *Baseline Pinch* and over the first 90 s compared to *Post Touch*. On ED16, ED17 and ED18, a significant increase in *Beak Distance* was observed over all time intervals *Post Pinch* compared to *Baseline Pinch* and *Post Touch*. The greatest increase in *Beak Distance* occurred during the first 30 s of *Post Pinch*. The group of ED18 embryos that received an injection of the local anesthetic lidocaine (ED18 w/Lido) did not exhibit reduced beak movements compared to same-age embryos that did not receive the local anesthetic (Figure S2). *Beak Distance* was still significantly increased in ED18 w/Lido in the first 30 s of *Post Pinch* ($p < 0.0001$).

Beak Angle results are displayed in the Supplementary Information (Figure S3). Briefly, *Beak Angle* showed a similar pattern of changes as *Beak Distance*. Additionally, significant increases in *Beak Angle* during *Post Pinch* were observed from ED15 onward.

3.2. Head Movements in Response to a Noxious Stimulus

The medial eye corner was tracked to analyze the head movements of chicken embryos. Changes were particularly observed on ED13 and ED16 to ED18 in the first 30 s of *Post Pinch*. On these days, the embryos showed a significant increase in head movements after *Pinch* compared to after *Touch* (ED13: $p = 0.0254$; ED16: $p = 0.0381$; ED17: $p = 0.026$; ED18: $p < 0.0001$) and during *Baseline Pinch* (ED13: $p = 0.0256$; ED16: $p = 0.0001$; ED17: $p < 0.0001$; ED18: $p < 0.0001$). At ED12, head movements increased significantly at 30–60 s after *Pinch* compared to those 30–60 s after *Touch* ($p = 0.0372$). At ED14, head movements also increased significantly in the first 30 s after the stimulus compared to those in the corresponding baseline period. These movements were observed after both stimuli (*Pinch*: $p = 0.0153$; *Touch*: $p = 0.0069$). In addition, a significant difference between head movements in response to *Pinch* and those in response to *Touch* was observed at 30–60 s after the stimulus ($p = 0.0175$). Head movements were significantly reduced in ED18 w/Lido embryos in the first 30 s of *Post Pinch* compared to those of ED18 embryos in the same period ($p < 0.0001$). Head movements on ED15 to ED18 are displayed in Figure 3, while data on ED9, ED12 to ED14 and ED18 w/Lido embryos are provided in the Supplementary Information (Figures S4 and S5).

3.3. Limb Movements in Response to a Noxious Stimulus

To track limb movements, the movements of the *Elbow*, *Metatarsus* and *Tarsus* (ED9) were analyzed. Significant differences in limb movements between *Baseline Pinch* and *Post Pinch* and between *Post Pinch* and *Post Touch* were observed only on ED18 (Figures S6 and S7). An increase in elbow movements was observed between *Baseline Pinch* and *Post Pinch* ($p = 0.0023$) as well as between *Post Pinch* and *Post Touch* ($p = 0.0096$) during the first 30 s after the stimulus. Regarding metatarsus movements, ED18 embryos showed a significant increase between *Baseline Pinch* and *Post Pinch* ($p < 0.0001$) as well as between *Post Pinch* and

Post Touch (p = 0.0002) during the first 30 s after the stimulus. For ED18 w/Lido embryos, no significant differences in limb movements were observed between *Baseline* and the first 30 s of *Post Stimulus*. There was also no significant difference between the ED18 embryos and the ED18 w/Lido embryos. Other significant changes in limb movements were observed at specific time intervals during development.

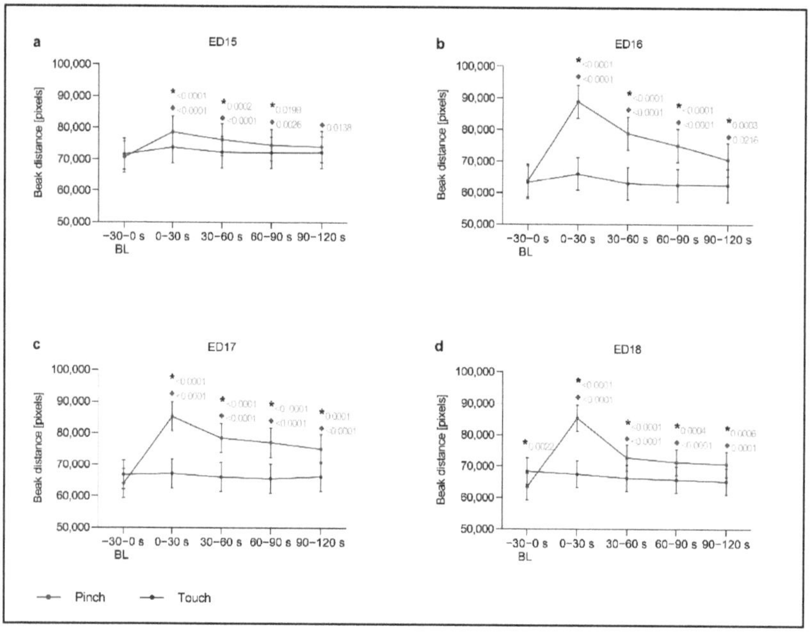

Figure 2. Beak Distance. This variable was defined as the distance between the upper and lower beak of embryos. It was measured at (**a**) ED15 (n = 16), (**b**) ED16 (n = 16), (**c**) ED17 (n = 16) and (**d**) ED18 (n = 15) before and after application of a control (*Touch*) or noxious stimulus (*Pinch*). The total distance in pixels across 30 s intervals (1500 frames) was evaluated. Plots show the estimated mean ±95% confidence intervals at the following 30 s intervals from Baseline (BL) to post-stimulation, with stimulation occurring at 0 s: −30–0, 0–30, 30–60, 60–90 and 90–120 s. Robust linear mixed effects were applied for all analyses. All contrasts (differences) between particular groups were assessed after model-fitting by the estimated marginal means with Tukey's p value correction for multiple comparisons. *Touch*: blue; *Pinch*: red. * Significant difference between *Pinch* and *Touch*; ♦ Significant difference from baseline. p values shown.

3.4. Characterization of Beak Movements in Response to a Noxious Stimulus

In particular, DLC analysis identified changes in beak movement during *Post Pinch* in embryos from ED15 to ED18. To characterize beak movements in further detail, manual observations were performed. The focus of the manual observations was on four behaviors: *Beak Shift*, *Mandibulation*, *Beak Opening* and *Wide Beak Opening*. An overview of the percentage of animals that exhibited each behavior at specific time intervals is shown in Table 2. In addition, the counts of each behavior are shown in Figures S8–S11.

Beak Opening was rarely displayed during *Baseline* and was observed in only 10.0% of animals from ED9 to ED18. *Beak Opening* was particularly rare on ED9 and ED12 to ED14. Before ED12, a maximum of 10.0% of animals exhibited this behavior within a single time interval; up to ED14, a maximum of 20.0% of animals exhibited this behavior within a single time interval. Starting from ED15, an increasing frequency (31.3%) of *Beak Opening* was observed after the application of the noxious stimulus. At ED16, 87.5% of embryos showed *Beak Opening* in the first 30 s of *Post Pinch*. Additionally, 50.0% of ED17 embryos and 62.5% of ED18 embryos showed this behavioral response to *Pinch*. During these days, at least twice as many embryos showed *Beak Opening* during *Post Pinch* as those during *Post Touch*.

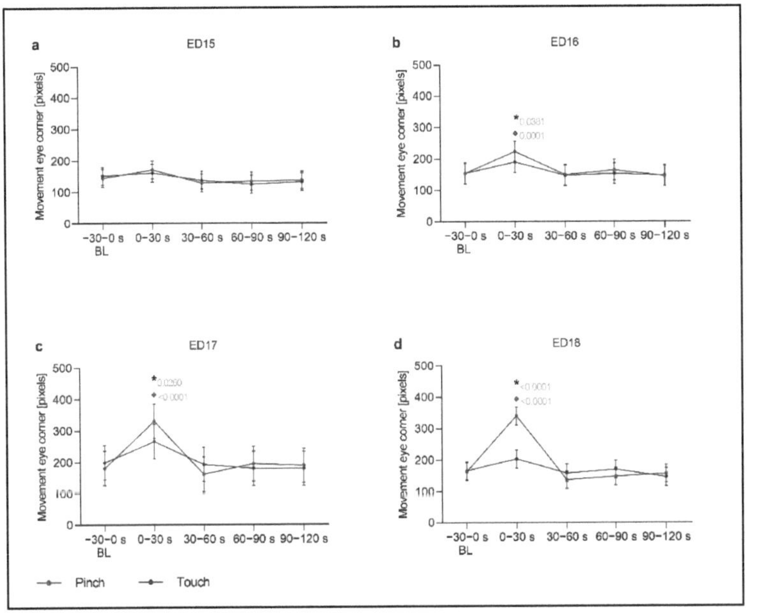

Figure 3. Eye Corner Movement. This variable was used to detect head movements of embryos at (**a**) ED15 ($n = 16$), (**b**) ED16 ($n = 16$), (**c**) ED17 ($n = 16$) and (**d**) ED18 ($n = 15$) before and after application of two stimuli (*Touch* and *Pinch*). The total distance in pixels across 30 s intervals (1500 frames) was evaluated. Plots show the estimated mean ±95% confidence intervals at the following 30 s intervals from Baseline (BL) to post-stimulation, with stimulation occurring at 0 s: −30–0, 0–30, 30–60, 60–90 and 90–120 s. Robust linear mixed effects were applied for all analyses. All contrasts (differences) between particular groups were assessed after model-fitting by the estimated marginal means with Tukey's p value correction for multiple comparisons. *Touch*: blue; *Pinch*: red. * Significant difference between *Pinch* and *Touch*; ♦ Significant difference from baseline. p values shown.

Table 2. Percentage of chicken embryos showing beak movements. Overview of the percentage of chicken embryos that showed beak movements (*Beak Shift*, *Mandibulation*, *Beak Opening* or *Wide Beak Opening*) during the 30 s before (*Baseline*) and 30 s after (*Post*) the stimulus.

		ED9 $n = 10$		ED12 $n = 10$		ED13 $n = 10$		ED14 $n = 16$		ED15 $n = 16$		ED16 $n = 16$		ED17 $n = 16$		ED18 $n = 16$		ED18 w/Lido $n = 5$	
Amount of embryos [%]		*Touch*	*Pinch*	*Touch*	*Pinch*	*Touch*	*Pinch*	*Touch*	*Pinch*	*Touch*	*Pinch*	*Touch*	*Pinch*	*Touch*	*Pinch*	*Touch*	*Pinch*	*Touch*	*Pinch*
Beak Shift	*Baseline*	0.0	0.0	0.0	0.0	10.0	30.0	18.8	25.0	31.3	0.0	18.8	25.0	25.0	31.3	25.0	6.3	40.0	40.0
	Post	0.0	0.0	30.0	30.0	20.0	20.0	18.8	31.3	31.3	25.0	18.8	18.8	25.0	18.8	31.3	6.3	20.0	60.0
Mandib-ulation	*Baseline*	20.0	30.0	40.0	30.0	10.0	50.0	12.5	12.5	37.5	12.5	62.5	25.0	43.8	37.5	31.3	37.5	80.0	80.0
	Post	30.0	20.0	50.0	60.0	50.0	40.0	56.3	56.3	62.5	81.3	68.8	93.8	62.5	87.5	68.8	87.5	80.0	60.0
Beak Opening	*Baseline*	10.0	0.0	10.0	10.0	10.0	10.0	6.3	0.0	0.0	6.3	0.0	6.3	6.3	6.3	0.0	0.0	0.0	20.0
	Post	0.0	0.0	00.0	10.0	10.0	20.0	0.0	12.5	12.5	31.3	31.3	87.5	18.8	50.0	18.8	62.5	0.0	20.0
Wide Beak Opening	*Baseline*	0.0	0.0	0.0	0.0	10.0	0.0	0.0	6.3	0.0	0.0	0.0	0.0	0.0	0.0	0.0	0.0	20.0	0.0
	Post	0.0	0.0	0.0	0.0	0.0	0.0	0.0	6.3	12.5	18.8	6.3	25.0	0.0	81.3	0.0	87.5	0.0	40.0

Wide Beak Opening, characterized by visible tongue movement, was observed only sporadically during *Baseline* on all developmental days. This behavior was observed in only one animal each on ED13, ED14 and in ED18 w/Lido embryos during baseline. Moreover, this specific beak movement was not observed during *Post Pinch* and *Post Touch* for ED9 to ED13 embryos and was observed only once during *Post Pinch* on ED14. On ED15 and ED16, this behavior was increasingly observed. A total of 18.8% (ED15) and 25.0% (ED16) of embryos exhibited *Wide Beak Opening* in the first 30 s of *Post Pinch*. A total of 81.3% and 87.5% of embryos on ED17 and ED18, respectively, showed more *Wide Beak Opening* in the first 30 s of *Post Pinch*. However, this behavior was never observed during *Post Touch* or corresponding baseline periods at these ages.

Beak Shift was observed from ED12 onward, but it did not appear to be associated with *Pinch*. *Mandibulation* was also observed across all embryonic days. Changes were observed in *Mandibulation* at all times in *Post Pinch* and *Post Touch* and regularly during both baseline periods.

Since *Beak Opening* and *Wide Beak Opening* were the most noticeable *Post Pinch* responses, the focus of comparisons with the additional control group that received local anesthetic (ED18 w/Lido) was on these two movements, as the application of lidocaine reduced these behaviors. In the ED18 w/Lido group, 40.0% of the embryos reacted with *Wide Beak Opening* to the noxious mechanical stimulus; in the ED18 embryos without a lidocaine injection, 87.5% exhibited this behavior. *Beak Opening* was observed in 20.0% of the ED18 w/Lido animals and 62.5% of the untreated ED18 embryos. Neither *Mandibulation* nor *Beak Shift* appeared to be associated with a specific reaction in any time interval, similar to embryos without lidocaine treatment. In other words, no noticeable increase or decrease in these behaviors was observed after a stimulus.

4. Discussion

In this exploratory study, we investigated the movements of chicken embryos in response to a noxious stimulus at different developmental stages. We used DeepLabCut, a Python-based markerless pose estimation software, as well as manual observations to determine their responses.

Recently, the use of artificial intelligence and deep learning systems in behavioral studies has increased, and the availability of free software such as DLC allows such techniques to be used by researchers with less sophisticated programming experience [28–30]. In our study, we trained a model to provide satisfactory accuracy of tracking individual body parts on each embryonic day. One of the major advantages of using the markerless pose estimation software DLC is that it enables unbiased analysis. Calculations of distances are not based on subjective perception by an observer and are therefore quantifiable and reliable. Therefore, deep learning systems in general and DLC in particular offer a means of detecting and classifying behaviors that may not be detectible to the naked eye. However, the DLC analysis did not allow us to distinguish between types of beak movements. Thus, for better differentiation of beak movements, we added manual observation of these movements and identified four different patterns.

Pain behavior in general is influenced by a variety of factors specific to the stimulus or the affected animal. For example, noxious agents can differ in duration (acute or chronic), source (somatic or visceral) and severity (mild to severe), each of which may provoke a different reaction [9,10,31]. Since behavioral responses vary extensively depending on the species and stimulus, any description is valid only for the specifically described case and cannot be transferred to another species without re-evaluation [10]. In our study, we applied an acute mechanical stimulus to the beak base of chicken embryos. The beak of chickens is known to be equipped with nociceptors [32] and therefore represents a pain-sensitive area [11]. The beak has also been reported as the region in chicken embryos where the earliest response to stimuli is observed [33]. Chumak observed reflex movements in the form of flexions of the head on day 7 of incubation in response to pinpricks in the beak region, describing reflexes provoked by external stimuli (isolated movements of the head or wing) and spontaneous voluntary movements (involving generalized head, trunk and limb movements) [33].

Nociceptive reflexes have evolved as protective mechanisms [34]. A noxious stimulus is transmitted via peripheral nociceptors to the spinal cord and transmitted to motor neurons, resulting in muscle contraction and thus the nociceptive reflex [34–36]. Chumak reported more specific responses, including increased defensive movements, in chicken embryos at ED14/15 but characterized these responses as reflexive [33]. Hamburger and Oppenheim reported that coordinated movements appear around ED17 [14]. Since our study was based solely on observations of movements by chicken embryos, a conclusion regarding whether the observed movements are reflexes or coordinated movements cannot be drawn.

We analyzed the movements of chicken embryos in response to a noxious stimulus applied to the beak from ED9 to ED18. Consistent with the assumption that a response to a stimulus is expected at the site of stimulus application, as was shown for well-innervated regions such as the beak [10], our DLC data for *Beak Angle* and *Beak Distance* showed the most noticeable changes after the stimulus. Both parameters, *Beak Angle* and *Beak Distance*, quantified beak movements. A significant increase in beak movements was detected immediately after *Pinch* from ED15 to ED18. As the increase in beak movements during *Post Pinch* was significant compared to those during *Baseline Pinch* and *Post Touch*, we assumed that the increase in beak movements was a reaction to the noxious stimulus and was not a random movement of the chicken embryos.

Further differentiation of the movements through manual observation revealed that *Beak Opening* (starting on ED16) and *Wide Beak Opening* (starting on ED17) were recurring movements in response to the noxious stimulus. Individual, slow beak openings have been described in connection with the penetration of the air sac membrane shortly before hatching, at the end of day 18 [14]. This description, however, does not match the rapid and clustered movements that we observed following the stimulus. Since these beak openings do not appear to be part of the typical behavior of chick embryos and markedly occurred only after a noxious stimulus, they may represent a nocifensive response by the embryo. Whether this can be interpreted as the presence of pain sensation remains unclear because an experience of pain presupposes consciousness [37], and no indications can be made about this in the context of this part of the study.

Hamburger and Oppenheim also described a behavior that they called beak clapping, which involves rapid opening and closing of the beak in sequences that occurred at irregular intervals [14]. The description and random occurrence of this behavior matches *Mandibulation* in our study. Likewise, the movement was randomly observed across time intervals and had no clear connection to any of the stimuli. However, a similar behavior was observed in adult chickens as a response to low atmospheric pressure stunning before slaughter [38]. In this case, the mandibulation was discussed as a possible sign of reduced welfare or a physiological reaction to hypoxia [38]. As in the other studies, the embryos in our study underwent stress from the opening of the egg, preparation and stimuli. Therefore, it is possible that *Mandibulation* is also a sign of stress in chicken embryos.

Application of the local anesthetic lidocaine did not yield a significant reduction in the beak movements of chicken embryos on ED18 according to the DLC analysis. However, in the manual observations, the application of lidocaine reduced the percentage of embryos that responded to stimuli with *Wide Beak Opening* and *Beak Opening* by approximately half. Furthermore, local anesthetics are known to be effective in birds [39–41] and can be used in chickens, e.g., for spinal anesthesia [42] or brachial plexus blockade [43]. However, there are no reliable empirical data regarding the mode of action of local anesthetics in chicken embryos. Additionally, we emphasize that only a small number of embryos were examined; thus, the results must be interpreted with caution. The inability of local anesthesia to reduce beak movements could also stem from the injection of lidocaine, which itself constitutes a noxious stimulus. In addition, numbness in the beak due to local anesthesia could have led to behavioral changes [44]. This is supported by the fact that head movements were significantly reduced by applying lidocaine to ED18 embryos.

Overall, stress could not be completely eliminated within the experimental setup; thus, its potential influence on behavior must be considered. The fenestrated egg does not represent a completely typical environment for the embryo because of the increased exposure to environmental influences, such as light. Additionally, the invasiveness of the preparation itself can induce stress, which is known to alter the behavior of birds [9]. We attempted to reduce external influences by standardizing the temperature and humidity during the experiments and adjusting them to match the typical incubation conditions as closely as possible. However, since direct access to the embryo was necessary for stimulation and the embryo had to be visible to assess responses, some stressors were unavoidable.

We were also interested in whether limb movements changed after the noxious stimulus; however, we did not detect any overarching pattern until ED17. Occasional significant differences in limb movements during *Post Pinch* compared to those during *Baseline Pinch* or *Post Touch* were inconsistent over several EDs or time intervals and are therefore likely due to random movements, which have been described previously in the literature [12–14,45–49]. Hamburger and Oppenheim stated that before ED15, the observed leg motility was not connected to any sensory input but appeared randomly due to autonomous cell discharges [15]. Wu et al. counted unilateral and bilateral simultaneous limb movements and found a maximum of movements between ED10 and ED13 for the former and two maxima on ED13 and ED17 for the latter [50]. In the present study, we detected a significant increase in elbow and metatarsal movements during the first 30 s of *Post Pinch* compared to those during the first 30 s of *Baseline Pinch* and *Post Touch* on only ED18, suggesting that these movements may represent an actual response to the noxious stimulus.

5. Conclusions

We observed the movements of chicken embryos from ED9 to ED18 before and after noxious stimulation. During *Post Pinch*, the observed movement changes in ED15 to ED18 embryos were most likely a response to the noxious mechanical stimulus and can therefore be interpreted as nocifensive behavior. The results of our current movement analysis in combination with the corresponding results of the cardiovascular changes [22] and the evaluation of the onset of physiological neuronal signals [23] in chicken embryos during this developmental period provide valuable information that enhances our understanding of the development of nociception and pain perception in chickens.

Supplementary Materials: The following supporting information can be downloaded at: https://www.mdpi.com/article/10.3390/ani13182859/s1, Figure S1: Beak Distance ED13/ED14; Figure S2: Beak Distance Lidocaine; Figure S3: Beak Angle; Figure S4: Movement Eye corner ED9/ED12-14; Figure S5: Movement Eye corner Lidocaine; Figure S6: Movement Metatarsus respectively Tarsus (ED9); Figure S7: Movement Elbow; Figure S8: Beak Shift; Figure S9: Mandibulation; Figure S10: Beak Opening; Figure S11: Wide Beak Opening; Table S1: Overview on the number of datasets from DLC included in the final analysis; Table S2: Statistical metrics. All MATLAB analysis code used in this study is available in a public GitHub repository: https://github.com/ondracej/dlcAnalysisEmbryo (accessed on 6 September 2023).

Author Contributions: Conceptualization, C.B., J.W., A.M.S., J.R., T.F., B.S.; methodology, C.B., J.W., A.M.S., J.R., S.C.S., L.W., S.K., M.A., T.F., B.S., H.P.; software, J.M.O.; validation, C.B., J.W., A.M.S.; formal analysis, J.M.O., Y.Z.; investigation, S.C.S., L.W., A.M.S., J.W., J.R.; resources, C.B., B.S.; data curation, S.C.S., J.W., A.M.S., J.M.O.; writing—original draft preparation, S.C.S.; writing—review and editing, S.C.S., J.W., A.M.S., J.R., J.M.O., S.K., M.A., T.F., B.S., C.B.; visualization, S.C.S., J.W., A.M.S.; supervision, C.B., T.F., B.S., H.P.; project administration, C.B.; funding acquisition, C.B., T.F., B.S. All authors have read and agreed to the published version of the manuscript.

Funding: This research was funded by the German Federal Ministry of Food and Agriculture (BMEL, https://www.bmel.de, accessed on 6 September 2023) based on a decision of the Parliament of the Federal Republic of Germany, granted by the Federal Office for Agriculture and Food (BLE, https://www.ble.de, accessed on 6 September 2023, grant number: 2821HS005).

Institutional Review Board Statement: According to Directive 2010/63/EU of the European Parliament and the German Animal Welfare Law, no ethical approval was required for the use of chicken embryos in the experiments. The experiments were performed in an AAALAC-certified animal facility. All experiments and the euthanasia of embryos were conducted in strict accordance with the institutional Guidelines for Care and Use of Laboratory Animals and under general animal welfare principles.

Informed Consent Statement: Not applicable.

Data Availability Statement: The data presented in this study are available on request from the corresponding author.

Acknowledgments: The authors thank the scientific advisory board including Michael Erhard, Wolf Erhardt, Harald Luksch, Heidrun Potschka, Hans Straka and Britta Wirrer. In addition, the authors thank Marie-Louise Schmid, Johannes Fischer and Hicham Sid for their helpful support.

Conflicts of Interest: The authors declare no conflict of interest. The funders had no role in the design of the study; in the collection, analyses or interpretation of data; in the writing of the manuscript or in the decision to publish the results.

References

1. Livingston, A. Pain and analgesia in domestic animals. In *Comparative and Veterinary Pharmacology*, 1st ed.; Cunningham, F., Elliott, J., Lees, P., Eds.; Springer: Berlin/Heidelberg, Germany, 2010; pp. 159–189. [CrossRef]
2. Gentle, M.J. Pain in birds. *Anim. Welf.* **1992**, *1*, 235–247. [CrossRef]
3. Paul-Murphy, J.R.; Hawkins, M. Bird-specific considerations: Recognizing pain behavior in pet birds. In *Handbook of Veterinary Pain Management*, 3rd ed.; Gaynor, J.S., Muir, W.W., Eds.; Elsevier Mosby: St. Louis, MO, USA, 2014; pp. 536–554. [CrossRef]
4. Douglas, J.M.; Guzman, D.S.-M.; Paul-Murphy, J.R. Pain in birds: The anatomical and physiological basis. *Vet. Clin. N. Am. Exot. Anim. Pract.* **2018**, *21*, 17–31. [CrossRef] [PubMed]
5. Raja, S.N.; Carr, D.B.; Cohen, M.; Finnerup, N.B.; Flor, H.; Gibson, S.; Keefe, F.J.; Mogil, J.S.; Ringkamp, M.; Sluka, K.A.; et al. The revised International Association for the Study of Pain definition of pain: Concepts, challenges, and compromises. *Pain* **2020**, *161*, 1976–1982. [CrossRef] [PubMed]
6. Julius, D.; Basbaum, A.I. Molecular mechanisms of nociception. *Nature* **2001**, *413*, 203–210. [CrossRef]
7. Zimmermann, M. Ethical considerations in relation to pain in animal experimentation. *Acta Physiol. Scandinavica. Suppl.* **1986**, *554*, 221–233.
8. Korbel, R.; Lierz, M. Vögel. In *Anästhesie und Analgesie beim Klein- und Heimtier mit Exoten, Labortieren, Vögeln, Reptilien, Amphibien und Fischen*, 2nd ed.; Erhardt, W., Baumgartner, C., Haberstroh, J., Eds.; Schattauer GmbH: Stuttgart, Germany, 2012; pp. 790–834.
9. Mikoni, N.A.; Guzman, D.S.-M.; Paul-Murphy, J. Pain recognition and assessment in birds. *Vet. Clin. N. Am. Exot. Anim. Pract.* **2023**, *26*, 65–81. [CrossRef]
10. Mikoni, N.A.; Guzman, D.S.-M.; Fausak, E.; Paul-Murphy, J. Recognition and assessment of pain-related behaviors in avian species: An integrative review. *J. Avian Med. Surg.* **2022**, *36*, 153–172. [CrossRef]
11. Gentle, M.J. Pain issues in poultry. *Appl. Anim. Behav. Sci.* **2011**, *135*, 252–258. [CrossRef]
12. Hamburger, V.; Balaban, M. Observations and experiments on spontaneous rhythmical behavior in the chick embryo. *Dev. Biol.* **1963**, *6*, 533–545. [CrossRef]
13. Hamburger, V.; Balaban, M.; Oppenheim, R.; Wenger, E. Periodic motility of normal and spinal chick embryos between 8 and 17 days of incubation. *J. Exp. Zool.* **1965**, *159*, 1–13. [CrossRef]
14. Hamburger, V.; Oppenheim, R. Prehatching motility and hatching behavior in the chick. *J. Exp. Zool.* **1967**, *166*, 171–203. [CrossRef] [PubMed]
15. Hamburger, V.; Wenger, E.; Oppenheim, R. Motility in the chick embryo in the absence of sensory input. *J. Exp. Zool.* **1966**, *162*, 133–159. [CrossRef]
16. Peters, J.J.; Vonderahe, A.R.; Powers, T.H. The functional chronology in developing chick nervous system. *J. Exp. Zool.* **1956**, *133*, 505–518. [CrossRef]
17. Eide, A.L.; Glover, J.C. Developmental dynamics of functionally specific primary sensory afferent projections in the chicken embryo. *Anat. Embryol.* **1997**, *195*, 237–250. [CrossRef] [PubMed]
18. Eide, A.L.; Glover, J.C. Development of the longitudinal projection patterns of lumbar primary sensory afferents in the chicken embryo. *J. Comp. Neurol.* **1995**, *353*, 247–259. [CrossRef]

19. Mathis, A.; Mamidanna, P.; Cury, K.M.; Abe, T.; Murthy, V.N.; Mathis, M.W.; Bethge, M. DeepLabCut: Markerless pose estimation of user-defined body parts with deep learning. *Nat. Neurosci.* **2018**, *21*, 1281–1289. [CrossRef]
20. Mathis, A.; Warren, R. On the inference speed and video-compression robustness of DeepLabCut. *bioRxiv* **2018**. [CrossRef]
21. Nath, T.; Mathis, A.; Chen, A.C.; Patel, A.; Bethge, M.; Mathis, M.W. Using DeepLabCut for 3D markerless pose estimation across species and behaviors. *Nat. Protoc.* **2019**, *14*, 2152–2176. [CrossRef]
22. Weiss, L.; Saller, A.M.; Werner, J.; Süß, S.C.; Reiser, J.; Kollmansperger, S.; Anders, M.; Potschka, H.; Fenzl, T.; Schusser, B.; et al. Nociception in Chicken Embryos, Part I: Analysis of Cardiovascular Responses to a Mechanical Noxious Stimulus. *Animals* **2023**, *13*, 2710. [CrossRef]
23. Kollmansperger, S.; Anders, M.; Werner, J.; Saller, A.M.; Weiss, L.; Süß, S.C.; Reiser, J.; Schneider, G.; Schusser, B.; Baumgartner, C.; et al. Nociception in Chicken Embryos, Part II: Embryonal Development of Electroencephalic Neuronal Activity *In Ovo* as a Prerequisite for Nociception. *Animals* **2023**, *13*, 2839. [CrossRef]
24. Spurlin, J., III; Lwigale, P. A technique to increase accessibility to late-stage chick embryos for in ovo manipulations. *Dev. Dyn.* **2013**, *242*, 148–154. [CrossRef] [PubMed]
25. Itoh, Y.; Suzuki, M.; Ogawa, A.; Munechika, I.; Murata, K.; Mizuno, S. Identification of the sex of a wide range of carinatae birds by PCR using primer sets selected from chicken EE0.6 and its related sequences. *J. Hered.* **2001**, *92*, 315–321. [CrossRef] [PubMed]
26. Insafutdinov, E.; Pishchulin, L.; Andres, B.; Andriluka, M.; Schiele, B. DeeperCut: A deeper, stronger, and faster multi-person pose estimation model. In Proceedings of the ECCV (European Conference on Computer Vision), Amsterdam, The Netherlands, 11–14 October 2016. [CrossRef]
27. He, K.; Zhang, X.; Ren, S.; Sun, J. Deep residual learning for image recognition. In Proceedings of the IEEE Conference on Computer Vision and Pattern Recognition, Las Vegas, NV, USA, 27–30 June 2016. [CrossRef]
28. Mathis, M.W.; Mathis, A. Deep learning tools for the measurement of animal behavior in neuroscience. *Curr. Opin. Neurobiol.* **2020**, *60*, 1–11. [CrossRef]
29. von Ziegler, L.; Sturman, O.; Bohacek, J. Big behavior: Challenges and opportunities in a new era of deep behavior profiling. *Neuropsychopharmacology* **2021**, *46*, 33–44. [CrossRef] [PubMed]
30. Hardin, A.; Schlupp, I. Using machine learning and DeepLabCut in animal behavior. *Acta Ethol.* **2022**, *25*, 125–133. [CrossRef]
31. Henke, J.; Tacke, S.; Erhardt, W. Analgesie. In *Anästhesie und Analgesie beim Klein- und Heimtier mit Exoten, Labortieren, Vögeln, Reptilien, Amphibien und Fischen*, 2nd ed.; Erhardt, W., Baumgartner, C., Haberstroh, J., Eds.; Schattauer GmbH: Stuttgart, Germany, 2012; pp. 383–431.
32. Kuenzel, W.J. Neurobiological basis of sensory perception: Welfare implications of beak trimming. *Poult. Sci.* **2007**, *86*, 1273–1282. [CrossRef]
33. Chumak, V.I. Dinamika reflektomykh reaktsii i vklyuchenie retseptornykh apparatov u embriona kuritsy (Dynamics of reflex reactions and initiation of receptor systems in the chick embryo). In *Voprosy Fiziologii i Patologii Tsentral'noi Nervnoi Sistemy Cheloveka i Zhivotnykli v Ontogeneze*; Sbornik: Moscow, Russia, 1961; pp. 63–68.
34. Antognini, J.F.; Carstens, E. In vivo characterization of clinical anaesthesia and its components. *Br. J. Anaesth.* **2002**, *89*, 156–166. [CrossRef]
35. Diener, M.; Gerstberger, R. Zentrales Nervensystem. In *Physiologie der Haustiere*, 5th ed.; von Engelhardt, W., Breves, G., Diener, M., Gäbel, G., Eds.; Enke: Stuttgart, Germany, 2015; pp. 131–145.
36. Willis, W.D.; Westlund, K.N. Neuroanatomy of the pain system and of the pathways that modulate pain. *J. Clin. Neurophysiol.* **1997**, *14*, 2–31. [CrossRef]
37. Diesch, T.J.; Mellor, D.J.; Johnson, C.B.; Lentle, R.G. Responsiveness to painful stimuli in anaesthetised newborn and young animals of varying neurological maturity (wallaby joeys, rat pups and lambs). In Proceedings of the AATEX (World Congress on Alternatives & Animal Use in the Life Sciences), Tokyo, Japan, 21–25 August 2007.
38. Mackie, N.; McKeegan, D.E.F. Behavioural responses of broiler chickens during low atmospheric pressure stunning. *Appl. Anim. Behav. Sci.* **2016**, *174*, 90–98. [CrossRef]
39. Glatz, P.C.; Murphy, L.B.; Preston, A.P. Analgesic therapy of beak-trimmed chickens. *Aust. Vet. J.* **1992**, *69*, 18. [CrossRef]
40. Hocking, P.M.; Gentle, M.J.; Bernard, R.; Dunn, L.N. Evaluation of a protocol for determining the effectiveness of pretreatment with local analgesics for reducing experimentally induced articular pain in domestic fowl. *Res. Vet. Sci.* **1997**, *63*, 263–267. [CrossRef] [PubMed]
41. Paul-Murphy, J.; Ludders, J.W. Avian analgesia. *Vet. Clin. N. Am. Exot. Anim. Pract.* **2001**, *4*, 35–45. [CrossRef] [PubMed]
42. Khamisabadi, A.; Kazemi-Darabadi, S.; Akbari, G. Comparison of anesthetic efficacy of lidocaine and bupivacaine in spinal anesthesia in chickens. *J. Avian Med. Surg.* **2021**, *35*, 60–67. [CrossRef] [PubMed]
43. Figueiredo, J.P.; Cruz, M.L.; Mendes, G.M.; Marucio, R.L.; Riccó, C.H.; Campagnol, D. Assessment of brachial plexus blockade in chickens by an axillary approach. *Vet. Anaesth. Analg.* **2008**, *35*, 511–518. [CrossRef] [PubMed]
44. Binshtok, A.M.; Bean, B.P.; Woolf, C.J. Inhibition of nociceptors by TRPV1-mediated entry of impermeant sodium channel blockers. *Nature* **2007**, *449*, 607–610. [CrossRef]
45. Provine, R.R.; Sharma, S.; Sandel, T.; Hamburger, V. Electrical activity in the spinal cord of the chick embryo, in situ. *Proc. Natl. Acad. Sci. USA* **1970**, *65*, 508–515. [CrossRef]
46. Provine, R.R. Development of between-limb movement synchronization in the chick embryo. *Dev. Psychobiol.* **1980**, *13*, 151–163. [CrossRef]

47. Sharma, S.; Provine, R.R.; Hamburger, V.; Sandel, T. Unit activity in the isolated spinal cord of chick embryo, in situ. *Proc. Natl. Acad. Sci. USA* **1970**, *66*, 40–47. [CrossRef]
48. Bekoff, A. Ontogeny of leg motor output in the chick embryo: A neural analysis. *Brain Res.* **1976**, *106*, 271–291. [CrossRef]
49. Bradley, N.S.; Solanki, D.; Zhao, D. Limb movements during embryonic development in the chick: Evidence for a continuum in limb motor control antecedent to locomotion. *J. Neurophysiol.* **2005**, *94*, 4401–4411. [CrossRef]
50. Wu, K.-C.; Streicher, J.; Lee, M.; Hall, B.; Müller, G. Role of motility in embryonic development I: Embryo movements and amnion contractions in the chick and the influence of illumination. *J. Exp. Zool.* **2001**, *291*, 186–194. [CrossRef] [PubMed]

animals

MDPI

Review

Pain Relief Interventions in Australian Livestock Husbandry: A Review of Animal Welfare and Pain Duration

Lee Metcalf [1,*], Sabrina Lomax [2], Dominique Van der Saag [1], Sanjay Garg [3] and Peter J. White [1]

1 School of Veterinary Science, University of Sydney, Camperdown, NSW 2006, Australia; dominique.van.der.saag@sydney.edu.au (D.V.d.S.); p.white@sydney.edu.au (P.J.W.)
2 School of Life and Environmental Sciences, University of Sydney, Camperdown, NSW 2006, Australia; sabrina.lomax@sydney.edu.au
3 UniSA Clinical & Health Sciences, University of South Australia, Adelaide, SA 5001, Australia; sanjay.garg@unisa.edu.au
* Correspondence: lee.metcalf@sydney.edu.au

Simple Summary: It is well established that animals feel pain akin to humans, although the expression of that pain is not as easy to perceive, especially considering that many species actively conceal or disguise pain, distress, or weakness. Current methods of husbandry practices used to improve welfare or production cause inherently painful tissue damage. Current interventions focus on immediate pain relief, but research indicates persistent pain behaviours post procedure, with pain experienced after routine husbandry procedures such as castration, tail docking, dehorning, and mulesing reported as lasting for days and sometimes weeks after the operation, affecting the animal's welfare and production performance. As livestock handlers, animal owners and veterinarians become better at recognising situations where pain and distress are experienced, efforts are increasing to improve pain mitigation methods. The challenges of avoiding multiple handling of livestock, or relying on owner compliance, may be found in developing long-acting pain relief solutions.

Abstract: In veterinary medicine and livestock production, ensuring good animal husbandry is vital for the physical and emotional wellbeing of animals under our care. Pain poses challenges for assessment and mitigation, especially in species unable to express pain overtly. This review examines current pain mitigation interventions in routine husbandry, focuses on the duration of pain after procedures and implications for animal welfare. Pain behaviours have been observed for days or weeks after regular husbandry procedures, and many studies have noted pain-related behaviour persisting until study finalisation, suggesting potential undocumented pain beyond study completion. Current products registered in Australia for pain mitigation in livestock primarily target immediate pain associated with procedures. The future of pain relief in livestock demands longer-acting solutions to address post-procedural pain adequately. Providing pain relief for at least 72 h post surgery is recommended, but current products require retreatment intervals to achieve this, posing practical challenges, especially in livestock. Methods of pain relief provision, such as voluntary consumption of medicated feed, transdermal medication delivery and long-acting formulations offer potential solutions for prolonged pain relief, with research ongoing in these areas. There is a need for further research and development of longer-acting pain relief to ensure optimal welfare of livestock.

Keywords: pain; sheep; cattle; livestock; analgesia; production animals

1. Introduction

Within veterinary medicine and livestock production, it is recognised that good animal husbandry is necessary to ensure the physical and emotional well-being of livestock and companion animals. The concept of "a life worth living" [1] places the responsibility on animal owners and veterinarians to ensure that there is a balance between positive and negative experiences in an animal's lifetime; that "suffering is somehow compensated for by pleasant experiences". There are some invasive husbandry procedures performed on livestock that are painful but considered necessary, either to ensure ongoing welfare for the animal or to facilitate efficient and safe management. These procedures result in physical injury to tissue, and in Australian livestock are often performed without any pre- or post-procedural anaesthesia or analgesia, although this is changing due to some Australian States having legislative requirements for pain relief for certain procedures or at certain ages [2,3]. Even in companion animals, owners are often provided with the choice as to whether post-surgical pain relief is provided, and the decision is frequently driven by cost or owner perception of animal pain [4].

Pain is an "aversive sensation and feeling associated with actual or potential tissue damage, or described in terms of such damage" [5]. There are different types of pain, which may be classified by the duration, the part of the body in which it is located, symptoms, syndromes, or mechanisms [6]. The importance and challenges of pain assessment in animals have been covered extensively in previous research [7–11]. Understanding the pathways and timing of the pain experience is necessary to determine the best method of pain mitigation in any species, whether or not they can express that pain.

It is interesting to note that in the early days of veterinary anaesthetics, anaesthesia was originally used for restraint rather than pain relief. The drugs initially used would induce paralysis but not necessarily provide any pain relief either during or after a procedure [12]. Some veterinarians and producers misunderstood that general anaesthesia or heavy sedation which chemically restrained the animal was not synonymous with analgesia and assumed that a lack of reaction to pain during and after surgery was due to an analgesic effect of the anaesthesia [8,13].

Practices such as branding and ear tagging/notching of livestock for identification purposes; dehorning and disbudding cattle and goats to prevent injury to other members of the herd, or handlers; castration and spaying of livestock and companion animals to prevent unwanted pregnancies and aggressive mating behaviour; and tail docking and mulesing of sheep to prevent fly strike, are all performed in Australia to improve management and production, and to ensure the holistic welfare of the animal. As an awareness of animal welfare expands, common practices are being assessed for necessity and alternatives, as well as the need for pain relief. Currently, less painful alternatives being investigated are either not viable and/or effective, or their integration may take several years or generations of breeding. It is, therefore, incumbent upon animal owners, producers, and veterinarians to ensure that animal welfare is maintained through the delivery of appropriate pain mitigation.

For production animals, consumers are increasingly demanding products such as wool, meat, milk and eggs that have been produced under proven welfare standards, including adequate pain relief [14]. There are still barriers towards the provision of pain mitigation on-farm, including cost, recognition of pain by producers, withdrawal periods, and entrenched generational farming practices [15]. Recent surveys of sheep and cattle producers have shown that only one-quarter of those surveyed are providing pain relief for routine procedures, with the most common objections from those producers not using pain relief being the time it takes, or that they do not believe it is necessary [16,17].

Over the last 27 years, the Five Domains Model for animal welfare assessment has been developed and updated to provide a way to evaluate the welfare of individuals or groups of animals [18] with particular emphasis on well-being and positive experience. The Five Domains indicate that the welfare of animals is associated with both mental and physical aspects and infers that animals should be provided with adequate nutrition, environment, the ability to behave naturally and receive adequate healthcare whilst ensuring that the animal's mental state is also protected. Good animal husbandry is necessary to ensure the health and wellbeing of livestock and the reality exists that some invasive husbandry procedures conducted for welfare or management purposes are painful. Studies in livestock have shown that the response of animals to pain is influenced by several different parameters, such as sex, age, body weight, prior experience and familiarity with the environment [19]. The emotional distress experienced during aversive procedures has also been demonstrated in livestock, with studies in calves showing an aversion to the location of a painful experience such as hot-iron disbudding [20] or preference for analgesia [21], and sheep displaying handler aversion for several months after a mulesing operation [22]. An important component of animal welfare, therefore, is appropriate pain relief before, during and after painful procedures, to ensure that the human-animal interaction is as stress-free as possible and that the restriction of behavioural interactions and negative experiences are minimized [23–25].

There has been a historical assumption [13,15] that neonates have less developed pain perception than older animals, and therefore procedures should be conducted as soon as possible after birth—this has even been the case with human neonates, even though the physiological markers of pain in humans are measurable from 26 weeks gestation [26]. Studies in lambs have shown that reaction to pain changes over time from birth, with one study showing an increase in electroencephalographic (EEG) response to castration as lambs increased in age from 1 day to 6 weeks [27]. A study of EEG responses of piglets that were tail-docked at either 2 or 20 days of age showed that the procedure appeared less acutely painful when performed soon after birth rather than at 20 days of age [28]. In contrast, a study of lambs [29] found that those animals castrated within a day of birth when compared with those castrated at 10 days of age, showed a higher pain response when tail docked at 3–5 weeks of age, leading to the conclusion that a "noxious stimulus" early in life (such as the pain associated with castration) can cause increased pain sensitivity later in life. This concept was further demonstrated in dairy calves disbudded at 3 days of age vs. 35 days of age [30]. The experience of pain has even been shown to be intergenerational—a study in sheep [31] found that ewes experiencing pain from tail docking or a simulated lipopolysaccharide (LPS) infection (*E.coli* LPS challenge) at 3–4 days of life showed higher levels of pain-related behaviour as adults when lambing. In addition, the LPS-treated ewes gave birth to lambs who also displayed a lower tolerance to pain at 2–3 days of age. There may therefore be a long-term and even trans-generational effect of pain experienced in neonates which would be worth further examination.

There is a lack of clarity and consistency for those in the industry when considering the legislative requirements for pain relief for livestock throughout the different states in Australia. The legal requirement for pain relief in mulesing of sheep, for example, ranges from support for the voluntary adoption of pain relief by the NSW government, with some technical assistance to find alternatives to mulesing [32], to the Victorian government's Prevention of Cruelty to Animals Regulations 2019 making the performance of mulesing without pain relief an offence [2]. Some industry bodies (for example, some wool buying groups) have taken a lead with regards to animal welfare, with certain requirements of their producers [33,34]; however, as up to 75% of the country's cattle and sheep farmers are not voluntarily providing pain relief for other routine procedures [16,17], it seems that until there is clear nationwide legislation with specific requirements, pain relief in the production space will remain inconsistent and often inadequate.

This review examines the current pain mitigation interventions in routine husbandry practice; of particular concern for animal welfare is the duration of pain following routine procedures, which highlights the need for more effective pain management strategies. This review examines the existing landscape, identifies gaps in available pain relief, and proposes avenues for future research to ensure the best pain relief and welfare standards in husbandry practices.

2. Search Methodology

A literature review was conducted by a search of CAB Abstracts via Web of Science (1910—present) and BIOSIS Previews via Web of Science (1926—present). Further databases were not included due to frequent overlap of articles across databases. Keywords included "pain", "chronic pain", "pain relief", "husbandry", "welfare"," long acting", "extended release", "sustained release".

To specifically address pain duration after husbandry, a search was conducted using PICO methodology:

PICO Elements		Search Terms	Boolean Operator
Patient/population	Cattle Sheep Pigs Goats	cattle OR cow* OR bovine* OR steer OR sheep OR ovine OR ram OR wether OR goat* OR kid* OR caprine OR lamb* OR pig* OR sow* OR porcine	-
Intervention	Castration Spay Mulesing Dehorning/disbudding Tail docking	Castrat* OR spay* OR spey* OR tail* OR mule* OR *horn* OR *bud*	AND
Comparison	-	-	-
Outcome	Long term pain	Pain* AND long* NEAR term	AND

(*) is a truncation symbol to search for all endings to a word.

In total, 150 articles were reviewed for inclusion and then some were excluded for the following reasons:

1. Studies that tracked pain up to 36 h only, as the currently available pain mitigation products provide relief for up to 36 h;
2. Studies that were for surgical procedures not considered as routine husbandry (such as orthopaedic surgery);
3. Articles that were not in English;
4. Articles that were reviews rather than original studies.

A small selection of hand-picked information found using the standard literature review search method was also included, resulting in a total of 33 studies included in the review and presented in Table 1.

To expressly search for current pain mitigation, a review of two veterinary drug handbooks [35,36] was conducted to identify common analgesic and anti-inflammatory drug classes (Table 2). Personal knowledge of the authors was used to identify common off-label use in Australian Practice.

To identify those products registered specifically for post-surgical pain relief in livestock, a search was conducted of the Public Chemical Registration Information System (PubCRIS) database of the Australian Pesticides and Veterinary Medicines Authority (APVMA) [37]. The search terms used were the active constituents of interest, based on the results of Table 2. Each search result ("Product List") based on the active constituent was exported as a CSV file to Microsoft Excel, and the registered host/pest and claim identified and sorted alphabetically. Where several identical products were found (generics), the label of the first registered product (based on registration date) was reviewed.

Products included for review were then identified based on the host alias of "beef", "beef calf", "bos indicus", "bos taurus", "bovine", "buffalo", "bull", "bullock", "calf", "cow", "heifer", "steer", "lamb", "sheep", "ewe", "ram", "swine", "pig", "gilt", "sow", "weaner" and "wether", and the pest alias of "inflammation" and "pain". The labels of those included products were reviewed on PubCRIS to identify the relevant particulars (Table 3). Products which provide general anaesthesia were not included, as these are not used in the context of routine on-farm husbandry in Australia.

Table 1. Summary of studies showing pain timeframes post routine husbandry.

Species, Breed	Age	Procedure, Method	No. of Animals (Per Group)	Acute Pain Relief Received *	Study Duration (Days)	Duration of Pain (Days)	Parameters Measured as Indicator of Pain	Parameters Indicating Long Term Pain **	Ref
Goat, Saanen	9–14 days	Disbudding, cautery vs. caustic paste vs. liquid nitrogen vs. clove oil injection vs. sham	50 (10)	None	42	42	Cortisol, haptoglobin up to 24 h. Skin surface temperature, Average Daily Gain (ADG), lesion measurements, lying time, head and body shaking, head scratching, self grooming, feeding	Prolonged healing (note behaviour was tracked for 24 h, ADG for 7 days).	[38,39]
Cattle, Holstein or Jersey	24–38 days	Disbudding, cautery	24 (13 test and 11 control)	Local anaesthetic (lignocaine block) vs. placebo (saline), all received meloxicam 1 mg/kg per os (PO)	11	11	Ethogram of behaviour (head scratch, tub, shake, ear flick, tail flick, buck/jump, grooming, transition to lying)	Behavioural—head shake and ear flick	[40]
Cattle, Holstein or Jersey	3 days or 35 days	Disbudding, cautery vs. sham handling	48 (12)	Local anaesthetic (lignocaine block) for all animals, meloxicam 1 mg/kg PO for disbudded calves only	63	63	Pressure (algometer), infrared thermography, ADG, wound healing	Wound healing, pressure/pain sensitivity	[30]
Cattle, Holstein or Jersey	21–28 days	Disbudding, cautery vs. sham handling	44(11)	Lignocaine block vs. placebo (saline), then meloxicam 1 mg/kg PO for disbudded calves	21	21	Pressure algometry, behaviour indicative of conditioning either 6 h or 20 days post surgery, testing a preference for analgesia	Pressure algometry, behaviour indicative of conditioning showing a preference for analgesia	[21]
Cattle, Holstein	7 days and 28 days	Disbudding, cautery vs. sham handling	30 (10)	Perineural 2% lignocaine, meloxicam 0.5 mg/kg intravenous (IV)	105	105	Visual analogue scale, quantitative sensory testing (pressure-pain threshold), mechanical allodynia, withdrawal reflexes	Behavioural signs, trigeminal hyperalgesia and allodynia	[41]
Cattle, Holstein-Friesian	4–5.5 weeks	Disbudding, cautery vs. sham handling	46 (6–8)	Placebo (saline cornual injection) vs. lignocaine 2% cornual nerve block vs. lignocaine 2% cornual nerve block with meloxicam 0.5 mg/kg IV	3 (75 h)	3 (75 h)	Play behaviour, wound sensitivity via von Frey monofilaments	Wound sensitivity via von Frey monofilaments	[42]
Cattle, Holstein	16–20 weeks	Dehorning, scoop with thermocautery	12 (6)	Saline vs. meloxicam 0.5 mg/kg IV	10	10	Cortisol, substance P, activity and behaviour, heart rate, ADG	ADG	[43]
Cattle, Angus or Hereford	Newborn or weaning (214 days)	Castration, surgical	62 (15–16)	Nil vs. meloxicam 1 mg/kg PO	>300 days	7	Activity via accelerometer (7 days), ADG	Activity via accelerometer (for 7 days)	[44]
Cattle, Hereford X or Black Angus	37–59 days	Castration, surgical vs. sham handling	158 (52–54)	Placebo (saline intramuscular (IM) injection) vs. Meloxicam 0.5 mg/kg IM vs. no injection for sham animals	14	14	Hair cortisol concentration (HCC), lying time, ADG	HCC	[45]

Table 1. *Cont.*

Species, Breed	Age	Procedure, Method	No. of Animals (Per Group)	Acute Pain Relief Received *	Study Duration (Days)	Duration of Pain (Days)	Parameters Measured as Indicator of Pain	Parameters Indicating Long Term Pain **	Ref
Cattle, Holstein	166 ± 0.4 days	Castration, surgical vs. positive control (previous castrates (steers)) vs. negative control (left intact (bulls)).	132 (44)	Flunixin 3 mg/kg IM at 0 and 48 h	126	10	Physical activity measured by pedometer, meal size and duration, lying time	Lying time (5 days), feed intake (10 days) and physical activity (10 days)	[46]
Cattle, Angus cross	2 months	Castration, surgical vs. band vs. sham handling	132 (24)	Placebo (lactated ringers solution) injected subcutaneously (SC) vs. meloxicam 0.5 mg/kg SCn	62	62	ADG and bodyweight (BW), pressure on wound, scrotal temperature, wound swelling, wound healing, behaviour: suckling, lying, standing, walking, head turn, lesion licking, foot stamp, tail flick, proximity to dam	BW and ADG, pressure on wound, wound swelling, proximity to dam	[47]
Cattle, Angus or Angus x	1 week vs. 2 months vs. 4 months	Castration, surgical vs. band vs. sham handling	108 (11–12)	None stated	69	35	Salivary cortisol and HCC, Substance P and Haptoglobin, wound temperature and healing, weigh gain, body temperature, pain behaviour, lying time, stride length	ADG at weaning (surgical castration), swelling (band castration)	[48]
Cattle, Angus x	7–8 days	Castration, surgical vs. band vs. sham handling	72 (12)	Placebo lactated Ringer's solution injection SC vs. meloxicam 0.5 mg/mL SC	56	56	HCC, haptoglobin, serum amyloid A, scrotal swelling, scrotal temperature, wound healing, stride length, behaviour, body weight, body temperature	Inflammation (banded group), HCC	[49]
Cattle, Angus	Not stated—BW ~300 kg.	Castration, surgical	48 (12)	Placebo ring block of lactated Ringer's solution vs. lidocaine 2% + epinephrine ring block vs. meloxicam 0.5 mg/kg SC and placebo ring block vs. meloxicam 0.5 mg/kg SC and lidocaine 2% + epinephrine ring block	28	3	Salivary cortisol, haptoglobin, scrotal temperature, stride length, visual analogue score.	Haptoglobin	[50]
Cattle, Angus x Hereford	25 ± 2 days	Castration, surgical	48 (24)	Placebo (saline) IV vs. flunixin 1.1 mg/kg IV. Lignocaine ring block (3 mL) used on all animals.	63	21–35	Healing and inflammation, wound surface temperature, ADG, Substance P, Lying behaviour	Inflammation (peaked at day 3), healing score	[51]
Cattle, Ayshire	5–7 days	Castration, surgical vs. rubber ring vs. Burdizzo vs. combination Burdizzo and ring vs. control (no castration)	40 (8)	None stated	51	42	Plasma cortisol, behaviour, lesion score	Rubber ring group showed wound directed behaviours, abnormal standing, high lesion scores	[52]

Table 1. *Cont.*

Species, Breed	Age	Procedure, Method	No. of Animals (Per Group)	Acute Pain Relief Received *	Study Duration (Days)	Duration of Pain (Days)	Parameters Measured as Indicator of Pain	Parameters Indicating Long Term Pain **	Ref
Cattle, Holstein	28 days	Castration, surgical vs. rubber ring	21 (10 and 11)	Lignocaine 2% local anaesthetic and meloxicam 0.5 mg/mL SC for all calves	56	56	Wound healing, inflammation, weight gain, feed intake, lying time, wound-directed behaviours	Rubber ring group showed lower weight gain after rubber ringing, scrotal inflammation, wound-directed behaviours, reduced lying	[53]
Cattle, Holstein	4–5 months	Castration, surgical vs. rubber ring	60 (15)	Placebo (saline) vs. meloxicam 1 mg/kg PO	3	3	Substance P, heart rate, cortisol, lying time, tail movements, observed painful behaviour, swelling (inflammation)	Lying time, observed painful behaviour, swelling	[54]
Cattle, Simmental or Simmental x Red Holstein	21–28 days	Castration, rubber ring vs. Burdizzo vs. sham handling	70 (10–15)	Placebo (saline) local infiltration 10 mL vs. lignocaine 2% local infiltration 10 mL	90	90	Expression of pain during castration, serum cortisol (to 72 h), behaviour, posture, scrotal condition including palpation, histology	Reaction to local palpation (up to 50 days), abnormal standing (up to 90 days) (rubber ring group)	[55]
Cattle, Holstein	Adult (lactating)	Tail docking, rubber ring vs. control	64 (16)	None vs. caudal epidural anaesthetic lignocaine 2% 4 mL	6	6	Tail movement and position, posture, milk production, feed intake	Tail movement and position, posture	[56]
Cattle, Holstein	12 months	Tail docking, rubber ring vs. undocked control	164 (133 + 31 control)	None stated	Tested at 6.2 ± 1.9 years of age after docking <12 months old	Tested at 6.2 ± 1.9 years of age	Pressure, thermal and pinprick sensitivity tests	Pressure, heat and cold sensitivity, pinprick sensitivity test	[57]
Pig, not stated	9 or 17 weeks	Tail docking, surgical—2/3rd removed vs. 1/3rd removed vs. sham handling	108 (12–23)	None	112	56	Mechanical Nociceptive Thresholds (MNT)	Mechanical Nociceptive Thresholds (MNT)	[58]
Pig, Landrace x Large white	2 days	Tail docking, clip vs. cautery vs. control	120 (40)	None	21 weeks	N/A	Histology of tail at slaughter	Histology showing evidence of neuroma formation indicative of neuropathic pain.	[59]
Pig, Landrace/Large White x synthetic sire line	3 days	Tail docking, cautery	16 (4)	None	112	112	Examination of tail stump at 1, 4, 8 and 16 weeks post amputation for histopathological changes (healing, neuroma formation	Traumatic neuromata after 28 days and ongoing past 16 weeks (112 days).	[60]
Pig, Landrace/Large White x synthetic sire line	3 or 63 days	Tail docking, amputation vs. sham handling	96 (8)	3 days old: none. 63 days old: meloxicam 0.2 mg/kg IM	112	112	Examination at 1, 8 and 16 weeks for changes in gene expression, traumatic neuroma development and inflammation	Changes in gene expression associated with both inflammatory pain and neuropathic pain	[61]

Table 1. *Cont.*

Species, Breed	Age	Procedure, Method	No. of Animals (Per Group)	Acute Pain Relief Received *	Study Duration (Days)	Duration of Pain (Days)	Parameters Measured as Indicator of Pain	Parameters Indicating Long Term Pain **	Ref
Pig, Piétrain x Hypor	2–8 days	Castration, surgical	186 (95 and 91)	CO_2 anaesthesia vs. none	8	6	Behaviour: general (suckling, socialisation, movement, suckling) specific pain related (huddling, trembling, spasms, scratching, tail wagging), posture, isolation	Pain-related behaviours, walking frequency, lying, sucking, interaction	[62]
Sheep, White Swiss Mountain	>10 to 24 weeks	Castration, surgical vs. Burdizzo vs. rubber ring vs. sham handling	70 (10)	Lidocaine 2% 4 mg/kg infiltration injection vs. bupivacaine 0.5% 1.5 mg/kg infiltration injection	30	21	Response to pain during castration, cortisol levels up to 48 h, food intake day of castration, behaviours and postures, lesions, palpation, bodyweight measurements, histology	Local palpation, average daily gain	[63]
Sheep, White Swiss Mountain and x Charolais	2–7 days	Castration, ring vs. Burdizzo vs. sham handling	70 (11–12)	Placebo (saline) vs. lidocaine 4 mg/kg infiltration	90	21	Response to pain during castration, cortisol levels up to 48 h, behaviours and postures, lesions, palpation, bodyweight measurements, histology	Active behaviour (especially the rubber ring lambs), scrotal swelling, palpation (9 days). Lesions were present >21 days.	[64]
Sheep, breed not specified	1 week vs. 4–6 weeks	Castration, rubber ring vs. combined Burdizzo/ring vs. sham handling	30 (6)	None stated	4 (castration day 2)	3	Moving (including play), eating, standing, lying and abnormal postures	Play behaviour, reduced lying, and abnormal posture	[65]
Sheep, breed not specified	45 days	Tail docking, cautery iron vs. sham handling	50 (25)	Lignocaine 2% 2 mL injected locally prior to docking)	90	90	Infra-red thermography, Mechanical nociceptive threshold, inflammation, histopathology	Mechanical nociceptive threshold, inflammation (significance to day 30), histopathology (moderate to marked fibrosis of the epineurial and perineurial connective tissue, nerve proliferation)	[66]
Sheep, Merino	10–12 weeks	Mulesing, Sodium lauryl sulfate (SLS) injection vs. surgical vs. sham handling	32 (10–11)	Topical local anaesthetic as a wound dressing for surgically mulesed group	42	7	Haematology, cortisol, haptoglobin, β-endorphin, rectal temperature, body weight, standing postures, ADG	ADG, haptoglobin	[67]
Sheep, Merino	6–7 months	Mulesing, surgical vs. Sham	20 (10)	None stated	113	112	Wound healing, Paddock observations of behaviour (lying, grazing), arena observations of handler aversion, cortisol and β-endorphin, growth rate	Wound healing (by day 22), handler aversion (up to day 112), weight gain (day 14)	[22]
Sheep, Merino	10–12 weeks	Mulesing, surgical vs. intradermal injection SLS vs. skin clip vs. none (control)	44 (11)	None	25	25	Plasma cortisol, haptoglobin, weight, gait	In surgical mulesing: decreased weight gain (to day 25), lower feed intake (to day 15), higher cortisol levels (to Day 7), higher haptoglobin (to day 14)	[68]

* The use of general anaesthesia that was reversed after the procedure is not included; ** Parameters indicating long term pain were experienced to the length of time in the "Duration of pain" column, unless otherwise specified.

Table 2. Analgesic and anti-inflammatory medications available for prescription in Australian Veterinary Practice [35–37,69,70].

Drug Type	Use	Schedule in Australia	Common Side Effects	Example Generic Molecules in this Class
Opioid	Analgesia, sedation, strong pain relief	8	Bradycardia, respiratory depression, sedation, constipation, tolerance	Methadone, butorphanol, buprenorphine, tramadol *, morphine *
NSAID	Analgesia & anti-inflammatory, chronic and acute	4, 5	Renal & hepatic toxicity, mild and transient vomiting, soft stool, inappetance, lethargy, gastrointestinal erosions/ulcerations	Meloxicam, ketoprofen, flunixin, tolfenamic acid, carprofen, grapiprant, other coxibs
Corticosteroid	Anti-inflammatory and immunosuppression	4	Hepatopathy, hyperlipidaemia, diabetes, delayed wound healing, immunosuppression leading to infection, GI ulceration. Use with NSAIDs can lead to increased risk of GI injury	Dexamethasone, prednisolone, prednisone
α_2 Agonist	Sedation, muscle relaxation & analgesia	4	Profound sedation, vomiting, startle behaviour, bradycardia, respiratory depression, hypothermia	Clonidine, detomidine, dexmedetomidine, medetomidine, xylaxine
Local Anaesthetic	Pain blocking/prevention	4, 5	CNS stimulation in large doses	Lignocaine, procaine, bupivacaine, prilocaine, mepivacaine
Other therapies and off-label products	Sedation, potentiation, analgesia	Various	Sedation (except paracetamol)	Diazepam, gabapentin *, paracetamol ** cannabidiol *

* Not registered in Australia for animals but used in veterinary practice. Gabapentin off-label use is widespread and very common for the treatment of neuropathic pain as well as a sedative/anxiolytic in companion animals [70]. Cannabidiols are gaining traction as a pain relief option in companion animals [71] and may be scripted for various purposes, including chronic pain, under state-by-state regulations in Australia [72]. ** Paracetamol is only registered by the APVMA in Australia as an anti-pyretic in piglets [37], but is commonly used off-label for analgesia [69,73].

Table 3. Summary of registered Australian products with specific claims for surgical pain in cattle and sheep [37].

Product (Brand if Applicable)	Prescription or OTC	Drug Class	Duration of Action *	Claim (Associated with Surgical Pain)
Lignocaine 2%, Prilocaine 2% (cattle only)	Prescription	Local anaesthesia pre-procedure	1–4 h	Infiltration anaesthesia and nerve block
Bupivacaine 0.4%, lignocaine 4%, adrenaline, cetrimide (Tri-Solfen)	OTC	Local anaesthesia post-procedure	After 30 s and up to 4 h	Topical local anaesthesia and antiseptic spray for castration, mulesing and tail docking in lambs, and castration and dehorning or disbudding in calves.
Lignocaine 2% (sheep only) (Numocaine for Numnuts device)	OTC	Local anaesthesia peri-procedure	Up to 3 h	Local anaesthestic injection via Numnuts applicator for tail docking and castration via rubber rings in sheep
Meloxicam 0.5% injection (cattle only) 2% injection, 4% injection (cattle only)	Prescription	NSAID	No duration of action specified on the label	Cattle—to assist in the control of pain particularly that after heat cautery dehorning in young cattle. It is recommended that a cornual nerve block anaesthesia is used in conjunction for dehorning. Sheep: As a single dose for alleviation of pain and inflammation pain in sheep more than 14 days old.
Flunixin 5% (cattle only)	Prescription	NSAID	24–36 h	Suppression of post-operative swelling and lameness
Meloxicam 1% (Buccalgesic, Butec)	OTC	NSAID	No duration of action specified on the label	Oral Transmucosal NSAID for alleviation of pain in lambs after mulesing, tail docking and castration, and in conjunction with a cornual block in calves for disbudding and dehorning, and in conjunction with a local anaesthetic for castration to enhance pain relief and minimise tissue damage and distress.
Meloxicam 1.5% oral (Meloxi-care)	Prescription	NSAID	No duration of action specified on the label	For the reduction of pain and inflammation associated with band or surgical castration administer orally two hours before the painful procedure.

* Claimed on label.

3. Pain Duration after Routine Husbandry

To understand the length of time that pain is experienced after a surgical procedure, it is necessary to consider the physiological mechanism underlying the type of pain. Pain itself is a protective mechanism, as it signals for potential or actual tissue damage, and ensures that an animal (if able) moves away from or avoids further injury [74]. Pain that is induced by surgical procedures should and can be pre-empted and mitigated to an appropriate degree.

During the initial phase of an injury, nociceptors are activated, nerve fibres deliver the sensation of pain to the brain, and the response causes the body to flinch or move away from the pain source. The tissue damage at the site causes the release of inflammatory and other mediators, which initially activate the nociceptors, and persistent pain sensitises those nociceptors [75] leading to longer-term pain.

Inflammation at a local level is a tissue stress response by the body's immune system, whereby damaged tissue, and infected or necrotic cells are identified and removed [76] and the healing process is initiated. The immune and vascular response of inflammation, which includes the formation and release of prostaglandins, involves (at a tissue level) redness, swelling, heat, and pain at the site of injury [77]. Damage from injury is detected by both tissue-resident macrophages and nociceptors at the injury site. Inflammatory mediators are responsible for inflammatory pain, while prostaglandins can enhance the sensitivity of nociceptors by lowering their threshold for activation, thus increasing the pain sensation [78]. While the inflammatory response is vital for healing [77], inflammatory pain can be intense and lead to an abnormally heightened sensitivity to pain (hyperalgesia), pain experienced from usually non-painful stimuli (allodynia), and sustained or increased pain perception (sympathetically maintained pain) [6].

The bulk of research performed to date regarding pain mitigation in livestock has focused on the acute, immediate pain experienced during a procedure and in the following 2–8 h. However, there have been several studies in animals showing that post-procedural pain lasts for longer than the first few hours, with neuropathic or inflammatory pain being postulated as the likely cause [79]. An example of this longer-lasting pain has been established after rubber ringing (ischaemic amputation) for tail docking and castration. The constrictive rubber ring leads to ischaemic necrosis of the tissue, which ultimately sloughs away, making the procedure bloodless but intensely painful, with significant behaviours indicative of severe pain, such as rolling, writhing and abnormal standing shown for at least 4 h after the ring is placed [80], then other observations such as reduced playing and lying, wound-directed behaviours and swelling, as well as atypical postures and abnormal walking seen for several days afterwards [52,81]. This is particularly interesting in the context of the Australian production industry, given that in recent Australian industry surveys of 2003 sheep producers and 803 beef producers, it was reported that 98% of male lambs and 85% of male calves owned by the producers surveyed are castrated with rubber rings, with only a quarter of these receiving any form of pain relief [16,17]. Studies in other procedures commonly performed, such as surgical castration, tail docking, dehorning and others have been shown to cause pain for days or weeks afterwards [40,46,53].

A selection of studies that collected pain data for more than 3 days, and that variously investigated aspects of routine husbandry methods in livestock is presented in Table 1; while many were not specifically designed to do so, the studies illustrate that observations of pain have been made for days or weeks following these procedures.

Pain behaviours have been observed for days or weeks post procedure, and many of the studies seen in Table 1 were still observing pain-related behaviour on the last day of recording. It is therefore possible that pain continued undocumented after these studies were completed, and this limits conclusions as to the true extent or duration of pain experienced. There are limited studies specifically designed to evaluate the duration of pain, but rather the focus of much research into pain relief has been intended to compare procedure methods, ages, or acute pain relief, so the assessments of pain in studies have not been specifically designed to detect longer lasting or inflammatory pain [42,43]. Some techniques used for measurement of pain (such as palpation or pressure) tend to induce pain, so that the animal may have been relatively pain-free without interference, thus confounding the interpretation of persistent pain. It is also of note that in the studies presented in Table 1, animals that were provided with acute pain relief still experienced observable pain for days and sometimes weeks afterwards [41,46,53].

These findings of longer-term post-procedural pain are not unexpected if considered in the context of the human experience. A human patient who has experienced amputation or major abdominal surgery is routinely provided with significant pain relief for several days post injury or surgery, since it is understood that surgery will cause the release of inflammatory mediators, which activate nociceptors, and if the post-surgical pain is persistent, the nociceptors become sensitised. Prolonged inflammatory states leading to this sensitisation can cause changes to the nociceptors that can lead to chronic pain pathophysiology [75]. Examples of this in humans who have undergone what may be considered equivalent surgeries have been reported, with one study [82] relating that up to 50% of patients who have had an amputation of a limb or digit will experience pain for at least 6 months, while in another study [83] 32% of hysterectomy patients still reported pain after 6 months. It is therefore highly likely that animals who have experienced amputation of tail or horns, or spaying/castration, may experience pain for a similar period.

If the inflammatory response (and therefore pain) is resolved during normal wound healing, the central nervous system (CNS) will revert to normal activity, thus avoiding long-term chronic pain caused by changes to the nociceptors via inflammation [75]. Extrapolation of this concept to non-human mammals demonstrates a need for pain mitigation in animals that decreases the inflammatory response for a longer period than the acute peri-or post-operative phase if long-term or chronic pain is to be avoided.

4. The Current State of Pain Mitigation

In animals, the medications available for pain relief or pain mitigation are limited, and the products available over the counter to owners are even fewer.

Different classes of drugs act in different ways upon the body, and the ideal analgesic targets the cause or mechanism of the pain [84], and that allows animals to achieve functionality and normal behaviour as soon as possible. This may mean that the method of pain relief changes throughout the injury and healing, or that multi-modal pain relief is required.

There are six broad categories of drugs, (Table 2), that are used in the treatment of pain and inflammation: opioids, non-steroidal anti-inflammatory drugs (NSAIDs), α_2-agonists, local anaesthetics, corticosteroids, and "others". These others such as non-opioid analgesics and antipyretics, or tranquilisers or anticonvulsants whose primary purpose may not be analgesia but may act as an adjunct to known analgesics.

Pain mitigation after non-routine or major surgery in animals is usually tailored to the specific animal and circumstance, and as per Table 2, there are several options available for appropriate multi-modal post-surgical pain relief.

The fact that most research has concentrated on the immediate pain associated with procedures is reflected in the current products registered in Australia for the mitigation of pain in cattle and sheep. These include a topical local anaesthetic (Tri-Solfen Wound Anaesthetic and Antiseptic Solution, Dechra Veterinary Products Ltd. (Somersby, NSW, Australia), specialised local anaesthetic in a device (for rubber ringing only) (Numnuts device with lignocaine), and oral trans-mucosal meloxicam (Butec OTM, Troy Laboratories Ltd., Glendenning, NSW, Australia) available over the counter, or injectable local anaesthetic or NSAIDs available from a veterinarian [37]. Other livestock species such as pigs and goats have fewer registered products available and generally require a veterinary prescription for off-label use. Local anaesthetic products are effective for approximately 1–4 h post procedure depending on the dose rate, molecule and/or combination used and NSAIDs, depending upon the product, will be effective for 4 to 36 h [36]. A summary of

Australian products registered for post-husbandry pain relief in cattle and sheep is shown in Table 3.

When considering the context of routine livestock husbandry on the farm, or the conduct of minor surgical procedures in the clinic, the requirement for a universal approach to pain mitigation in a large number of animals accounts for several parameters in addition to efficacy: availability (over-the-counter vs. prescription), practicality (single dose), and ease of application for non-veterinarian users. Another consideration is the selection of an active pharmaceutical ingredient (API) that has an established safety and efficacy profile across various species, including the food safety aspect in production species, and that is economically viable. Finally, there is the requirement for the animal to be able to function normally whilst under treatment, which precludes many of the drugs that affect the central nervous system. One of the major signs of recovery in livestock is the ability to "mother up" and/or graze effectively, as well as move from the point of treatment to the paddock as soon as practicable after treatment.

The NSAID group of products meets many of the above requirements, and there are several NSAIDs registered in livestock in Australia that meet the efficacy and safety criteria, providing relief relatively quickly after the first dose whilst still allowing the animal to be ambulant and functional (Table 3). In their current form, however, they do not provide pain relief for an adequate period when considering the length of pain duration experienced by animals after procedures (Table 1). NSAIDs are known to have some general contraindications, most of which are relevant in an older or debilitated population; such as renal, hepatic, cardio or pulmonary insufficiencies or dysfunctions, animals that are pregnant, or animals on concomitant systemic corticosteroids or other NSAIDs, or that are dehydrated [37,85].

5. Future Directions: Longer-Term Pain Relief

The need for the provision of pain relief for an appropriate length of time in animals undergoing surgery or painful routine procedures is becoming increasingly recognised [86]. It is a recommendation that pain relief should be provided for at least 72 h post surgery [87], but with current registered products available to veterinarians and owners, retreatment at hourly or daily intervals is required to achieve this level of pain mitigation. When an animal is hospitalised post surgery, this can be easily achieved. However, most animals are discharged on the same day or within 24 h of surgery, and for livestock especially, which undergo routine procedures on-farm, the stress to the animals of re-handling (mustering, physical separation and restraint, the risk of re-injury from handling and restraint, and needle sticks) that would be needed to re-treat, can negatively affect the animal's welfare. Another important consideration when using a product that provides sustained pain relief is the time to onset of action. An ideal product for long-acting pain relief is one which has rapid onset of analgesia, and then maintains this over a sustained period without requiring re-treatment.

One method of providing medication that does not require rehandling is in-feed medication, where the animals voluntarily consume nutritional supplements in the form of licks or blocks that contain the drug of interest, allowing the provision of medication over several days or even weeks, in a non-invasive manner. This has been reported for self-medication of endo parasiticides in wildlife and zoo animals [88], and sheep [89]. There is work underway with pain relief medication and voluntary consumption of NSAIDs through medicated feed or supplements has been trialled with carprofen in chickens with lameness [90,91] flunixin in cattle [92] and sheep [93,94] and meloxicam in cattle [95]. It has been found that this method provides an ongoing level of pain relief for animals and that future research in this area is warranted. Some of the challenges for livestock dosing include ensuring appropriate palatability so that voluntary consumption of adequate but not excessive medication is achievable, and managing accurate dose rates for medications of this type, especially considering the differences in pharmacokinetics of the oral route in monogastric versus ruminant species, as well as determining the withholding periods for meat and/or milk. In addition, maintenance of the drug product potency in this format where exposure to heat, moisture and UV radiation may destabilise the API needs to be evaluated.

The application of transdermal medication has long been an option for providing pain relief in humans, and a buprenorphine solution for cats delivered as a low volume topical dose to the unclipped dorsal cervical skin provides extended plasma buprenorphine concentrations and opioid physiological effects [96]. A study in cattle using transdermally-delivered ketoprofen as a back-line pour-on compared with conventional intramuscular administration [97] showed that the transdermal formulation was slightly superior in terms of overall drug exposure, giving rise to the possibility of transdermal delivery of longer-acting NSAIDs (such as meloxicam) providing a greater duration of action, although this concept is unproven. A study in sheep using transdermal patches of fentanyl [98] in order to decrease the need for post-operative handling showed promise, providing up to 72 h of pain relief; however, fentanyl as a drug product would not be feasible for large-scale use in livestock, given the practicality of an adherent patch on fleece or hair, which would need to be clipped or shorn to allow adherence to the skin, and the possibility of animals removing and ingesting the patch during self or social grooming leading to toxicity. In addition, the controlled scheduling that is used to prevent potential abuse and misuse, and consequent difficulty of procurement adds another layer of complexity. A recent and promising development in the delivery of transdermal pain relief is lignocaine-impregnated elastrator bands (developed by Chinook Contract Research Inc., Canada) that have been tested in calves and lambs [99–101], with results showing effective levels of lignocaine in tissue for 3 to 7 days, although the sloughing of the scrotum and testes tended to be slower when compared with conventional bands in a larger trial of lambs [102].

Another method of extending the pain relief available to animals, especially livestock where re-handling would exacerbate stress, is to develop an extended-release (ER) or sustained release (SR) pain relieving medication, in the form of an anaesthetic, analgesic or NSAID that is dosed once at the time of surgery.

There is currently a liposomal encapsulated bupivacaine injectable suspension under investigation for the extension of the duration of local anaesthesia, and in dogs undergoing cranial cruciate ligament rupture surgery, an intra-thecal injection of the sustained release bupivacaine at the time of wound closing has shown promising results with some pain relief still present up to 72 h after surgery [86]. In a subsequent study of dogs undergoing similar surgeries, the animals receiving liposomal encapsulated bupivacaine injection were less likely to require rescue analgesia and required lower amounts of opioids than the dogs that received conventional bupivacaine [103]. Another novel formulation of bupivacaine involving sucrose acetate isobutyrate, a highly viscous sugar that has also been used for the sustained release of drugs, has been tested as a cornual nerve block when disbudding calves. The level of anaesthesia was prolonged (8–36 h) when compared with a lignocaine cornual block (0.5–1.5 h) [104], or bupivacaine cornual block (4 h) [105].

A compounded sustained release formulation of injectable buprenorphine (an opioid analgesic) has been tested in sheep [106] and guinea pigs [107,108] and has been shown to provide a steady state of the minimum threshold for therapeutic benefit for 72 h in sheep and up to 48 h in guinea pigs. This may be an option for veterinarian use; however, it is not an option for livestock owners due to the scheduling constraints of buprenorphine, which is a strictly controlled drug in Australia and other countries.

From a practical standpoint, to allow owner-treatment of livestock, an NSAID provides a practical solution to longer-acting pain relief. A sustained release injectable formulation of meloxicam in a polymer-based matrix has been trialed in sheep [87], but the formulation provided only 48 to 60 h at a presumed therapeutic level of meloxicam and requires further investigation. Although studies have shown that pain from routine husbandry such as castration, tail docking and dehorning lasts for weeks to months [55,57,60,61], the main pain indicators that impede normal function such as walking, eating and socialization generally persist for at least 3 to 7 days [43,62,109,110]. A sustained-release formulation of a well-characterized drug product with a good safety profile such as meloxicam, which provided a therapeutic level of pain relief for 72 to 96 h, would substantially improve the welfare and productivity of animals that undergo routine husbandry on the farm.

A further option for longer-term pain relief in livestock is the use of some of the NSAID APIs that are currently registered in humans or companion animals only. The NSAID mavacoxib, a long-acting COX-2 inhibitor, has a half-life in dogs of more than 2 weeks [111], so investigation of its efficacy in livestock may lead to an efficacious and longer-acting pain relief product.

The safety concerns with regards to the use of NSAIDs are generally focused on animals with renal, hepatic, pulmonary or cardiac insufficiencies or dysfunction, with gastrointestinal disease, or that are being treated with concomitant systemic corticosteroids or NSAIDs [85]. For the target population species indicated for this project, the animals are young and generally in good health, and not being treated with other NSAIDs or steroids. One issue that must be considered is that in animals that have experienced trauma with active haemorrhage or blood loss, the use of NSAIDs is contraindicated [85], which may preclude certain procedures, such as mulesing, from being treated with some NSAID (especially those that include significant COX-1 inhibition) sustained release formulations.

6. Conclusions

As livestock handlers, owners and veterinarians become better at recognising situations where pain and distress are experienced, they should strive to improve methods of pain mitigation. Inflammatory pain post surgery is a well-established concept and demonstrates a requirement for mitigating the inflammatory response post surgery, ideally for at least 5 to 7 days.

All the NSAIDs currently available for use in veterinary practice in Australia provide relief from inflammatory pain and have been shown to meet the appropriate safety criteria in many species; however, current products require frequent retreatment to provide an adequate period of pain mitigation, posing practical difficulties, especially for livestock. The challenge is providing a solution that allows a single dose to provide relief for at least a week or longer, to animals that are only handled once (at the time of surgery), or for which repeated doses are not viable. Currently, there are no commercially available, registered anti-inflammatory solutions for livestock available in Australia or globally, that will provide an adequate level of pain mitigation for an extended period. Potential solutions being researched include in-feed dosing via voluntary consumption of medicated feed, transdermal medication delivery, and extended-release formulations. Continued research into the development of extended-release formulations for pain mitigation in livestock is warranted to provide better animal welfare now and in the future.

Author Contributions: Conceptualisation, writing original draft, preparation: L.M. Review and editing: D.V.d.S., S.L., S.G. and P.J.W. All authors have read and agreed to the published version of the manuscript.

Funding: This research received no external funding.

Institutional Review Board Statement: Not applicable.

Informed Consent Statement: Not applicable.

Data Availability Statement: No new data were created or analysed in this review; therefore, data sharing is not applicable.

Conflicts of Interest: The authors declare no conflicts of interest.

References

1. Yeates, J. Is 'a life worth living' a concept worth having? *Anim. Welf.* **2011**, *20*, 397–406. [CrossRef]
2. Victorian State Government. Prevention of Cruelty to Animals Regulations 2019. Available online: https://www.legislation.vic.gov.au/as-made/statutory-rules/prevention-cruelty-animals-regulations-2019 (accessed on 2 May 2024).

3. Tasmanian Government. Animal Welfare (Cattle) Regulations. 2023. Available online: https://www.legislation.tas.gov.au/view/whole/html/inforce/2023-03-08/sr-2023-006 (accessed on 28 May 2024).
4. Steagall, P.V.; Monteiro, B.P.; Ruel, H.L.M.; Beauchamp, G.; Luca, G.; Berry, J.; Little, S.; Stiles, E.; Hamilton, S.; Pang, D. Perceptions and opinions of Canadian pet owners about anaesthesia, pain and surgery in small animals. *J. Small Anim. Pract.* **2017**, *58*, 380–388. [CrossRef] [PubMed]
5. Raja, S.N.; Carr, D.B.; Cohen, M.; Finnerup, N.B.; Flor, H.; Gibson, S.; Keefe, F.J.; Mogil, J.S.; Ringkamp, M.; Sluka, K.A.; et al. The revised International Association for the Study of Pain definition of pain: Concepts, challenges, and compromises. *Pain* **2020**, *161*, 1976–1982. [CrossRef] [PubMed]
6. Yam, M.F.; Loh, Y.C.; Tan, C.S.; Adam, S.K.; Manan, N.A.; Basir, R. General Pathways of Pain Sensation and the Major Neurotransmitters Involved in Pain Regulation. *Int. J. Mol. Sci.* **2018**, *19*, 2164. [CrossRef] [PubMed]
7. Prunier, A.; Mounier, L.; Le Neindre, P.; Leterrier, C.; Mormède, P.; Paulmier, V.; Prunet, P.; Terlouw, C.; Guatteo, R. Identifying and monitoring pain in farm animals: A review. *Animal* **2013**, *7*, 998–1010. [CrossRef]
8. Short, C.E. Fundamentals of pain perception in animals. *Appl. Anim. Behav. Sci.* **1998**, *59*, 125–133. [CrossRef]
9. National Research Council (US). 1. Pain in Research Animals: General Principles and Considerations. Available online: https://www.ncbi.nlm.nih.gov/books/NBK32655/ (accessed on 24 May 2024).
10. McLennan, K.M. Why Pain Is Still a Welfare Issue for Farm Animals, and How Facial Expression Could Be the Answer. *Agriculture* **2018**, *8*, 127. [CrossRef]
11. Lizzaraga, I.; Chambers, J.P. Use of analgesic drugs for pain management in sheep. *N. Z. Vet. J.* **2012**, *60*, 87–94. [CrossRef]
12. Downing, R. Pain Management and the Human-Animal Bond. In *Handbook of Veterinary Pain Management*; Gaynor, J.S., Muir, W.W., III, Eds.; Elsevier: St. Louis, MO, USA, 2015; pp. 3–9.
13. Monteiro, B.P.; Lascelles, B.D.X.; Murrell, J.; Robertson, S.; Steagall, P.V.M.; Wright, B. 2022 WSAVA guidelines for the recognition, assessment and treatment of pain. *J. Small Anim. Pract.* **2023**, *64*, 177–254. [CrossRef]
14. Doughty, A.K.; Coleman, G.J.; Hinch, G.N.; Doyle, R.E. Stakeholder Perceptions of Welfare Issues and Indicators for Extensively Managed Sheep in Australia. *Animals* **2017**, *7*, 28. [CrossRef]
15. Van Dyke, R.; Connor, M.; Miele, A. An Investigation into the Perceptions of Veterinarians towards Perioperative Pain Management in Calves. *Animals* **2021**, *11*, 1882. [CrossRef] [PubMed]
16. Sheep Sustainability Framework. On-Farm Insights from the National Producer Survey. October 2022. Available online: https://www.sheepsustainabilityframework.com.au/globalassets/sheep-sustainability/media/ssf-on-farm-insights-report-web-25oct2022.pdf (accessed on 23 October 2023).
17. Sloane, B.; Walker, L. Final Report: MLA Project Proof Beef. Report Code E.SUS.0005. Available online: https://www.mla.com.au/research-and-development/reports/2023/e.sus.0005--mla-project-proof-beef/ (accessed on 23 October 2023).
18. Mellor, D.J.; Beausoleil, N.J.; Littlewood, K.E.; McLean, A.N.; McGreevy, P.D.; Jones, B.; Wilkins, C. The 2020 Five Domains Model: Including Human-Animal Interactions in Assessments of Animal Welfare. *Animals* **2020**, *10*, 1870. [CrossRef] [PubMed]
19. Small, A.; Fisher, A.D.; Lee, C.; Colditz, I. Analgesia for Sheep in Commercial Production: Where to Next? *Animals* **2021**, *11*, 1127. [CrossRef] [PubMed]
20. Ede, T.; Lecorps, B.; von Keyserlingk, M.A.G.; Weary, D.M. Calf aversion to hot-iron disbudding. *Sci. Rep.* **2019**, *9*, 5344. [CrossRef] [PubMed]
21. Adcock, S.J.J.; Tucker, C.B. Conditioned place preference reveals ongoing pain in calves 3 weeks after disbudding. *Sci. Rep.* **2020**, *10*, 3849. [CrossRef] [PubMed]
22. Fell, L.R.; Shutt, D.A. Behavioural and hormonal responses to acute surgical stress in sheep. *Appl. Anim. Behav. Sci.* **1989**, *22*, 283–294. [CrossRef]
23. Colditz, I.; Paull, D.; Lee, C. Social transmission of physiological and behavioural responses to castration in suckling Merino lambs. *Appl. Anim. Behav. Sci.* **2012**, *136*, 136–145. [CrossRef]
24. Guesgen, M.J.; Beausoleil, N.J.; Stewart, M. Effects of early human handling on the pain sensitivity of young lambs. *Vet. Anaesth. Analg.* **2013**, *40*, 55–62. [CrossRef] [PubMed]
25. Guesgen, M.J.; Beausoleil, N.J.; Minot, E.O.; Stewart, M.; Stafford, K.J. Social context and other factors influence the behavioural expression of pain by lambs. *Appl. Anim. Behav. Sci.* **2014**, *159*, 41–49. [CrossRef]
26. Chen, J.S.; Kandle, P.F.; Murray, I.V.; Fitzgerald, L.A.; Sehdev, J.S. Physiology, Pain. Available online: https://www.ncbi.nlm.nih.gov/books/NBK539789/ (accessed on 21 January 2024).
27. Johnson, C.B.; Sylvester, S.P.; Stafford, K.J.; Mitchinson, S.L.; Ward, R.N.; Mellor, D.J. Effects of age on the electroencephalographic response to castration in lambs anaesthetized with halothane in oxygen from birth to 6 weeks old. *Vet. Anaesth. Analg.* **2009**, *36*, 273–279. [CrossRef]
28. Kells, N.J.; Beausoleil, N.J.; Chambers, J.P.; Sutherland, M.A.; Morrison, R.S.; Johnson, C.B. Electroencephalographic responses of anaesthetized pigs (Sus scrofa) to tail docking using clippers or cautery iron performed at 2 or 20 days of age. *Vet. Anaesth. Analg.* **2017**, *44*, 1156–1165. [CrossRef] [PubMed]
29. McCracken, L.; Waran, N.; Mitchinson, S.; Johnson, C.B. Effect of age at castration on behavioural response to subsequent tail docking in lambs. *Vet. Anaesth. Analg.* **2010**, *37*, 375–381. [CrossRef] [PubMed]
30. Adcock, S.J.J.; Tucker, C.B. The effect of disbudding age on healing and pain sensitivity in dairy calves. *J. Dairy Sci.* **2018**, *101*, 10361–10373. [CrossRef] [PubMed]

31. Clark, C.; Murrell, J.; Fernyhough, M.; O'Rourke, T.; Mendl, M. Long-term and trans-generational effects of neonatal experience on sheep behaviour. *Biol. Lett.* **2014**, *10*, 20140273. [CrossRef] [PubMed]
32. Saunders, D. Reference D22/50766 Response to e-Petition. Available online: https://www.sheepcentral.com/wp-content/uploads/2022/11/Government-response-Mandatory-mulesing-phase-out-by-2030.pdf (accessed on 4 May 2024).
33. Scott, A. NZM Alerts Farmers to Plan for Docking Pain Relief. Available online: https://www.farmersweekly.co.nz/news/nzm-alerts-farmers-to-plan-for-docking-pain-relief/ (accessed on 20 May 2023).
34. Textile Exchange. Responsible Wool Standard 2.2 RAF-101a-V2.2-2021.10.01. Available online: https://textileexchange.org/app/uploads/2020/08/RAF-101a-V2.2-Responsible-Wool-Standard.pdf (accessed on 24 February 2024).
35. Plumb, D. *Plumb's Veterinary Drug Handbook*, 9th ed.; Wiley-Blackwell: Hoboken, NJ, USA, 2018.
36. Papich, M.G. *Saunders Handbook of Veterinary Drugs: Small and Large Animal*, 4th ed.; Elsevier, Inc.: Maryland Heights, MO, USA, 2016.
37. Australian Pesticides and Veterinary Medicines Authority (APVMA). PubCRIS. Available online: https://www.apvma.gov.au/registrations-and-permits/search-registered-chemical-products-and-permits (accessed on 19 April 2024).
38. Hempstead, M.N.; Waas, J.R.; Stewart, M.; Cave, V.M.; Sutherland, M.A. Evaluation of alternatives to cautery disbudding of dairy goat kids using behavioural measures of post-treatment pain. *Appl. Anim. Behav. Sci.* **2018**, *206*, 32–38. [CrossRef]
39. Hempstead, M.N.; Waas, J.R.; Stewart, M.; Cave, V.M.; Sutherland, M.A. Evaluation of alternatives to cautery disbudding of dairy goat kids using physiological measures of immediate and longer-term pain. *J. Dairy Sci.* **2018**, *101*, 5374–5387. [CrossRef]
40. Adcock, S.J.J.; Cruz, D.M.; Tucker, C.B. Behavioral changes in calves 11 days after cautery disbudding: Effect of local anesthesia. *J. Dairy Sci.* **2020**, *103*, 8518–8525. [CrossRef]
41. Castoni, D.; Mirra, A.; Suter, M.R.; Gutzwiller, A.; Spadavecchia, C. Can disbudding of calves (one versus four weeks of age) induce chronic pain? *Physiol. Behav.* **2019**, *199*, 47–55. [CrossRef] [PubMed]
42. Mintline, E.M.; Stewart, M.; Rogers, A.R.; Cox, N.R.; Verkerk, G.A.; Stookey, J.M.; Webster, J.R.; Tucker, C.B. Play behavior as an indicator of animal welfare: Disbudding in dairy calves. *Appl. Anim. Behav. Sci.* **2013**, *144*, 22–30. [CrossRef]
43. Coetzee, J.F.; Mosher, R.A.; KuKanich, B.; Gehring, R.; Robert, B.; Reinbold, J.B.; White, B.J. Pharmacokinetics and effect of intravenous meloxicam in weaned Holstein calves following scoop dehorning without local anesthesia. *BMC Vet. Res.* **2012**, *8*, 153. [CrossRef]
44. Brown, A.C.; Powell, J.G.; Kegley, E.B.; Gadberry, M.S.; Reynolds, J.L.; Hughes, H.D.; Carroll, J.A.; Burdick Sanchez, N.C.; Thaxton, Y.V.; Backes, E.A.; et al. Effect of castration timing and oral meloxicam administration on growth performance, inflammation, behavior, and carcass quality of beef calves. *J. Anim. Sci.* **2015**, *93*, 2460–2470. [CrossRef]
45. Creutzinger, K.C.; Stookey, J.M.; Marfleet, T.W.; Campbell, J.R.; Janz, D.M.; Marqués, F.J.; Seddon, Y.M. An investigation of hair cortisol as a measure of long-term stress in beef cattle: Results from a castration study. *Can. J. Anim. Sci.* **2017**, *97*, 499–509. [CrossRef]
46. Devant, M.; Marti, S.; Bach, A. Effects of castration on eating pattern and physical activity of Holstein bulls fed high-concentrate rations under commercial conditions. *J. Anim. Sci.* **2012**, *90*, 4505–4513. [CrossRef] [PubMed]
47. Gellatly, D.; Marti, S.; Pajor, E.A.; Melendez, D.M.; Moya, D.; Janzen, E.D.; Yang, X.; Milani, M.R.M.; Schwartzkopf-Genswein, K.S. Effect of a single subcutaneous injection of meloxicam on chronic indicators of pain and inflammatory responses in 2-month-old knife and band-castrated beef calves housed on pasture. *Livest. Sci.* **2021**, *244*, 104305. [CrossRef]
48. Marti, S.; Meléndez, D.M.; Pajor, E.A.; Moya, D.; Heuston, C.E.M.; Gellatly, D.; Janzen, E.D.; Schwartzkopf-Genswein, K.S. Effect of band and knife castration of beef calves on welfare indicators of pain at three relevant industry ages: II. Chronic pain1. *J. Anim. Sci.* **2017**, *95*, 4367–4380. [CrossRef] [PubMed]
49. Marti, S.; Melendez, D.M.; Pajor, E.A.; Moya, D.; Gellatly, D.; Janzen, E.D.; Schwartzkopf-Genswein, K.S. Effect of a single dose of subcutaneous meloxicam prior to band or knife castration in 1-wk-old beef calves: II. inflammatory response and healing. *J. Anim. Sci.* **2018**, *96*, 4136–4148. [CrossRef] [PubMed]
50. Meléndez, D.M.; Marti, S.; Janzen, E.D.; Moya, D.; Gellatly, D.; Pajor, E.A.; Schwartzkopf-Genswein, K.S. Effect of lidocaine and meloxicam on indicators of pain and distress after knife castration in weaned beef calves. *J. Anim. Sci.* **2017**, *95*, 5–6. [CrossRef]
51. Mintline, E.M.; Varga, A.; Banuelos, J.; Walker, K.A.; Hoar, B.; Drake, D.; Weary, D.M.; Coetzee, J.F.; Stock, M.L.; Tucker, C.B. Healing of surgical castration wounds: A description and an evaluation of flunixin. *J. Anim. Sci.* **2014**, *92*, 5659–5665. [CrossRef]
52. Molony, V.; Kent, J.E.; Robertson, I.S. Assessment of acute and chronic pain after different methods of castration of calves. *Appl. Anim. Behav. Sci.* **1995**, *46*, 33–48. [CrossRef]
53. Nogues, E.; von Keyserlingk, M.A.G.; Weary, D.M. Pain in the weeks following surgical and rubber ring castration in dairy calves. *J. Dairy Sci.* **2021**, *104*, 12881–12886. [CrossRef]
54. Olson, M.E.; Ralston, B.; Burwash, L.; Matheson-Bird, H.; Allan, N.D. Efficacy of oral meloxicam suspension for prevention of pain and inflammation following band and surgical castration in calves. *BMC Vet. Res.* **2016**, *12*, 102. [CrossRef] [PubMed]
55. Thüer, S.; Mellema, S.; Doherr, M.G.; Wechsler, B.; Nuss, K.; Steiner, A. Effect of local anaesthesia on short- and long-term pain induced by two bloodless castration methods in calves. *Vet. J.* **2007**, *173*, 333–342. [CrossRef]
56. Tom, E.M.; Duncan, I.J.H.; Widowski, T.M.; Bateman, K.G.; Leslie, K.E. Effects of Tail Docking Using a Rubber Ring with or Without Anesthetic on Behavior and Production of Lactating Cows. *J. Dairy Sci.* **2002**, *85*, 2257–2265. [CrossRef] [PubMed]
57. Troncoso, R.J.; Herzberg, D.E.; Meneses, C.S.; Müller, H.Y.; Werner, M.P.; Bustamante, H. Mechanical/thermal sensitivity and superficial temperature in the stump of long-term tail-docked dairy cows. *PeerJ* **2018**, *6*, e5213. [CrossRef] [PubMed]

58. Di Giminiani, P.; Edwards, S.A.; Malcolm, E.M.; Leach, M.C.; Herskin, M.S.; Sandercock, D.A. Characterization of short- and long-term mechanical sensitisation following surgical tail amputation in pigs. *Sci. Rep.* **2017**, *7*, 4827. [CrossRef]
59. Kells, N.J.; Beausoleil, N.J.; Johnson, C.B.; Sutherland, M.A.; Morrison, R.S.; Roe, W. Comparison of neural histomorphology in tail tips from pigs docked using clippers or cautery iron. *Animal* **2017**, *11*, 1222–1227. [CrossRef]
60. Sandercock, D.A.; Smith, S.H.; Di Giminiani, P.; Edwards, S.A. Histopathological Characterization of Tail Injury and Traumatic Neuroma Development after Tail Docking in Piglets. *J. Comp. Pathol.* **2016**, *155*, 40–49. [CrossRef] [PubMed]
61. Sandercock, D.A.; Barnett, M.W.; Coe, J.E.; Downing, A.C.; Nirmal, A.J.; Di Giminiani, P.; Edwards, S.A.; Freeman, T.C. Transcriptomics Analysis of Porcine Caudal Dorsal Root Ganglia in Tail Amputated Pigs Shows Long-Term Effects on Many Pain-Associated Genes. *Front. Vet. Sci.* **2019**, *18*, 314. [CrossRef]
62. Van Beirendonck, S.; Driessen, B.; Verbeke, G.; Geers, R. Behavior of piglets after castration with or without carbon dioxide anesthesia. *J. Anim. Sci.* **2011**, *89*, 3310–3317. [CrossRef]
63. Melches, S.; Mellema, S.C.; Doherr, M.G.; Wechsler, B.; Steiner, A. Castration of lambs: A welfare comparison of different castration techniques in lambs over 10 weeks of age. *Vet. J.* **2007**, *173*, 554–563. [CrossRef]
64. Mellema, S.C.; Doherr, M.G.; Wechsler, B.; Thueer, S.; Steiner, A. Influence of local anaesthesia on pain and distress induced by two bloodless castration methods in young lambs. *Vet. J.* **2006**, *172*, 274–283. [CrossRef] [PubMed]
65. Thornton, P.D.; Waterman-Pearson, A.E. Behavioural Responses to Castration in Lambs. *Anim. Welf.* **2002**, *11*, 203–212. [CrossRef]
66. Larrondo, C.; Bustamante, H.; Paredes, E.; Gallo, C. Long-term hyperalgesia and traumatic neuroma formation in tail-docked lambs. *Anim. Welf.* **2019**, *28*, 443–454. [CrossRef]
67. Colditz, I.G.; Paull, D.R.; Lee, C.; Fisher, A.D. Physiological and behavioural effects of intradermal injection of sodium lauryl sulfate as an alternative to mulesing in lambs. *Aust. Vet. J.* **2010**, *88*, 483–489. [CrossRef] [PubMed]
68. Hemsworth, P.H.; Barnett, J.L.; Karlen, G.M.; Fisher, A.D.; Butler, K.L.; Arnold, N.A. Effects of mulesing and alternative procedures to mulesing on the behaviour and physiology of lambs. *Appl. Anim. Behav. Sci.* **2009**, *117*, 20–27. [CrossRef]
69. Bello, A.M.; Dye, C. Current perceptions and use of paracetamol in dogs among veterinary surgeons working in the United Kingdom. *Vet. Med. Sci.* **2023**, *9*, 679–686. [CrossRef] [PubMed]
70. Di Cesare, F.; Negro, V.; Ravasio, G.; Villa, R.; Draghi, S.; Cagnardi, P. Gabapentin: Clinical Use and Pharmacokinetics in Dogs, Cats, and Horses. *Animals* **2023**, *13*, 2045. [CrossRef] [PubMed]
71. Yu, C.H.J.; Rupasinghe, H.P.V. Cannabidiol-based natural health products for companion animals: Recent advances in the management of anxiety, pain, and inflammation. *Res. Vet. Sci.* **2021**, *140*, 38–46. [CrossRef]
72. Australian Pesticides and Veterinary Medicines Authority (APVMA). Cannabis in Veterinary Chemical Products. Available online: https://www.apvma.gov.au/resources/chemicals-news/cannabis-veterinary-chemical-products (accessed on 28 May 2024).
73. Mercer, M.A.; McKenzie, H.C.; Byron, C.R.; Pleasant, R.S.; Bogers, S.H.; Council-Troche, R.M.; Werre, S.R.; Burns, T.; Davis, J.L. Pharmacokinetics and clinical efficacy of acetaminophen (paracetamol) in adult horses with mechanically induced lameness. *Equine Vet. J.* **2023**, *55*, 524–533. [CrossRef]
74. Anderson, D.E.; Muir, W.W. Pain Management in Cattle. *Vet. Clin. N. Am. Food Anim. Pract.* **2005**, *21*, 623–635. [CrossRef]
75. Vascopoulos, C.; Lema, M. When does acute pain become chronic? *Br. J. Anaesth.* **2010**, *105*, i65–i85. [CrossRef] [PubMed]
76. Ferrero-Miliani, L.; Nielsen, O.H.; Andeson, P.S.; Girardin, S.E. Chronic inflammation: Importance of NOD2 and NALP3 in interleukin-1B generation. *Clin. Exp. Immunol.* **2006**, *147*, 227–235. [CrossRef] [PubMed]
77. Chen, L.; Deng, H.; Cui, H.; Fang, J.; Zuo, Z.; Deng, J.; Li, Y.; Wang, X.; Zhao, L. Inflammatory responses and inflammation-associated diseases in organs. *Oncotarget* **2018**, *9*, 7204–7218. [CrossRef] [PubMed]
78. Medzhitov, R. Inflammation 2010: New Adventures of an Old Flame. *Cell* **2010**, *140*, 771–776. [CrossRef] [PubMed]
79. Small, A.; Fisher, A.D.; Lee, C.; Colditz, I. *Project Final Report: Gap Evaluation of Pain Alleviation Research*; CSIRO Project No ON-00550; Australian Wool Innovation: Sydney, Australia, 2020.
80. Lester, S.J.; Mellor, D.J.; Holmes, R.J.; Ward, R.N.; Stafford, K.J. Behavioural and cortisol responses of lambs to castration and tailing using different methods. *N. Z. Vet. J.* **1996**, *44*, 45–54. [CrossRef]
81. Mellema, S.C.; Doherr, M.G.; Wechsler, B.; Thuer, S.; Steiner, A. Influence of local anaesthesia on pain and distress induced by bloodless castration methods in young lambs. Einfluss der Lokalanasthesie auf Schmerz und Stress verursacht durch unblutige Kastrationsmethoden beim jungen Lamm. *SAT Schweiz. Arch. Tierheilkd.* **2007**, *149*, 213–225. [CrossRef] [PubMed]
82. Kehlet, H.; Jensen, T.S.; Woolf, C.J. Persistent postsurgical pain: Risk factors and prevention. *Lancet* **2006**, *367*, 1618–1625. [CrossRef] [PubMed]
83. Katz, J.; Cohen, L. Preventive Analgesia Is Associated with Reduced Pain Disability 3 Weeks but Not 6 Months after Major Gynecologic Surgery by Laparotomy. *Anesthesiology* **2004**, *101*, 169–174. [CrossRef] [PubMed]
84. Muir, W.W., III. Overview of Drugs Administered to Treat Pain. In *Handbook of Pain Management*, 3rd ed.; Gaynor, J.S., Muir, W.W., III, Eds.; Elsevier: Amsterdam, The Netherlands, 2015; pp. 113–141.
85. Budsberg, S.C. Nonsteroidal Anti-Inflammatory Drugs. In *Handbook of Veterinary Pain Management*, 3rd ed.; Gaynor, J.S., Muir, W.W., III, Eds.; Elsevier: Amsterdam, The Netherlands, 2015; pp. 142–160.
86. Lascelles, B.D.X.; Kirkby Shaw, K. An extended release local anaesthetic: Potential for future use in veterinary surgical patients? *Vet. Med. Sci.* **2016**, *2*, 229–238. [CrossRef] [PubMed]
87. Dunbar, M.L.; Walkowiak, K.J.; Faustich, J.S.; Rendahl, A.K.; Graham, M.L. Preliminary Evaluation of Sustained-release Compared with Conventional Formulations of Meloxicam in Sheep (*Ovis aries*). *J. Am. Assoc. Lab. Anim. Sci.* **2019**, *58*, 339–345. [CrossRef]

88. Gradé, J.; Tabuti, J.; Van Damme, P. Four footed pharmacists: Indications of self-medicating livestock in Karamoja, Uganda. *Econ. Bot.* **2009**, *63*, 29–42. [CrossRef]
89. Fishpool, F.J.; Kahn, L.P.; Tucker, D.J.; Nolan, J.V.; Leng, R.A. Voluntary intake of a medicated feed block by grazing sheep is increased by gastrointestinal nematode infection. *Anim. Prod. Sci.* **2012**, *52*, 1136–1141. [CrossRef]
90. Danbury, T.C.; Weeks, C.A.; Waterman-Pearson, A.E.; Kestin, S.C.; Chambers, J.P. Self-selection of the analgesic drug carprofen by lame broiler chickens. *Vet. Rec.* **2000**, *146*, 307–311. [CrossRef]
91. Siegel, P.B.; Gustin, S.J.; Katanbaf, M.N. Motor ability and self-selection of an analgesic drug by fast-growing chickens. *J. Appl. Poult. Res.* **2011**, *20*, 249–252. [CrossRef]
92. Odensvik, K. Pharmacokinetics of flunixin and its effect on prostaglandin F2 alpha metabolite concentrations after oral and intravenous administration in heifers. *J. Vet. Pharmacol. Ther.* **1995**, *18*, 254–259. [CrossRef]
93. Marini, D.; Colditz, I.G.; Hinch, G.; Petherick, J.C.; Lee, C. Self-administration by consumption of flunixin in feed alleviates the pain and inflammation associated with castration and tail docking of lambs. *Appl. Anim. Behav. Sci.* **2017**, *188*, 26–33. [CrossRef]
94. Marini, D.; Colditz, I.G.; Lee, C. Can Lambs in Pain Identify Medicated Feed? *Front. Anim. Sci.* **2021**, *2*, 741631. [CrossRef]
95. Van der Saag, D.; Garling, E.; White, P.; Govendir, M.; Kimble, B.; Lomax, S. Developing a self-medication strategy for practical delivery of long-lasting analgesia to cattle. In Proceedings of the 8th International Conference on the Assessment of Animal Welfare at the Farm and Group Level, Cork, Ireland, 16–19 August 2021; Wageningen Academic: Wageningen, The Netherlands, 2021; p. 81.
96. Freise, K.J.; Reinemeyer, C.; Warren, K.; Lin, T.-L.; Clark, T.P. Single-dose pharmacokinetics and bioavailability of a novel extended duration transdermal buprenorphine solution in cats. *J. Vet. Pharmacol. Ther.* **2022**, *45*, S31–S39. [CrossRef]
97. Mills, P.C.; Ghodasara, P.; Satake, N.; Alawneh, J.; Fraser, B.; Kopp, S.; McGowan, M. A Novel Transdermal Ketoprofen Formulation Provides Effective Analgesia to Calves Undergoing Amputation Dehorning. *Animals* **2020**, *10*, 2442. [CrossRef]
98. Christou, C.; Oliver, R.A.; Rawlinson, J.; Walsh, W.R. Transdermal fentanyl and its use in ovine surgery. *Res. Vet. Sci.* **2015**, *100*, 252–256. [CrossRef]
99. Saville, J.W.; Ross, J.A.; Trefz, T.; Schatz, C.; Matheson-Bird, H.; Ralston, B.; Granot, O.; Schmid, K.; Terry, R.; Allan, N.D.; et al. Development and Field Validation of Lidocaine-Loaded Castration Bands for Bovine Pain Mitigation. *Animals* **2020**, *10*, 2363. [CrossRef]
100. Ross, J.A.; Roche, S.M.; Beaugrand, K.; Schatz, C.; Hammad, A.; Ralston, B.J.; Hanson, A.M.; Allan, N.; Olson, M. Assessment of the Effective Tissue Concentrations of Injectable Lidocaine and a Lidocaine-Impregnated Latex Band for Castration in Calves. *Animals* **2024**, *14*, 977. [CrossRef]
101. Ross, J.A.; Roche, S.M.; Beaugrand, K.; Schatz, C.; Hammad, A.; Ralston, B.J.; Hanson, A.M.; Allan, N.; Olson, M. Assessment of the Pharmacokinetics and Pharmacodynamics of Injectable Lidocaine and a Lidocaine-Impregnated Latex Band for Castration and Tail Docking in Lambs. *Animals* **2024**, *14*, 255. [CrossRef]
102. Roche, S.M.; Ralston, B.J.; Olson, B.; Sharpe, B.D.; Schatz, C.; Beaugrand, K.; Ross, J.A.; Broomfield, M.A.; Allan, N.; Olson, M. Efficacy of a Lidocaine-Impregnated Elastrator Band for Castration and Tail Docking in Lambs. *Animals* **2024**, *14*, 1403. [CrossRef]
103. Reader, R.C.; McCarthy, R.J.; Schultz, K.L.; Volturo, A.R.; Barton, B.A.; O'Hara, M.J.; Abelson, A.L. Comparison of liposomal bupivacaine and 0.5% bupivacaine hydrochloride for control of postoperative pain in dogs undergoing tibial plateau leveling osteotomy. *J. Am. Vet. Med. Assoc.* **2020**, *256*, 1011–1019. [CrossRef]
104. Venkatachalam, D.; Kells, N.; Chambers, P.; Jacob, A.; Ward, N.; Singh, P. Pharmacokinetics and efficacy of a novel long-acting bupivacaine formulation for cornual nerve block in calves. *Front. Vet. Sci.* **2022**, *9*, 1060951. [CrossRef]
105. McMeekan, C.M.; Mellor, D.J.; Stafford, K.J.; Bruce, R.A.; Ward, R.N.; Gregory, N.G. Effects of local anaesthesia of 4 to 8 hours duration on the acute cortisol response to scoop dehorning in calves. *Aust. Vet. J.* **1998**, *76*, 281–285. [CrossRef]
106. Walkowiak, K.J.; Graham, M.L. Pharmacokinetics and Antinociceptive Activity of Sustained-Release Buprenorphine in Sheep. *J. Am. Assoc. Lab. Anim. Sci.* **2015**, *54*, 763–768.
107. Smith, B.J.; Wegenast, D.J.; Hansen, R.J.; Hess, A.M.; Kendall, L.V. Pharmacokinetics and Paw Withdrawal Pressure in Female Guinea Pigs (*Cavia porcellus*) Treated with Sustained-Release Buprenorphine and Buprenorphine Hydrochloride. *J. Am. Assoc. Lab. Anim. Sci.* **2016**, *55*, 789–793.
108. Zanetti, A.S.; Putta, S.K.; Casebolt, D.B.; Louie, S.G. Pharmacokinetics and Adverse Effects of 3 Sustained-release Buprenorphine Dosages in Healthy Guinea Pigs (*Cavia porcellus*). *J. Am. Assoc. Lab. Anim. Sci.* **2017**, *56*, 768–778.
109. Väisänen, M.A.; Tuomikoski, S.K.; Vainio, O.M. Behavioral alterations and severity of pain in cats recovering at home following elective ovariohysterectomy or castration. *J. Am. Vet. Med. Assoc.* **2007**, *231*, 236–242. [CrossRef] [PubMed]
110. Väisänen, M.; Vainio, O.; Oksanen, H. Postoperative signs in 96 dogs undergoing soft tissue surgery. *Vet. Rec.* **2004**, *155*, 729–733. [PubMed]
111. Cox, S.R.; Lesman, S.P.; Boucher, J.F.; Krautmann, M.J.; Hummel, B.D.; Savides, M.; Marsh, S.; Fielder, A.; Stegemann, M.R. The pharmacokinetics of mavacoxib, a long-acting COX-2 inhibitor, in young adult laboratory dogs. *J. Vet. Pharmacol. Ther.* **2010**, *33*, 461–470. [CrossRef] [PubMed]

animals

MDPI

Review

An Exploration of Analgesia Options for Australian Sheep

Shari Cohen [1,*], Emily Foss [2], Thierry Beths [2] and Gabrielle C. Musk [3]

1 Animal Welfare Science Centre, University of Melbourne, Parkville, Melbourne, VIC 3010, Australia
2 Melbourne Veterinary School, University of Melbourne, Parkville, Melbourne, VIC 3010, Australia; emilypagefoss@gmail.com (E.F.); thierry.beths@unimelb.edu.au (T.B.)
3 School of Human Sciences, University of Western Australia, Perth, WA 6009, Australia; gabrielle.musk@uwa.edu.au
* Correspondence: shari.cohen@unimelb.edu.au

Simple Summary: Sheep may undergo a variety of painful husbandry and disease processes in their lifetime, which negatively impact their welfare. These procedures can cause considerable pain that may be unalleviated due to a lack of pain relief options across many different settings such as farm, clinical, and biomedical contexts. The choice of pain relief may be restricted due to licensing requirements (e.g., Australian regulations) or lack of known effectiveness. In a biomedical setting, a variety of potential pain relief options have been used but not validated for pain relief or safety (human residues or sheep welfare). A review of the farm, veterinary, and biomedical literature was undertaken to identify important gaps in sheep analgesia, pain management, and potential options for pain relief to promote better sheep welfare across these industries.

Abstract: During their lifetime, sheep undergo many painful husbandry and disease processes. Procedures undertaken on the farm, such as tail docking, castration, and mulesing, all cause considerable pain. In addition, sheep may experience painful diseases and injuries that require treatment by veterinary practitioners, and in biomedical research, sheep may undergo painful experimental procedures or conditions. It is important due to ethics, animal welfare, social licence, and, at times, legal requirements for farmers, veterinary practitioners, and researchers to provide pain relief for animals in their care. While there is a heightened awareness of and a greater interest in animal welfare, there remain few licensed and known analgesia options for sheep within Australia. A literature review was undertaken to identify currently known and potential future options for analgesic agents in sheep in farm and biomedical settings. Non-steroidal anti-inflammatories, opioids, local anaesthetics, α_2 adrenoreceptor agonists, and NMDA receptor antagonists are some of the more common classes of analgesic drugs referred to in the literature, but few drugs are registered for use in sheep, with even fewer proven to be effective. Only six analgesic product formulations, namely, lignocaine (e.g., Numocaine®), Tri-Solfen®, ketamine, xylazine, and meloxicam (oral transmucosal and injectable formulations), are currently registered in Australia and known to be efficacious in some types of painful conditions in sheep. The gap in knowledge and availability of analgesia in sheep can pose risks to animal welfare, social licence, and research outcomes. This article presents a summary of analgesic agents that have been used in sheep on farms and in clinical veterinary and biomedical research settings along with details on whether their efficacy was assessed, doses, routes of administration, indication for use, and pain assessment techniques (if any) used. The outcome of this research highlights the challenges, gaps, and opportunities for better analgesia options in sheep.

Keywords: analgesia; sheep; pain; ovine

1. Introduction

Sheep in Australian meat and wool production enterprises undergo painful husbandry and disease processes throughout their life. Most lambs are 'marked' between 4 and 12 weeks of age [1]. Surgical or painful procedures undertaken at this time may include earmarking, tail docking, castration, and/or mulesing. These procedures cause considerable pain with impacts on animal welfare, especially if performed without any analgesia [1]. Sheep also experience painful conditions such as shearing cuts, mastitis, foot abscesses, dystocia, and flystrike, for which they may or may not be treated by a veterinarian or farmer. The lack of administration of pain relief for painful husbandry practices entrenched within Australia's sheep farming industry is waning in public acceptability [2]. Phasing out these procedures or at least providing analgesia is a practice more commonly being advocated for by both industry and the public. The Australian Wool Innovation (AWI) industry organisation in 2017 released a Merino Husbandry Practices Survey, which reported that up to 85% of lambs were likely to receive some form of pain relief when mulesed. AWI also reported that up to 42% of producers used pain relief for tail docking and castration [3]. A 2018 Meat and Livestock Australia (MLA) survey found that up to 39% of producers would be willing to use pain relief for marking if it were available and effective [4]. There is an increasing demand from local and global retail brands as well as industry markets for more ethical, higher-welfare-produced wool and meat from producers committed to using pain relief. Markets and retailers typically grant a price premium to more ethical, higher-welfare products, offering producers greater financial benefits with greater market access for their products. According to the Australian Wool Exchange (AWEX), data reveal that wool from sheep treated with pain relief receives a premium that often offsets the cost of any pain relief administered [5,6].

In biomedical research, various procedures including orthopaedic, reproductive, cardiac, and abdominal surgeries are performed on sheep [7–9]. The use of pain relief in these procedures can ensure better animal welfare and higher ethical standards, promote the Three Rs, and minimise potential impacts on research outcomes. In addition, researchers, institutions, and animal ethics committees are working under Australian legislative requirements published by the National Health and Medical Research Council Code of Practice for the Care and Use of Animals for Scientific Purposes (2013 and 2018) to consider and manage pain and distress. These codes also require any choice of analgesic regimen to be consistent with current best veterinary or medical practice, appropriate for the species and life stage of the animal, and compatible with the purpose and aims of the project [10]. In Australia, the type and dose of pain relief given to sheep in biomedical trials can include licensed and unlicensed drugs, with the latter often extrapolated from veterinary drug use in other species or from human medicine [11]. The literature provides an array of analgesic agents at various doses administered for various conditions to sheep. However, many of these analgesic agents have not been investigated for safety (for sheep or in meat) or efficacy, and in some of these publications, methods of pain assessment are not disclosed. Even if pain assessment in sheep is performed via sheep-specific and generic parameters [12], this does not ensure that the analgesic choice selected is effective, appropriate, or safe. This issue poses potential animal welfare concerns and risks confounding experimental work due to unmitigated pain or side effects of these therapies [13–16]. To achieve best practice in pain relief, research, and sheep management, further research is needed to ensure that preventative and multi-modal analgesic regimes are fit for purpose.

There are also additional ethical responsibilities, societal demands, and potential legal requirements of veterinary practitioners, farmers, and researchers to provide adequate pain relief to animals in their care. Heightened public awareness and interest in animal welfare are key drivers to ensure that appropriate pain relief is administered to farm and experimental animals. Increasing societal concern for animal welfare is reflected in the public statement of the Royal Society for the Prevention of Cruelty to Animals (RSPCA) that 'all future systems must identify and adopt humane husbandry and management practices that do not cause pain, suffering or distress to animals. In the interim, best practice pain relief must be used' [17]. Specifically for those working in the Australian sheep industry, the Australian Animal Welfare Standards and Guidelines for Sheep state that lambs must have analgesia for many common painful husbandry procedures from 6 months of age onwards [18]. When lambs are under 6 months old, pain relief is not required but still recommended. Additionally, livestock South Australia (an industry body), Victorian state regulations, and Tasmanian state regulations all mandate pain relief for mulesing and recommend it for all other invasive procedures from various ages [19]. The future sustainability of the sheep industry will likely require further investment, development, and formal experimental trials of suitable products for safe administration and effective analgesia.

Non-steroidal anti-inflammatory drugs (NSAIDs), opioids, local anaesthetics, α_2 adrenoreceptor agonists, and N-*methyl*-D-aspartate (NMDA) receptor antagonists are classes of analgesic drugs reported in the literature. Depending on the national jurisdiction (e.g., the European Union), analgesia options may be different, limited, or unavailable [20]. In Australia, there are only six analgesic formulations registered (also known as 'licensed') for use in sheep: lignocaine (2%); Tri-Solfen® (lignocaine hydrochloride 40.6 g/L, bupivacaine hydrochloride 4.2 g/L, adrenaline (as acid tartrate) 24.8 mg/L, and cetrimide 5 g/L); ketamine (as hydrochloride 100 mg/mL); xylazine (as hydrochloride 20 mg/mL); and oral transmucosal and injectable formulations of meloxicam (20 mg/mL) [21]. Their product registration is as follows: lignocaine is a local anaesthetic registered for use since 1998; Tri-Solfen® was registered in 2011 and is a topical anaesthetic and antiseptic solution; xylazine is an α_2 adrenoreceptor agonist registered since 1998 [22]; ketamine is an NMDA receptor antagonist registered since 1994 [22]; meloxicam, a non-steroidal anti-inflammatory drug (NSAID), has been registered in its injectable form since 2016;the oral transmucosal formulation, known as Buccalgesic® and, more recently, Butec®, is the most recent analgesic drug to be registered for sheep, receiving approval in 2017 [22]. The paucity of effective, registered (permitted), and available products for sheep analgesia poses animal welfare concerns and limits best practice across all jurisdictions in the wider sheep industry.

The aim of this review of analgesic agents used in Australian sheep on farms and in veterinary clinics and biomedical research settings is to identify the possible large array of known and potential analgesic drugs. There are potentially far more future analgesic options that could be available or viable to alleviate pain in sheep if further research, appropriate pain assessment, and safe registration are undertaken. The intent of this review is to offer a starting point to highlight these options as well as promote, encourage, and improve sheep analgesia and welfare across biomedical, veterinary, and farming enterprises.

2. Materials and Methods

A structured approach to the review was undertaken, as outlined in Figure 1. The electronic literature databases CAB Direct and PubMed were searched from 2010 to March 2022 for the following key terms: analgesia, local anaesthetic, pain relief, opioid, NSAID, ovine, sheep, lamb, ewe, and ram. Further databases were not included in the search due to frequent overlap of articles across databases. Only full-text articles in English or translated into English were included, as the authors' primary language is English, and non-English articles could not be confirmed to match the information presented in the abstract or used to extract additional information required for review. The criteria for article inclusion required publications to include the analgesic dose, route given, and purpose for analgesic use in sheep either on a farm or in a biomedical research setting. Confirmation and evaluation of pain assessment was not a criterion for inclusion, as the review sought to outline both potential and known options for sheep analgesia rather than assess analgesic effectiveness. The quality and impact factors of journals were not included or used as a criterion for inclusion or exclusion due to the exploratory nature of the review. Two hundred and forty-two (242) articles were found to meet the criteria for inclusion and downloaded into Endnote X9 (Clarivate, Philadelphia, PA, USA). A small selection of hand-picked known information on sheep analgesia methods found using the standard literature review search method were also included.

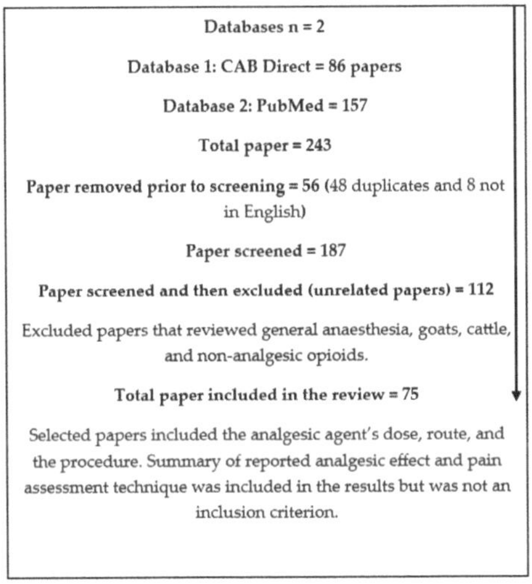

Figure 1. Database review process.

Papers that were deemed unrelated and therefore excluded were those that focused primarily on general anaesthesia, non-target species (goats or cattle), or non-analgesic opioids. A total of 75 articles met the criteria for review. The results were categorised into five tables by drug class. The analgesic drug classes were NSAIDs, opioids, local anaesthetics, α_2 adrenoreceptor agonists, and other miscellaneous drugs (e.g., paracetamol and ketamine). Details of drug action, dose, route, indication, summary of analgesic effect, pain assessment method used, and the number of sheep involved in the study were included. The details on drug 'action' highlight the pharmacokinetic differences between drugs within their class. The 'dose', 'route', and 'summary of analgesic effect' sections show the variation in these methods of administration between studies. The 'indication' for use lists any painful or potentially painful procedures or disease states experienced by sheep. The use of a 'pain assessment method' was the assessment tool or constellation of indicators used to identify pain to determine if any pain assessment method was used. The effectiveness of the method used to identify pain was not assessed, as this was outside the scope of the paper. The 'number of sheep' was included to show study size.

In several studies, analgesics were administered as part of a surgical anaesthesia protocol and were not the sole focus of the study. These study designs could cause interpretation difficulties, as the primary purpose was not to study analgesic effect. Only information on the reported analgesic agent or regimen was recorded, as the intent of the review was to identify drugs being used for analgesic purposes in sheep.

3. Results

The results demonstrated that a far greater number of analgesic drugs and/or regimens (32) have been used for analgesia in sheep than the six currently licensed formulations available in Australia. Multiple studies (21) attempted to utilise multimodal analgesia techniques. Three studies used analgesic drugs for a disease process rather than a procedure.

3.1. Non-Steroidal Anti-Inflammatory Drugs (NSAIDs)

Seven NSAIDs were identified in the reviewed literature: ketorolac, meloxicam, flunixin, diclofenac, ketoprofen, carprofen, and phenylbutazone (Table 1). Of these seven, only meloxicam is registered for use in sheep in Australia. Meloxicam was also the most common NSAID used and was utilised in three different multimodal NSAID combinations. The multimodal NSAID combinations were meloxicam with lignocaine, meloxicam with Tri-Solfen®, and flunixin with lignocaine. The table below outlines the literature reviewed.

Table 1. NSAIDs used for analgesia in sheep.

Drug	Action	Dose	Route	Indication	Summary of Analgesic Effect	Pain Assessment Method	Sheep (n)	Reference
Ketorolac	Nonselective COX-1 and COX-2 inhibitor	30 mg/sheep SID for 7 days postoperatively	Intravenous	Open heart surgery.	Not recorded.	None.	10	[7]
Meloxicam *	Selective COX-2 inhibitor	1.0 mg/kg	Oral	Surgical mulesing and hot-knife tail docking.	Slower to provide effective analgesia than Tri-Solfen®. Superior analgesia was seen when Tri-Solfen® and Buccalgesic® were used together.	Pain avoidance and postural behaviour, cortisol, haematology, and haptoglobin were used.	24	[23]
		1.0 mg/kg postoperatively	Oral	Laparotomy.	Provided similar analgesia to flunixin. Pain was not eliminated.	Sheep grimace scale, behaviour, blood drug concentration, infrared thermography, pressure mat gait analysis, mechanical nociceptive threshold, and vocalization were used.	12	[8]
		1.0 mg/kg	Oral	Hot-knife tail docking and surgical mulesing.	Analgesia evident at the 2 h observation. Pain was not eliminated. Best analgesia was seen when Tri-Solfen® and Buccalgesic® were used together.	Lamb behaviour was observed.	20	[24]
		1.0 mg/kg	Oral	Hot-iron tail docking and knife castration.	Provided substantial analgesia on the day of marking. Some analgesia evident the following morning.	Time to mother up and behaviours were observed.	30	[25]
		1.0 mg/kg	Oral	Tail docking and ring castration.	Reduced lamb mortality between marking and weaning.	Pain-related behaviour, average daily growth, and feed intake were measured.	78	[26]
		1.0 mg/kg	Intravenous	Forelimb pain.	Provided some pain relief.	Leucocyte count, neutrophil/lymphocyte ratio, haptoglobin, force plate pressure, skin temperature, and daily feed intake were measured.	10	[27]
		1.0 mg/kg	Subcutaneous	Sterile acute inflammation in forelimb.	This dose of meloxicam provided significant analgesic benefits to sheep.	Pain-related variables and inflammation-related variables were measured.	12	[28]
		1.0 mg/kg SID preoperatively and every 48 h postoperatively	Oral	Laser ablation of abscess.	Not recorded.	No.	1	[29]
		1.0 mg/kg SID for 10 days	Oral	No procedure. Trial for meat withdrawal intervals	Provides potential analgesia but not for longer than 24 h.	No.	27	[30]
		0. 5 mg/kg SID	Subcutaneous	No procedure.	Analgesic effect not recorded.	No.	6	[31]
		0.5 mg/kg postoperatively	Intramuscular	Rumen fistulation.	Effect not recorded.	No.	13	[32]
		1.0 mg/kg	Subcutaneous	Mulesing.	Minimal to no analgesia.	Behavioural responses were observed.	20	[33]
		15 min preoperatively, 1.0 mg/kg at mulesing	Subcutaneous	Mulesing.	Minimal to no analgesia.	Behavioural responses were observed.	20	[33]

Table 1. *Cont.*

Drug	Action	Dose	Route	Indication	Summary of Analgesic Effect	Pain Assessment Method	Sheep (n)	Reference
		1.0 mg/kg preoperatively	Intramuscular	Ring castration and hot iron docking.	Meloxicam had no analgesic effect.	Behavioural indicators of pain were recorded.	15	[34]
		1.0 mg/kg on day 1 and day 4 postpartum	Oral	Post-partum.	Not recorded.	No.	19	[35]
		Not recorded	Subcutaneous around scrotum	Castration.	Provided partial analgesia for ring castration.	Behaviour, plasma haptoglobin, cortisol, rectal temperature, haematology, and behaviour were recorded.	12	[36]
		0.5 mg/kg postoperatively	Intravenous	Elective laparoscopy.	Not recorded.	The UNESP-Botucatu composite scale was used.	48	[37]
Meloxicam * and lignocaine *	Selective COX-2 inhibitor and local anaesthetic	0.5 mg/kg meloxicam + 2 mL 2% lidocaine/sheep	Subcutaneous + intra-testicular injection	Castration.	Minimal analgesia.	Electroencephalography, behavioural observations, and eye temperature were recorded.	8	[38]
		1.0 mg/kg + 1 mL 2% lidocaine/sheep preoperatively	Intramuscular + subcutaneous into scrotal neck, spermatic cords, and tail	Ring castration and hot iron docking.	Some indication that meloxicam improved lignocaine's analgesic effect but did not fully alleviate pain.	Behavioural indicators of pain were recorded.	15	[34]
		5 mL of 2% lidocaine + 2% meloxicam/sheep	Administered together. Diluted in 5 mL saline, then injected SC into scrotal neck, spermatic cords, and tail	Ring castration and hot iron docking.	Analgesic effects were similar to those of the two drugs when administered separately, but the treatment did not fully alleviate pain.	Behavioural indicators of pain were recorded.	15	[34]
Meloxicam (sustained release)	Selective COX-2 inhibitor	1.5 mg/kg	Subcutaneous	No procedure.	Not measured.	No.	6	[31]
		3 mg/kg	Subcutaneous	No procedure.	Not measured	No.	6	[31]
Meloxicam and Tri-Solfen® *	Selective COX-2 inhibitor + (local anaesthetic + sympathomimetic + antiseptic)	1.0 mg/kg 15 min preoperatively + 8–10 mL/sheep	Subcutaneous + topical (on the mulesed area and tail-docking wound)	Ear marking, castration, tail docking, and mulesing.	No analgesia evident in lambs 1.5 h after the procedures	QBA method.	30	[39]
		1.0 mg/kg 15 min preoperatively + 8–10 mL/sheep	Subcutaneous meloxicam administered 15 min before mulesing and Tri-Solfen® applied after mulesing	Mulesing.	Provided analgesia in the first 6 h post-mulesing.	Behavioural responses were observed.	20	[33]
		1.0 mg/kg + lambs 5–10 kg 6 mL, 11–15 kg 8 mL, 16–20 kg 10 mL, >20 kg 12 mL	Oral + sprayed onto wounds	Hot knife tail docking and surgical mulesing.	Provided analgesia, but pain was not eliminated.	Lamb behaviour was observed.	20	[24]

Table 1. *Cont.*

Drug	Action	Dose	Route	Indication	Summary of Analgesic Effect	Pain Assessment Method	Sheep (n)	Reference
		1.0 mg/kg + lambs 5–10 kg 6 mL, 11–15 kg 8 mL, 16–20 kg 10 mL, >20 kg 12 mL	Oral + sprayed onto wounds	Hot knife tail docking and surgical mulesing.	Provided analgesia to surgical mulesing.	Behaviour, cortisol, and postures were recorded.	24	[23]
Flunixin	Nonselective COX-1 and COX-2 inhibitor	1.1 mg/kg on day 1	Intramuscular	Foot rot.	NSAID had no significant effect on recovery from lameness.	No.	16	[40]
		2.2 mg/kg postoperatively	Intravenous	Laparotomy.	Provided similar analgesia to meloxicam. Pain was not eliminated.	Sheep grimace scale, behaviour, blood drug concentration, infrared thermography, pressure mat gait analysis, mechanical nociceptive threshold, and vocalization were used.	12	[8]
		1.0 mg/kg every 24 h	Intravenous	Orchiectomy.	Moderate reduction in pain	Pain was assessed.	6	[41]
		4.0 mg/kg	Oral	Turpentine injection was used as a painful stimulus.	Minimal analgesia was seen.	Pain was assessed.	10	[42]
		4.0 mg/kg	Oral in feed	No procedure.	No.	No.	9	[43]
		5.0 mg/kg	Subcutaneous around scrotum	Castration.	Provided partial analgesia.	Behaviour, cortisol, rectal temperature, haematology, and plasma haptoglobin were recorded.	12	[36]
		1.1 mg/kg every 12 h	Intravenous	Retropharyngeal abscess and tracheostomy.	No.	No.	1	[29]
		1.1 mg/kg SID	Intravenous	Post orchiectomy analgesia.	Effect not recorded.	No.	10	[44]
Flunixin and lignocaine (2%)	Nonselective COX-1 and COX-2 inhibitor + local anaesthetic	1.1 mg/kg + 2% lidocaine at 2.5 mL + 5 mL	Intramuscular Subcutaneous (spermatic cords and scrotal neck)	Burdizzo castration.	Analgesic effect for up to 3 days post-castration.	Multiparametric: behaviour, inflammation, ANS, HPA, andoxidative stress.	24	[45]
		1.0 mg/kg 1 h preoperatively, then every 24 h for 2 days postoperatively + 2 mg/kg preoperatively	Intravenous + intrafunicular	Orchiectomy.	Reduced pain and distress preoperatively and postoperatively.	Pain was assessed.	6	[41]
Diclofenac (1%)	Selective COX-1 and COX-2 inhibitor	Placed around tracheostomy site	Topical (gel)	Tracheostomy.	No.	No.	1	[29]
Ketoprofen	Nonselective COX-1 inhibitor	3.0 mg/kg	Intravenous and intramuscular	No procedure.	Not recorded.	No.	6	[46]
		8.0 mg/kg	Oral	Turpentine injection was used as a painful stimulus.	Minimal analgesia.	Pain was assessed.	10	[42]

Table 1. *Cont.*

Drug	Action	Dose	Route	Indication	Summary of Analgesic Effect	Pain Assessment Method	Sheep (n)	Reference
		3.0 mg/kg for 3 days	Intramuscular	Polyarthritis caused by *Erysipelothrix rhusiopathiae.*	No.	No.	7	[47]
Carprofen	Selective COX-2 inhibitor	8.0 mg/kg	Oral	Turpentine injection was used as a painful stimulus	Achieved putative therapeutic concentrations within 2 h, but little evidence of therapeutic efficacy was seen.	Pain was assessed.	10	[42]
Phenylbutazone	Nonselective COX-1 and COX-2 inhibitor	1.0 g/sheep the day before and the day of the procedure and for 3 days postoperatively	Oral	Stifle surgery.	Effective analgesia.	Behavioural and physiological parameters were recorded.	30	[48]

* = registered for use in Australia.

A total of 11/28 studies did not report pain assessment methods. Routes of administration across drugs included the following: ketorolac–intravenous; meloxicam–intravenous, subcutaneous, transmucosal, and intramuscular; flunixin–intravenous, subcutaneous, intramuscular, and oral; diclofenac–topical; ketoprofen–intravenous, intramuscular, and oral; carprofen–oral; and phenylbutazone–oral.

3.2. Opioids

Seven opioids were identified: tramadol, buprenorphine, morphine, methadone, fentanyl, remifentanil, and oxycodone (Table 2). No opioid is currently registered for use in sheep in Australia. Fentanyl was the most used opioid and was found in six studies, and three multimodal combinations were reviewed. These were tramadol/lignocaine, buprenorphine/ketamine, and methadone/bupivacaine. The table below outlines the literature reviewed.

A total of seven out of twenty-two studies did not report pain assessment methods. Routes of administration included the following: tramadol–intravenous, intramuscular, transdermal, subcutaneous, and epidural; buprenorphine–intravenous, intramuscular, and subcutaneous (SR-only); morphine–intravenous, intramuscular, and epidural; methadone–intravenous and epidural; fentanyl–intravenous and transdermal; remifentanil–intravenous; and oxycodone–epidural.

3.3. Local Anaesthetics

The use of five local anaesthetics were identified: lignocaine, bupivacaine, levobupivacaine, procaine, and ropivacaine (Table 3). Of these, lignocaine is the only local anaesthetic registered for use in sheep in Australia. Lignocaine was also the most studied local anaesthetic, including eight multimodal combinations: lignocaine/xylazine, lignocaine/morphine, lignocaine/adrenalin, lignocaine/tramadol, bupivacaine/morphine, bupivacaine/lignocaine, bupivacaine/methadone, and bupivacaine/fentanyl and Tri-Solfen®. The table below outlines the literature reviewed.

A total of two out of twenty-nine studies did not report pain assessment methods. Routes of administration included lignocaine–intra-tissue, epidural, subcutaneous, paravertebral, intramuscular, and nerve blocks; bupivacaine–epidural, paravertebral, and nerve blocks; levobupivacaine–epidural; procaine–intra-tissue and subcutaneous; ropivacaine–epidural and nerve block; and Tri-Solfen®–topical.

3.4. α_2 Adrenoreceptor Agonists

Five α_2 adrenoreceptor agonists were identified: clonidine, xylazine, medetomidine, dexmedetomidine, and detomidine (Table 4). Xylazine is the only α_2 adrenoreceptor agonist registered for use in sheep in Australia. Medetomidine was the most used α_2 adrenoreceptor agonist, including two multimodal combinations. The latter were clonidine/lignocaine/buprenorphine, and dexmedetomidine/lignocaine. The table below outlines the literature reviewed.

Table 2. Opioids used for analgesia in sheep.

Drug	Action	Dose	Route	Indication	Summary of Effect	Pain Assessment Method	Sheep (n)	Reference
Tramadol	Weak μ agonist + serotonin reuptake inhibitor	4 and 6 mg/kg	Intravenous	Use of a mechanical nociceptive threshold (MNT) device.	Antinociceptive effects were not detected.	Physiological parameters, blood samples, and mechanical nociceptive threshold (MNT) values were recorded.	6	[49]
		1 mg/kg	Lumbosacral epidural	Needle pricks were used as a painful stimulus.	Analgesia lasted for 318.6 ± 5.08 min and began at 14.29 ± 1.24 min.	Pain was assessed in study.	7	[50]
		2 mg/kg	Epidural	Postoperative caesarean section analgesia.	Analgesia up to 8 h.	Adaptation of the UNESP-Botucatu One-Dimensional Scale for Post-Operative Pain Evaluation in Bovine was recorded.	2	[51]
		100.0 mg/sheep	Intravenous	Postoperative analgesia.	Not recorded.	No.	10	[7]
Tramadol (5%) and lignocaine (2%)	Weak μ agonist and serotonin reuptake inhibitor + local anaesthetic	2 mg/kg and 2 mg/kg	Lumbosacral	Laparo-ovariectomy.	No beneficial effect over epidural injection of lignocaine alone. Duration of analgesia was 133 ± 19.5 min.	Pain was assessed in study.	10	[52]
Buprenorphine	Partial μ and κ agonist, δ antagonist	10 μg/kg	Epidural	No procedure.	Not recorded.	No.	14	[53]
		0.01 mg/kg q 8 h for 48 h beginning 1 h before anaesthesia induction.	Intravenous bolus	Instrumentation of the foetus.	Acceptable postoperative analgesia.	Physiologic variables and behavioural were recorded.	6	[54]
Buprenorphine (Slow release/long acting 72 h)	Partial μ and κ agonist, δ antagonist	0.27 mg/kg	Intramuscular	A thermal portable device was used to assess SRB-induced antinociception.	Well-tolerated analgesic. Plasma concentrations increased; the thermal withdrawal time declined.	SRB-induced antinociception.	4	[55]
		4 mg/sheep pre- and postoperatively.	Subcutaneous	Third-degree flame skin burn and smoke inhalation.	Not recorded.	None.	11	[56]
Buprenorphine and Ketamine	Partial μ and κ agonist, δ antagonist + NMDA receptor antagonist	10 μg/kg + 1 mg/kg 30 min later. Then Ketamine at 5 mg/kg/h.	Intravenous	Experimental intervertebral disk nucleotomy.	Prevented increases in HR and MAP during surgery.	Cardiovascular response to noxious stimulation.	18	[57]
Morphine	μ agonist	0.1 mg/kg post operatively	Epidural	Caesarean section.	Analgesia up to 6 h.	Adaptation of the UNESP-Botucatu One-Dimensional Scale for Post-Operative Pain Evaluation in Bovine was used.	3	[51]
		0.1 mg kg	Thoracic epidural	No procedure.	Average duration of analgesia was 45 min.	Pain was assessed in study.	6	[58]
		0.5 mg/kg	Intramuscular (preoperative)	Stifle surgery.	Effective analgesia.	Behavioural and physiological parameters were recorded.	30	[48]
		0.1 mg/kg every 4 h post-op.	Not recorded	Laparotomy and hysterectomy.	Not recorded.	Pain scores were recorded.	6	[59]

Table 2. *Cont.*

Drug	Action	Dose	Route	Indication	Summary of Effect	Pain Assessment Method	Sheep (n)	Reference
		0.2 mg/kg post-op	Intravenous	Elective laparoscopy.	Not recorded.	The UNESP-Botucatu composite scale was used to assess acute postoperative abdominal pain.	48	[37]
Methadone	μ agonist + NMDA antagonist	0.3 mg/kg	Intravenous	Experimental intervertebral disk nucleotomy.	Prevented increases in HR and MAP during surgery.	Cardiovascular response to noxious stimulation.	18	[57]
		0.3 mg/kg	Lumbosacral epidural	No procedure.	Duration of analgesia was 220 min.	Pain scored by deep application of muscle pricks.	6	[60]
Methadone and Bupivacaine	μ agonist and NMDA antagonist+ local anaesthetic	0.15 mg/kg + 0.25 mg/kg	Lumbosacral epidural	No procedure.	Duration of analgesia was 180 min.	Pain scored by application of deep muscle pricks.	6	[60]
Fentanyl	μ agonist κ agonist	2 μg/kg followed by 10 μg/kg/h	Intravenous	Experimental intervertebral disk nucleotomy.	Prevented increases in HR and MAP during surgery.	Cardiovascular response to noxious stimulation.	18	[57]
		2 μg/kg/h	Transdermal patch foreleg and thorax	No procedure.	Provided sufficient analgesia if applied 3–6 h before painful event. Foreleg patch provided faster and longer lasting analgesia.	Measured blood levels to assess if fentanyl plasma concentrations had reached the minimum analgesia level for opioid-naïve humans of 0.6–1.5 ng/mL. Physiological parameters and behaviour were observed.	12	[61]
		2 patches: 100 ug/kg/h and 50 ug/kg/h placed 1 day pre-operatively	Transdermal patch on both forelimbs	Stifle surgery.	Effective analgesia.	Behavioural and physiological parameters were recorded.	30	[48]
		2 μg/kg/h	Transdermal patch on foreleg	Orthopaedic surgery.	Minimum dose rate of 2 μg/kg was required for analgesia.	Measured blood levels to assess fentanyl plasma concentrations. Physiological parameters and behaviour were observed.	8	[62]
		2.0 μg/kg loading dose followed by 2.5 μg/kg/hr infusion	Intravenous	Abdominal surgery.	Not recorded.	No.	10	[13]
		2 μg/kg/hr	Transdermal patch	Abdominal surgery.	Not recorded.	No.	10	[13]
		1.4 ± 0.2 μg kg/hour	Transdermal patch	Laparotomy and hysterectomy.	Not recorded.	No.	10	[14]
		75 μg/hour patch	Transdermal patch	Laparotomy and hysterectomy.	No analgesia noted.	Thermal and mechanical thresholds were measured.	8	[63]
		2 μg of /kg/h	Transdermal patch	Surgery for instrumentation of the foetus.	Acceptable postoperative analgesia.	Physiologic variables and behavioural changes indicative of pain were assessed.	6	[54]
Remifentanil	μ agonist	0.33 μg/kg/min for 1 h	Intravenous continuous infusion	Caesarean section.	Not recorded.	No.	7	[15]

Table 2. *Cont.*

Drug	Action	Dose	Route	Indication	Summary of Effect	Pain Assessment Method	Sheep (n)	Reference
Oxycodone	µ agonist	0.1 mg/kg infusion then 0.05 mg/kg/h for five days.	Epidural	Laparotomy.	Not recorded.	No.	10	[16]
		Initial 0.4 mg/kg bolus followed by 0.2 mg/kg boluses BID for five days.	Epidural	Laparotomy.	Not recorded.	No.	10	[16]

Table 3. Local anaesthetics used for analgesia in sheep.

Drug	Action	Dose	Route	Indication	Summary of Effect	Pain Assessment Method	Sheep (n)	Reference
Lignocaine (2%) *	Local anaesthetic	2 mg/kg	Intrafunicular	Orchiectomy.	No analgesia.	Pain was assessed in study.	6	[41]
		30 mg/site	Numnuts® device injection at ring site	Castration and tail docking with rubber rings.	Provided analgesia during the acute pain response.	Time to mother up, acute pain-related behaviours and postures were recorded.	50	[64]
		1.5 mL/site	Numnuts® device injection at ring site	Tail docking with rubber rings.	Abolished abnormal behaviours and signs of pain however some evidence of residual discomfort remained.	Pain-related behaviours were recorded.	10	[65]
		1.5 mL/site	Numnuts® device injection at ring site	Ring castration and tail docking.	Early onset but short-lived analgesia.	Active pain avoidance behaviours were recorded.	56	[66]
		8 mL/sheep	4-point regional nerve block distal to the fetlock	Single distal limb lameness.	Resulted in anaesthesia of the distal limb.	A pressure algometer was used to quantify analgesia.	18	[67]
		9 mL total. 3 mL per paravertebral nerve.	Paravertebral	Nociceptive stimuli.	Durations of analgesia was 65 ± 18 min.	Nociceptive effects were recorded.	6	[68]
		1.2 mg/kg	Lumbosacral epidural	No procedure.	Antinociceptive effects were observed up to 60 min.	Anti-nociceptive effects were recorded.	6	[69]
		2 mL/sheep	Subcutaneous–Metacarpi block	Nociceptive stimuli.	Analgesia of the Metacarpi was limited to 60 min.	Nociceptive threshold was measured.	4	[70]
		2.86 mg/kg	Lumbosacral epidural	Needle pricks.	Duration of analgesia 54.43 ± 3.28 min	Analgesia tested by recording response to sharp needle pricks.	7	[50]
		2 mL/sheep	Intra-testicular injection	Castration.	Not recorded.	Electroencephalography was used.	8	[71]
		1 mL/sheep	Subcutaneous into scrotal neck, spermatic cords, and tail prior to procedure	Ring castration and hot iron docking.	Reduced acute pain to some degree.	Behavioural indicators of pain were recorded.	15	[34]

Table 3. *Cont.*

Drug	Action	Dose	Route	Indication	Summary of Effect	Pain Assessment Method	Sheep (n)	Reference
		2 mL/sheep	Proximal paravertebral block	Caesarean section.	Not recorded.	No.	5	[51]
		2.5 mL + 5 mL/sheep	Subcutaneous (spermatic cords and scrotal neck)	Burdizzo castration.	Some analgesia within the first 2 h.	Behaviour, inflammation, ANS, HPA, and oxidative stress were recorded.	24	[45]
		Not recorded	Subcutaneous and Intramuscular inverted L block	Surgical placement of rumen fistula.	Not recorded.	Analgesia tested by recording response to sharp needle pricks.	13	[32]
		5 mg/kg	Brachial plexus block	Pin pricks and skin pinching with haemostats.	Produced forelimb analgesia within 11.3 min. Mean duration of analgesia was 100 min.	Responses to aversive pin pricks and skin pinches were recorded.	9	[72]
		5 mg/kg	Brachial plexus block	Pin pricks and skin pinching with haemostats.	Provided analgesia for 100 ± 38 min.	Responses to aversive pin pricks and skin pinches were recorded.	7	[73]
		5 mg/kg + 0.05 mg kg	Brachial plexus block	Pin pricks and skin pinching with haemostats.	Produced forelimb analgesia within 7 min. Mean duration of analgesia was 186.8 min.	Responses to aversive pin pricks and skin pinches were recorded.	9	[72]
Lignocaine and Xylazine	Local anaesthetic + α2 agonist	3.9 mg/kg + 0.05 mg/kg	Lumbosacral epidural	No procedure.	Provided prolonged anaesthesia that may contribute to pain relief in the immediate post-operative period.	Pain scoring system was used.	6	[74]
		5 mg/kg + 0.1 mg/kg	Brachial plexus block	Pin pricks and skin pinching with haemostats.	Mean duration of analgesia to brachial plexus was 103 ± 35 min. Produced forelimb analgesia within 11 min. Mean duration of analgesia was 133.2 min.	Responses to aversive pin pricks and skin pinches were recorded.	7	[73]
Lignocaine and Morphine	Local anaesthetic + μ agonist	5 mg/kg + 5 μg mL	Brachial plexus block	Pin pricks and skin pinching with haemostats.	Rapid onset analgesia with short duration of action. Duration of analgesia was 119.4 ± 52.5 min.	Responses to aversive pin pricks and skin pinches were recorded	9	[72]
Lignocaine and Adrenalin	Local anaesthetic + sympathomimetic	4 mg/kg	Lumbosacral epidural	Laparo-ovariectomy.	Not suitable for medium to long-term surgery.	Pain was assessed in study.	10	[52]
		9 mL total. 3 mL per paravertebral area.	Paravertebral	Nociceptive stimuli.	Durations of analgesia 95 ± 46 min.	Nociceptive stimuli response was recorded.	6	[68]
		4.2 mg/kg + 5 μg/mL	Lumbosacral epidural	Pin pricks and skin pinching with haemostats.	Provided prolonged anaesthesia that may contribute to pain relief immediately postoperatively.	Pain scoring was used.	6	[74]
Lignocaine and Tramadol	Local anaesthetic + Weak μ agonist + serotonin	5 mg kg + 1 mg/kg	Brachial plexus block	Pin pricks and skin pinching with haemostats.	Mean duration of sensory block was 79 ± 28 min.	Response to aversive pin pricks and skin pinches were recorded.	7	[73]

Table 3. *Cont.*

Drug	Action	Dose	Route	Indication	Summary of Effect	Pain Assessment Method	Sheep (n)	Reference
	reuptake inhibitor	2.46 mg/kg + 1 mg/kg	Lumbosacral epidural	Pin pricks with needles.	Rapid onset of perineal and cutaneous analgesia 5.58 ± 0.40 min and prolonged duration 100.7 ± 4.80 min.	Needle prick response was recorded.	7	[50]
		0.5 mg/kg	Femoral nerve or the sciatic nerve block.	Surgery on the femorotibial joint.	No clear benefit of nerve block.	Physiological and behavioural measures were recorded.	15	[48]
		1.5 mL/site	Numnuts® device injection into ring site	Rubber ring castration and tail docking.	More sustained analgesia than only lidocaine.	Active pain avoidance behaviours were recorded.	32	[66]
Bupivacaine (0.75%)	Local anaesthetic	1.2 ± 0.1 mg/kg.	Lumbosacral epidural	Pin pricks and skin pinching with haemostats.	Prolonged anaesthesia that might contribute to pain relief in the postoperative period.	Pain scoring was used.	6	[74]
Bupivacaine (0.5%)	Local anaesthetic	9 mL total. 3 mL per paravertebral nerve.	Paravertebral	Nociceptive stimuli.	Produces a longer duration of analgesia than lidocaine with or without epinephrine.	Nociceptive response to stimuli.	6	[68]
		1.25 mg/kg	Brachial plexus block	Pin pricks and skin pinching with haemostats.	Mean duration of sensory block was 335 ± 134 min.	Responses to aversive pin pricks and skin pinches were recorded.	7	[73]
		2 mL/site	Subcutaneous–Metacarpal block	Nociceptive stimuli.	Duration of anaesthesia 110.0 ± 47.26 min. Lasted for 120 min, and the best analgesia was between 60 and 120 min.	Nociceptive threshold was measured.	4	[70]
		0.5 mg/kg	Lumbosacral epidural	Deep muscle needle pricks.	Duration of analgesia was 240 min.	Response was scored after deep muscle pricks.	6	[60]
		0.5 mg/kg	Thoracic epidural	No procedure.	Average duration of analgesia was 60 min.	Pain was assessed in study.	6	[58]
Bupivacaine (0.25%)	Local anaesthetic	0.5 mg/kg	Lumbosacral epidural	Painful stimulus.	Duration of analgesia was 240 min.	Response to painful stimulus was recorded.	6	[11]
Bupivacaine and Morphine	Local anaesthetic + μ agonist	0.25 mg/kg + 0.05 mg kg	Thoracic epidural	No procedure.	Average duration of analgesia was 140 min.	Pain was assessed in study.	6	[58]
Bupivacaine and Lignocaine	Local anaesthetics	1 mL + 11 mL/sheep	Subcutaneous metacarpal ring block	Nociceptive stimuli.	Anaesthesia lasted twice as long than with lignocaine alone. Onset of analgesia was slower than bupivacaine alone.	Nociceptive threshold was measured.	4	[70]
Bupivacaine and Methadone	Local anaesthetic + μ agonist and NMDA antagonist	0.25 mg/kg + 0.3 mg/kg	Lumbosacral epidural		Duration of analgesia was 240 min.	Response to a painful stimulus was recorded.	6	[11]
Bupivacaine and Fentanyl	Local anaesthetic + μ agonist and κ agonist	0.25 mg/kg + 0.002 mg/kg	Lumbosacral epidural	Painful stimulus..	Duration of analgesia was 180 min.	Response to a painful stimulus was recorded.	6	[11]
Levobupivacaine	Local anaesthetic	0.05 mg/kg	Lumbosacral epidural	Deep muscle needle pricks.	30 ± 5 min of local anaesthesia.	Response to deep muscle pricks were recorded.	6	[75]

Table 3. *Cont.*

Drug	Action	Dose	Route	Indication	Summary of Effect	Pain Assessment Method	Sheep (n)	Reference
		0.15 mg/kg	Lumbosacral epidural	Deep muscle needle pricks.	145 ± 27 min of local anaesthesia.	Response to deep muscle pricks were recorded.	6	[75]
		0.25 mg/kg	Lumbosacral epidural	Deep muscle needle pricks.	290 ± 18 min of local anaesthesia.	Response to deep muscle prick was used as a painful stimulus.	6	[75]
Procaine (5%) and adrenalin (0.002%)	Local anaesthetic + sympathomimetic	0.3 mL/ lamb (2- to 3-day-old lambs) at time of procedure	Subcutaneous injection into Spermatic cords	Castration with rubber rings.	Produced acute analgesia for visceral pain.	Active behavioural responses and postures of the lambs were recorded.	8	[76]
		1.5 mL/site (75 mg per site)	Numnuts® device injection into ring site	Castration and tail docking using rubber rings.	More sustained and quicker onset of analgesia than lidocaine.	Active pain avoidance behaviours were recorded.	17	[66]
Ropivacaine (0.5%)	Local anaesthetic	10 mL/sheep	Block of the femoral and sciatic nerves under ultrasound guidance	Tibial osteotomy.	Analgesia for an average of 6 h.	Grimace scale, pain scoring, heart rate, respiratory rate, and mean blood pressure were recorded.	12	[9]
		10 mL/sheep	Epidural	Tibial osteotomy.	Analgesia for an average of 8 h.	Grimace scale, pain scoring, heart rate, respiratory rate, and mean blood pressure were recorded.	13	[9]
Tri-Solfen® * (Lignocaine, bupivacaine, adrenalin and cetrimide)	Local anaesthetic + sympathomimetic + antiseptic	Single spray of 1.5 mL/sheep applied to lesions	Topical	Treatment of Orf virus lesions.	Not recorded.	No.	11	[77]
		0.5 mL/kg	Topical spray onto wound	Mulesing + hot-iron tail docking.	Significant analgesia for at least 24 h after mulesing.	Body weight, behavioural responses, assessment of skin and wound sensitivity, and time to mother up and to feed were measured.	20	[78]
		Lambs 5–10 kg: 6 mL 11–15 kg: 8 mL 16–20 kg: 10 mL >20 kg: 12 mL	Topical spray onto wounds	Surgical mulesing and hot-knife tail docking.	Provided analgesia. Pain was not eliminated.	Lamb behaviour was observed.	20	[23]
		Lambs 5–10 kg 6 mL; 11–15 kg 8 mL; 16–20 kg 10 mL; >20 kg 12 mL	Spray onto wounds	Surgical mulesing and hot-knife tail docking.	Provided rapid-onset analgesia.	Pain avoidance behaviour, cortisol concentrations and postural behaviour were recorded.	24	[23]

* = registered for use in Australia.

Table 4. α_2 Adrenoreceptor agonists used for analgesia in sheep.

Drug	Action	Dose	Route	Indication	Summary of Effect	Pain Assessment Method	Sheep (n)	Reference
Clonidine, Lignocaine (2%) and Buprenorphine	α_2 agonist + local anaesthetic + Partial μ and κ agonist and δ antagonist	2 μg/kg + 2 mg/kg + 300 μg	Intrathecal	Spinal anaesthesia for orthopaedic surgery.	Addition of clonidine produces a faster onset and a long-lasting analgesia compared to lidocaine and buprenorphine combination.	Presence of reflexes were assessed, and ataxia was scored.	20	[79]
Xylazine *	α_2 agonist	0.4 mg/kg	Intramuscular	Skin and muscle needle pricks.	Xylazine has a mild analgesic effect on sheep during deep sedation.	Skin and muscle pricks were used as a painful stimulus.	5	[80]
		0.2 mg/kg	Intravenous	No procedure.	Produced skin analgesia and medium to deep degree of sedation.	Analgesic effects were recorded.	8	[81]
Medetomidine	α_2 agonist	15 μg/kg	Intravenous	No procedure.	Not recorded.	No.	4	[82]
		15 μg/kg	Oral	No procedure.	Not recorded.	No.	4	[82]
		6 μg/kg	Intravenous	No procedure.	No analgesia was achieved after administration. Produced light to medium sedation.	Pain scoring was performed.	8	[81]
		3 μg/kg/h	Intraperitoneal (continuous infusion) postoperatively	Laparotomy and hysterectomy.	Provided analgesia for 24 h after surgery.	Pain was assessed.	6	[59]
		3 μg kg/hour	Intraperitoneal via osmotic pump	Laparotomy and hysterotomy.	May have a role in providing post-operative analgesia.	Thermal and mechanical thresholds were recorded.	8	[63]
Dexmedetomidine	α_2 agonist	2.5 μg/kg	Lumbosacral epidural	Nociceptive stimuli.	Inferior antinociceptive effects compared to dexmedetomidine and lignocaine combination.	Anti-nociceptive effects were measured.	6	[69]
		1 μg/kg/h for 3 h	Intravenous	No procedure.	No recorded.	No.	1	[83]
Dexmedetomidine and Lignocaine	α_2 agonist + local anaesthetic	2.5 μg/kg + 1.2 mg/kg	Lumbosacral epidural	Nociceptive stimuli.	Prolonged analgesia	Anti-nociceptive effects were measured.	6	[69]
Detomidine	α_2 agonist	40 μg/Kg	Intravenous	No procedure	No analgesia was achieved after administration. Produced light to medium sedation.	Analgesic effects were recorded.	8	[81]

* = registered for use in Australia.

A total of 2/8 studies did not report pain assessment methods. Routes of administration included the following: clonidine–intrathecal; xylazine–intravenous and intramuscular; medetomidine–intravenous, oral, and intraperitoneal; dexmedetomidine–intravenous and epidural; and detomidine–intravenous.

3.5. Other Analgesia

In the recent literature, the use of fourteen analgesic drugs or drug combinations in sheep were identified outside of the drug classes in Tables 1–4. These were metamizole, ketamine, racemic ketamine, magnesium sulphate, proglumide, diltiazem, nifedipine, verapamil, L-AP_3, D L-AP_3, salicylic acid, paracetamol, and amitriptyline (Table 5). Of these fourteen, ketamine was the only drug registered for use in sheep in Australia. There were two multimodal combinations: ketamine/lignocaine and ketamine/magnesium sulphate. The table below outlines the literature reviewed.

A total of four out of seven studies did not report pain assessment methods. Routes of administration included the following: metamizole–intravenous; ketamine/racemic ketamine–subarachnoid and epidural; magnesium sulphate–epidural; proglumide–intracerebroventricular; diltiazem–intracerebroventricular; nifedipine–intracerebroventricular; verapamil–intracerebroventricular; L-AP_3–intracerebroventricular; D L-AP_3–intracerebroventricular; salicylic acid–intravenous and oral; paracetamol–intravenous and oral; and amitriptyline–intravenous, epidural, and intrathecal.

Table 5. NMDA Receptor Agonists and Other Analgesics in sheep.

Drug	Action	Dose	Route	Indication	Summary of Effect	Pain Assessment Method	Sheep (n)	Reference
Metamizole	Non opioid analgesic	1000 mg/sheep SID	Intravenous	Post-op analgesia.	Effect not recorded.	No.	10	[7]
Ketamine *	NMDA receptor antagonist	2.5 mg/kg	Lumbosacral epidural	Deep muscle needle pricks.	41 ± 7 min of analgesia.	Deep muscle pricks were used as a painful stimulus.	6	[84]
Racemic ketamine and Lignocaine (2%)	NMDA receptor antagonist + local anaesthetic	3.0 mg kg + 1.5 mg kg	Subarachnoid	Bilateral orchiectomy.	Produced surgical analgesia and recumbency.	Response to scrotal skin pricks recorded.	10	[44]
Ketamine and Magnesium Sulphate	NMDA receptor antagonists	2.5 mg/kg + 100 mg	Lumbosacral epidural	Deep muscle needle pricks.	115 ± 17 min of analgesia.	Response to deep muscle pricks recorded.	6	[84]
Magnesium sulphate	NMDA receptor antagonist	100 mg/sheep	Lumbosacral epidural	Deep muscle needle pricks.	29 ± 5 min of analgesia.	Response to deep muscle pricks recorded.	6	[84]
Proglumide	Inhibitor of Cholecystokinin	25 or 50 µg/kg	Intracerebroventricular	Mechanically induced duodenal distension.	Effective analgesic agent for duodenal pain.	Sheep behaviour, plasma catecholamines (CA), cortisol concentration, and clinical symptoms of visceral pain.	6	[85]
Diltiazem	Voltage-Dependent Calcium Channel Inhibitor	25 or 50 µg/kg	Intracerebroventricular	Mechanically induced duodenal distension.	Prevented nocifensive signs of behaviour and clinical symptoms, as well as increased plasma cortisol and catecholamine concentration in periphery and perhaps in CNS structures.	Sheep behaviour, plasma catecholamines (CA), cortisol concentration, and clinical symptoms of visceral pain.	6	[85]
Nifedipine	Voltage-Dependent Calcium Channel Inhibitor	25 or 50 µg/kg	Intracerebroventricular	Mechanically induced duodenal distension.	Provided peripheral analgesia and possibly CNS analgesia.	Sheep behaviour, plasma catecholamines (CA), cortisol concentration, and clinical symptoms of visceral pain.	6	[85]
Verapamil	Voltage-Dependent Calcium Channel Inhibitor	25 or 50 µg/kg	Intracerebroventricular	Mechanically induced duodenal distension.	Provided peripheral analgesia and possibly CNS analgesia.	Sheep behaviour, plasma catecholamines (CA), cortisol concentration, and clinical symptoms of visceral pain.	6	[85]
L-AP_3	Inhibitor of Metabotropic Glutaminergic Receptors ($mGluR_1$)	0.2, 0.4, and/or 0.8 mg total/sheep	Intracerebroventricular	Mechanically induced duodenal distension.	Worked as an analgesic and an antistress agent.	Sheep behaviour, plasma catecholamines (CA), cortisol concentration, and clinical symptoms of visceral pain.	6	[85]
DL-AP_3	Inhibitor of Metabotropic Glutaminergic Receptors ($mGluR_1$)	2, 4, and/or 8 mg total/sheep	Intracerebroventricular	Mechanically induced duodenal distension.	Worked as an analgesic and an antistress agent.	Sheep behaviour, plasma catecholamines (CA), cortisol concentration, and clinical symptoms of visceral pain.	6	[85]
Salicylic Acid	Monohydroxybenzoic acid, nonselective COX inhibitor	10, 50, 100, and 200 mg/kg	Intravenous	No procedure.	Not recorded.	No.	6	[86]
		100 and 200 mg/kg	Oral	No procedure.	Not recorded.	No.	6	[86]
Paracetamol	Non NSAID analgesic and anti-pyretic	10 mg/kg	Intravenous	Post-surgical analgesia.	Not recorded.	Undisclosed.	7	[87]
		15 mg/kg orally BID for 6 doses	Oral	Post-surgical analgesia.	Not recorded.	Undisclosed.	7	[87]
Amitriptyline	Tricyclic antidepressant	5 mg/sheep	Intravenous	No procedure.	Not recorded.	Undisclosed.	6	[88]
		10 mg/sheep	Intrathecal	No procedure.	Not recorded.	Undisclosed.	6	[88]
		50 mg/sheep	Epidural	No procedure.	Not recorded.	Undisclosed.	6	[88]

* = registered for use in Australia.

4. Discussion

4.1. NSAIDs

The mechanism of action of NSAIDs is to reduce the synthesis of prostaglandins by inhibiting cyclooxygenase (COX) enzymes in the arachidonic acid pathway [89]. NSAIDs have been shown to have anti-inflammatory, anti-pyretic, and analgesic effects. There was only one drug banned in Australia for use in livestock that was found in this review (phenylbutazone [90]) with the remainder of drugs either registered or potentially able to be used off licence/off label. Meloxicam remains the only NSAID registered for use in Australia for sheep and is available in transmucosal oral and injectable formulations. Both formulations were found in the studies reviewed. The use of 1.0 mg/kg dose of meloxicam was frequently used across all studies except for two studies which used a lower than recommended dose of 0.5 mg/kg [31,32]. An analgesic effect was not recorded when meloxicam was given at this lower dose, and it is unclear whether this lower dose would offer effective pain relief. Therefore, the use of 1.0 mg/kg remains the recommended dose based on the available literature. The timing of the administration of meloxicam varied. However, manufacturer guidelines state pain relief can be effective for up to 24 h. Most studies gave a single dose of meloxicam at the time of the painful procedure. Metacam® also has a broad claim for the alleviation of pain and inflammation which includes any conditions causing inflammation and pain in sheep [21]. It can therefore be prescribed to sheep with painful disease processes such as flystrike, mastitis, foot rot, and shearing cuts in addition to other painful conditions. Three studies recorded the use of an NSAID to alleviate a painful disease process rather than a procedure [27–29]. Meloxicam was used for post-partum analgesia although its analgesic effect was not recorded [35]. Flunixin was used for footrot analgesia but was found to have no significant effect on footrot induced lameness [40]. Ketoprofen was also given to reduce pain associated with polyarthritis caused by *Erysipelothrix rhusiopathiae* but its analgesic effect was not recorded [47].

While multimodal analgesia is currently recognised as best practice for lambs undergoing lamb marking in Australia [1], of the three studies that used multimodal analgesia, [26,32,91] only the combinations of Tri-Solfen® with meloxicam and lignocaine and meloxicam are registered for use in sheep (see Table 1). The combination of meloxicam and Tri-Solfen® provided some level of analgesia in most studies [23,24,33]. It should be remembered that Tri-Solfen® is only effective on open wounds. Therefore, it is thought to be suitable for mulesing and knife docking but not suitable for marking (castration and/or tail docking) with rubber rings or similar non-open-wound procedures [92]. As an alternative, the registered meloxicam and lignocaine combination can be used for rubber ring marking methods [34].

The use of drugs confirmed to provide analgesia in some types of painful procedures can be used to manage other painful disease processes on farm under veterinary supervision. Given the paucity of information and inconsistent numbers of formally assessed studies in sheep analgesia, this option may be feasible if there is clear communication with the sheep owner on the use of unlicensed products and a plan for the management of the animal in a farm context where withholding periods must be adhered to. Studies demonstrating NSAIDs are effective at relieving pain associated with naturally occurring diseases are limited, and future research should capitalise on opportunities to demonstrate efficacy. More research to assess the potential frequency, clinical analgesic effect, and refined dosing intervals is required to validate pain relief for both painful procedures and disease processes on farms. Additionally, to ensure withholding periods are appropriate with increased frequency or prolonged dosing regimens. Further research would be required and could be used to approve future prolonged drug dosing regimens across the wider sheep industry to offer more sustained pain relief and improve animal welfare. Drugs that may be of most interest for analgesic use individually or as part of multi-modal analgesia and/or research could include meloxicam, ketoprofen, flunixin, ketorolac, and carprofen.

4.2. Opioids

The opioids included in the review were full μ-opioid receptor agonists (morphine, tramadol, methadone, remifentanil, oxycodone), partial μ agonists (buprenorphine), and κ agonists (fentanyl). Opioid receptors are distributed in the periphery, spinal cord, and brain. Opioids combine reversibly with these receptors and alter the transmission and perception of pain. In addition to analgesia, opioids can cause side effects such as sedation, dysphoria, euphoria, and excitement [89]. All studies were for biomedical research procedures. No opioids are currently registered for use in sheep in Australia. Much of the information on opioid analgesia and pain relief validation methods in sheep is extrapolated from other species and human medicine. As evident in Table 2, there remains a large variation in doses, usage, and efficacy between studies. The use of opioids for analgesia in sheep should, therefore, be interpreted and used with care. More studies on the use of opioid dose, and frequency are required to review and confirm of analgesic effectiveness before assuming regimens are clinically suitable for sheep [11].

Fentanyl had the greatest number (6) of publications found in this review. Fentanyl was used intravenously and transdermally in the papers reviewed. Five of the studies using fentanyl patches assessed efficacy of its analgesic effect [48,54,57,61,62]. In the literature reviewed, only fentanyl patches were used transdermally. This finding contrasts with the use of transdermal analgesia, in small animal veterinary clinical practice where fentanyl patches as well as lidocaine and buprenorphine patches can be used for pain relief post-operatively in orthopaedic and laparotomy surgeries [93]. The multimodal combinations of tramadol/lignocaine, buprenorphine/ketamine, and methadone/bupivacaine were all validated for pain relief [52,57,60]. Unfortunately, due to the potential expense and possible risks of human abuse of opioids, it is unlikely opioids will become commonly available for pain relief in farming enterprises. Any potential registration of opioid drugs in sheep would also require the development of appropriate withholding periods to avoid any residues in animals intended for human consumption. However, the use of opioids for the treatment of more invasive and painful procedures is a likely important option in biomedical research. Given these animals do not typically enter the food chain there is minimal risk to human food safety and potentially lower opportunity for misuse as animals are typically held in a highly controlled and regulated environment. If opioids were found to be effective and registered for use in sheep, it would offer the opportunity for uplift and more multi-modal regimens in sheep undergoing painful procedures or conditions. Opioids that may offer the most potential for use or further exploration individually or as part of multi-modal analgesic options could be methadone, fentanyl, morphine, buprenorphine, oxycodone, and remifentanil. Sheep may then be routinely provided with a higher standard of pain relief more akin to small animal and human patients. Procedures such as fracture repair in stud sheep, caesarean sections, or other painful procedures could be performed with better analgesia and contribute to improved animal welfare.

4.3. Local Anaesthetics

Local anaesthetics block the transmission of nociceptive impulses in the periphery to the brain [94] to create a local anaesthetic effect in the area of injection and the surrounding tissues innervated by targeted nerves. Nearly all papers listed assessed the local anaesthetic for pain relief and recorded an analgesic effect. Local anaesthetics have been used in both biomedical research and on-farm. Lignocaine is currently the only single-agent local anaesthetic registered for use in sheep in Australia. This differs from other countries' requirements such as in the European Union where lignocaine is not available (versus procaine) for use production animals [20]. In all three studies that utilised a pre-calibrated 1.5 mL subcutaneous dose of lignocaine via the Numnuts® device, analgesic effects were confirmed when the device was correctly used [64–66].

All Tri-Solfen® studies were performed as part of farm studies. There were no studies of Tri-Solfen® use in a biomedical research setting. Interestingly, one study sprayed 1.5 mL Tri-Solfen® directly onto Orf virus lesions [77]. While pain relief was not confirmed in this trial, it could be tested in the future as an option for painful disease states with open wounds. More work should continue to adapt and where possible modify current registered products such as Tri-Solfen® and Numocaine® to promote best practice and maximise the opportunity for analgesia in sheep.

Due to the relatively fast onset of action and short duration of effect, local anaesthetics are often used as part of a multimodal analgesia regime. These types of drugs can also be combined with other more long-lasting analgesics. In the studies reviewed, only the meloxicam/lignocaine and lignocaine/xylazine combinations are registered for use in sheep. There were also several unregistered combinations used for pain relief with potential effectiveness across the literature. Combinations found were lignocaine/morphine, lignocaine, lignocaine/tramadol, bupivacaine/morphine, bupivacaine/lignocaine, bupivacaine/methadone and bupivacaine/fentanyl [11,46,48,50,52,58,66,68,70,72–74,76] Meloxicam/lignocaine combinations can be used for various lamb marking procedures (including rubber rings) whereas meloxicam/Tri-Solfen® combinations are only appropriate for open wound procedures such as mulesing and hot-knife tail docking. Lignocaine/xylazine combination is another option that can be administered into the epidural space for caesarean sections or laparotomies for use in veterinary or research procedures. Overall, ropivacaine, lignocaine, procaine, bupivacaine, and levobupivacaine all appear to be potentially viable options for local analgesia. Future trials could assess other combinations of local anaesthetics and/or classes of drugs (e.g., opioids) with meloxicam. The clinical importance of unregistered drug combinations could also be studied further and registered to offer greater options and potential effectiveness for pain relief on farms as well as in research settings.

4.4. α_2 Adrenoreceptor Agonists

α_2 adrenoreceptor agonists bind to α_2 adrenoreceptors on vascular smooth muscle, inducing contraction and vasoconstriction [95]. α_2 adrenoreceptor agonists are commonly used sedative agents in livestock, but have also demonstrated analgesic effects particularly at sub-sedative doses [1]. In veterinary clinical practice, they often form part of a pre-medication anaesthesia protocol due to their combined sedative and analgesic effects. Xylazine is can also be used for epidural anaesthesia in combination with lignocaine [74]. Xylazine is the only α_2 adrenoreceptor agonist currently registered for use in sheep in Australia. This contrasts with the literature reviewed which identified a range of α_2 adrenoreceptor agonists (clonidine, xylazine, medetomidine, dexmedetomidine and detomidine) being used in biomedical research settings. Many of these are yet to be formally trialled for effectiveness or administration/regimen optimised. Multiple studies across the literature also noted the common sedative effects of these drugs [80,81]. The majority of studies did not report use of α_2 adrenoreceptor agonists as a primary agent to treat painful procedures or conditions. However, in some studies it was administered to test analgesic properties via skin and muscle pricks, thermal or mechanical threshold. Appropriate dosing is key with these drugs as risks are associated with α_2 adrenoreceptor agonists used at higher doses in sheep such as pulmonary oedema and late gestation abortions [96]. Nonetheless, the use of α_2 adrenoreceptor agonists at smaller doses may prove to be a beneficial adjunct to pain management and/or as premedication for analgesic purposes. Further studies investigating α_2 adrenoreceptor agonists are required to assess timing and optimal dose for effective potential analgesic effect rather than anaesthetic effects across different dosing regimens.

4.5. NMDA Receptor Agonists and Other Drugs

Table 5 summarises drugs that were not classified into any of the previous categories. Ketamine was also reported in the literature in both veterinary and biomedical procedures as a general anaesthetic and analgesic. Both ketamine/lignocaine and ketamine/magnesium sulphate combinations were validated to provide analgesia [44,84]. Similar to most opioids, ketamine's highly regulated Schedule 8 classification in Australia and profound anaesthetics effects may make it more appropriate on farm for veterinary-only use and/or biomedical research settings [97]. However, unlike other potent analgesics (e.g., opioids) found to be used in this review, it is already registered for use in sheep in Australia. The benefit of this means it has immediate potential to be a used as an adjunct to pain relief at both higher and/or lower doses for painful conditions or when administered as part of an analgesic or anaesthetic regimen for painful procedures. Similar to other drugs found in this review, further studies are still required to evaluate effectiveness, dose rates, timing, frequency, and appropriate routes of administration.

A collection of 'other drugs' identified were found within a single biomedical research study investigating the voltage-dependent calcium channel inhibitors of diltiazem, nifedipine, verapamil, proglumide, L-AP$_3$ and DL-AP$_3$. In this particular study, all the drugs listed were thought to provide visceral analgesia in mechanically induced duodenal distension [85]. Therefore, these drugs may be useful for other types of painful visceral conditions. Salicylic acid, paracetamol, and amitriptyline were also used in other studies but without any analgesic assessment described. Additional research may demonstrate these drugs could be new options or novel applications for pain relief and animal welfare improvement in sheep or possibly other ruminants.

4.6. Limitations

This review was undertaken to identify potential analgesic drugs, combinations, regimens, and options used to (potentially) alleviate pain in sheep via the use of scientific databases and grey literature. It is recognised that although many of the drugs utilised may not have been comprehensively investigated or shown to successfully and consistently alleviated pain, the information collated provides a broad list of potential drugs candidates and starting points for drug regimens for future investigations. A key limitation of this study was in the search strategy utilised as it was not feasible to identify all analgesia studies in sheep using the presented search methods. The search strategy was intentionally limited to the use of target words and did not include all known synonyms. While this prevented a higher number of inappropriate or irrelevant results, it may have missed some research-only publications and did miss some of the known textbooks or online formularies which may have listed additional drugs and/or drug regimens [98–100].

Additionally, the search criteria omitted publications prior to 2010 and after March 2022, such the more recent use of mint terpenoid L-carvone in sheep [101]. Due to the lack of published studies specific to analgesia in sheep found in the search, and from authors' knowledge, a small selection of published and grey literature information that fit the criteria for inclusion was also included. It is important to note this study did not fully capture drugs registered in all other countries and did not include the most modern human analgesics developments, such as tapentadol [1]. Some of these drugs might be of value to explore when developing new studies testing analgesics in sheep. Finally, a full review of the analgesic agents and pain assessments strategies utilised in sheep was outside of the scope of this study. Therefore, there remains a wealth of further opportunities available for future publications and research to build upon this review.

4.7. General Discussion

The current estimated number of sheep in Australia is 74 million [102]. All of these animals will undergo painful husbandry procedures at some stage in their lifetime. Herein is an enormous opportunity and responsibility for farmers, researchers, animal ethics committees, and veterinarians to improve the welfare of millions of animals through better analgesic practices. Despite the obligation for the provision of analgesia for good animal welfare, only six commercial products (lignocaine (2%), Tri-Solfen® (lignocaine hydrochloride 40.6 g/L, bupivacaine hydrochloride 4.2 g/L, adrenaline (as acid tartrate) 24.8 mg/L and Cetrimide 5 g/L), ketamine (as hydrochloride 100 mg/mL), xylazine (as hydrochloride 20 mg/mL), and oral transmucosal and injectable formulations of meloxicam (20 mg/mL)) are registered to alleviate pain in Australian sheep. Only three multimodal combinations (meloxicam/lignocaine, meloxicam/Tri-Solfen®, xylazine/lignocaine) are registered despite the importance of multi-modal analgesia as part of best practices in analgesia for moderately to severely painful procedures [1]. In addition to physically painful conditions or procedures, sheep can also experience painful disease processes. However, only three studies were found that trialled the use of pain relief for a disease process rather than a procedure [40,47,77]. Assessing and validating analgesics for painful procedures and conditions is an essential requirement for good sheep welfare across farm, veterinary clinical and biomedical research settings. There has been minimally publicly available known interest across the meat, livestock, veterinary and biomedical research industries to seek registration of new or novel analgesics products in the last 5 years. The most recent analgesic drug registration for sheep was meloxicam (Buccalgesic® in 2017 [22] and Butec® in 2023 [103]) and there are no other types of drugs known to the authors at this time undergoing testing for registration purposes. Only one topically non-drug analgesic option using cooling via the device CoolSense [87] for mild pain has been studied, but it is yet to be registered for animal use. There is still much more work to explore and required, with many opportunities for collaboration across biomedical research, veterinary clinical and farming industries to bridge the gaps in sheep analgesics.

The aim of this review was to improve the health and welfare of sheep in farming, biomedical and veterinary practices by exploring potential opportunities for analgesics in the scientific literature against the currently approved drugs in Australia. Literature on Australian registered drugs were predominantly found in the context of farming while most of the use of unregistered drugs were found in the biomedical research context. The review demonstrates there is a far larger array of potentially effective analgesics in sheep in comparison to the few available registered products. The use of these non-registered drugs is permitted in many biomedical studies since these animals would not be allowed to and are highly unlikely to exit research facilities prior to humane killing or euthanasia. While this may be the case in Australia and other jurisdictions, this can contrast with other international regulations such those found in the United Kingdom may prevent the use of a more appropriate drugs (cascade system) or in Europe where the use of some analgesics may not be easily permitted even biomedical settings [20]. There is a high likelihood that out of all the non-traditional, unregistered or formally untested drugs described in this research that some may prove to be important alternatives or primary agents in alleviating on-farm, biomedical and veterinary clinical management of pain in sheep. Therefore, there is a need for more research into these and other analgesics to ensure the availability of suitably safe, tested, effective and registered analgesic products to promote better welfare and ensure refinement of research outcomes in sheep.

The information from the sheep biomedical literature shows that there is clearly a potential for improved sheep welfare and an opportunity to alleviate pain to a potentially greater extent and/or beyond the approved drugs in Australia. However, on review of these publications, the dosage and route of administration for many unregistered drugs were quite varied. There were also a reasonable number of publications across the different drugs class categories that did not state the method of pain assessment. This is of considerable concern as confirmation of pain relief post-administration of analgesics is foundational to good veterinary clinical practice. For articles that did state the method of pain assessment, it remains unknown if the methods were appropriate for the procedure and context or if any other indicators were utilised to assist in validations of pain relief. There were also concerns regarding the lack of information listed for the frequency and timing of pain relief administered. According to the ARRIVE guidelines [104] pain relief should be utilised where appropriate and disclosed within publications. Unfortunately, this lack of disclosure has been documented historically in other animal studies [105]. Other issues posing animal welfare risks and concerns include the potential low usage or at a minimum lack of disclosure of appropriate multi-modal pain relief for high impact procedures (e.g., orthopaedics). The appropriate use of multi-modal regimens should be further explored and could have improved from of the pain management regimens.

The extrapolation of analgesics across species for similar conditions can be useful in the absence of other more formal science-based evidence. Nonetheless, it is still vital to support and advocate for well-developed sheep-specific analgesic studies. While it is likely sheep will respond similarly to other small ruminants and mammals, there are well known examples in veterinary medicine where some forms of pain relief can be deadly in other species (e.g., cats and paracetamol [106]) or require significantly higher or lower dosages (e.g., meloxicam in cats [107] versus mice [108]). It is also crucial to ensure appropriate analgesic regimens are explored and suitable options identified for sheep during various life stages (e.g., pregnancy, lambs) and for a variety of painful conditions (e.g., mastitis, bloat, castration).

Given the relatively frequent of use of sheep as large animal models in biomedical studies, it is in the spirit of the three Rs and incumbent on animal ethics committees, researchers, the biomedical industry, and associated veterinarians, to consider if adjustments to experimental design could simultaneously capture, advance, and support better analgesic regimens (ancillary research) in sheep. Simple refinements such as ensuring the use of appropriate pain and animal welfare assessment methods as well as their inclusions in publications would be a great first simple step. These considerations should ideally be a prerequisite for animal ethics committee approval for any sheep undergoing potential painful procedures. Additionally, many of these biomedical studies collect and utilise blood as well as other tissue samples which may be able to be re-used or re-purposed for used in safety and food animal drug testing residue studies to inform withdrawal times for slaughter and safety. While this approach may not be suitable or possible in types of biomedical work with sheep, there are myriad of (lost) opportunities that can be captured to advance the knowledge, welfare, and management of pain management and analgesics in sheep. Without further consideration, advancement and focus on ideal pain regimes for sheep, both the biomedical and farming industries are unnecessarily exposed risks to public support (social licence [109]) as well as possible reduced production [26] and research outcomes [109].

This review highlighted a wide array of unregistered potential drugs and doses that could be useful in sheep. Many of these unregistered and/or minimally studied drugs and doses may have been administered under the assumptions that their mode of action and analgesia would be comparable to humans and other mammalian species. There is still a concern that the dosed, frequencies, and use of these drugs may not be optimal or appropriate. Many of the studies included in this review lacked detailed pain assessment strategies or other key animal welfare indicators to enhance validity. Further probable barriers when using pain relief in food-producing species include the potential for human risk of abuse with more potent analgesics (e.g., companion animals), costs, risk of residues in food and dosing frequencies for appropriate analgesia. There may also be challenges in the practicality, applicability, and appropriateness of when these medications would be suitable for farm, biomedical or veterinary clinical use. These studies should be undertaken to support and encourage the registration of analgesic formulations for sheep including those intended for human consumption. Further research and greater encouragement for collaboration across all sheep industries should be undertaken to improve animal welfare and research outcomes to better meet ethical, societal, and legal obligations.

5. Conclusions

Good animal welfare, industry, and veterinary practices dictate that pain relief must be administered to animals experiencing pain., Farmers, veterinarians, and researchers are expected and often required to provide best-practice pain relief to animals undergoing painful procedures and disease processes in their care, highlighting current gaps, challenges, and opportunities for better pain relief in sheep including dose rates, routes of administration, indication of use, and any pain assessment strategies utilised. Both current and possible future analgesia options are outlined with key agents identified for further research either as individual drugs or as part of a multimodal strategy to improve sheep analgesia and welfare. Further research should also focus on the assessment of the safety and efficacy of new drugs or new formulations of old drugs, food safety testing and registration of additional analgesic agents to alleviate pain and improve the welfare of sheep in Australia and worldwide across the farming, biomedical research, and veterinary industries.

Author Contributions: Conceptualisation S.C. and E.F. Methodology S.C., T.B. and G.C.M. Writing original draft preparation S.C. and E.F. writing review and editing G.C.M. and T.B. All authors have read and agreed to the published version of the manuscript.

Funding: This research received no external funding.

Institutional Review Board Statement: Not applicable.

Informed Consent Statement: This review was of literature approved by various animal ethics committees.

Data Availability Statement: The raw data supporting the conclusions of this article will be made available by the authors on request.

Conflicts of Interest: The authors declare no conflict of interest.

References

1. Small, A.; Fisher, A.D.; Lee, C.; Colditz, I. Analgesia for Sheep in Commercial Production: Where to Next? *Animals* **2021**, *11*, 1127. [CrossRef] [PubMed]
2. Hampton, J.O.; Jones, B.; McGreevy, P.D. Social License and Animal Welfare: Developments from the Past Decade in Australia. *Animals* **2020**, *10*, 2237. [CrossRef] [PubMed]
3. AWI. *Merino Husbandry Practices Survey—Final Report*; Australian Wool Innovation: Sydney, Australia, 2017. Available online: https://www.wool.com/globalassets/wool/sheep/research-publications/welfare/surveys/221027-2021-awi-merino-husbandry-practices-survey-final-report.pdf (accessed on 1 July 2023).
4. Howard, K.; Beattie, L. *A National Producer Survey of Sheep Husbandry Practices*; Meat and Livestock Australia: Sydney, NSW, Australia, 2018.
5. NSW Young Lawyers Animal Law Committee. *Inquiry into Prevention of Cruelty to Animals Amendment (Restrictions on Stock Animal Procedures) Bill 2019*; NSW Young Lawyers Animal Law Committee: Sydney, NSW, Australia, 2020.
6. AWI. *Premiums and Discounts for Mulesing Status*; Australian Wool Innovation: Sydney, Australia, 2018. Available online: https://www.wool.com/globalassets/wool/sheep/research-publications/welfare/wool-market/btb-june2018-premiums-discounts-mulesing-status.pdf (accessed on 1 July 2023).
7. Al Hussein, H.; Al Hussein, H.; Sircuta, C.; Cotoi, O.S.; Movileanu, I.; Nistor, D.; Cordos, B.; Deac, R.; Suciu, H.; Brinzaniuc, K.; et al. Challenges in Perioperative Animal Care for Orthotopic Implantation of Tissue-Engineered Pulmonary Valves in the Ovine Model. *Tissue Eng. Regen. Med.* **2020**, *17*, 847–862. [CrossRef] [PubMed]
8. Viscardi, A.V.; Reppert, E.J.; Kleinhenz, M.D.; Wise, P.; Lin, Z.; Montgomery, S.; Daniell, H.; Curtis, A.; Martin, M.; Coetzee, J.F. Analgesic Comparison of Flunixin Meglumine or Meloxicam for Soft-Tissue Surgery in Sheep: A Pilot Study. *Animals* **2021**, *11*, 423. [CrossRef] [PubMed]
9. Stenger, V.; Zeiter, S.; Buchholz, T.; Arens, D.; Spadavecchia, C.; Schüpbach-Regula, G.; Rohrbach, H. Is a Block of the Femoral and Sciatic Nerves an Alternative to Epidural Analgesia in Sheep Undergoing Orthopaedic Hind Limb Surgery? A Prospective, Randomized, Double Blinded Experimental Trial. *Animals* **2021**, *11*, 2567. [CrossRef]
10. NHMRC. *Australian Code for the Care and Use of Animals for Scientific Purposes*; Commonwealth of Australia: Canberra, Australia, 2013.
11. DeRossi, R.; Pagliosa, R.C.; Carvalho, A.Q.d.; Macedo, G.G.; Hermeto, L.C. Fentanyl and methadone used as adjuncts to bupivacaine for lumbosacral epidural analgesia in sheep. *Vet. Rec.* **2017**, *180*, 96. [CrossRef]
12. Gigliuto, C.; De Gregori, M.; Malafoglia, V.; Raffaeli, W.; Compagnone, C.; Visai, L.; Petrini, P.; Avanzini, M.A.; Muscoli, C.; Viganò, J.; et al. Pain assessment in animal models: Do we need further studies? *J. Pain Res.* **2014**, *7*, 227–236. [CrossRef]
13. Heikkinen, E.M.; Voipio, H.M.; Laaksonen, S.; Haapala, L.; Räsänen, J.; Acharya, G.; Erkinaro, T.; Haapsamo, M.; Hautajärvi, H.; Kokki, H.; et al. Fentanyl Pharmacokinetics in Pregnant Sheep after Intravenous and Transdermal Administration to the Ewe. *Basic Clin. Pharmacol. Toxicol.* **2015**, *117*, 156–163. [CrossRef] [PubMed]
14. Musk, G.C.; Catanchin, C.S.M.; Usuda, H.; Woodward, E.; Kemp, M.W. The uptake of transdermal fentanyl in a pregnant sheep model. *Vet. Anaesth. Analg.* **2017**, *44*, 1382–1390. [CrossRef] [PubMed]
15. Coonen, J.B.; Marcus, M.A.; Joosten, E.A.; van Kleef, M.; Neef, C.; van Aken, H.; Gogarten, W. Transplacental transfer of remifentanil in the pregnant ewe. *Br. J. Pharmacol.* **2010**, *161*, 1472–1476. [CrossRef]
16. Kinnunen, M.; Kokki, H.; Hautajärvi, H.; Lantto, J.; Räsänen, J.; Voipio, H.M.; Kokki, M. Oxycodone concentrations in the central nervous system and cerebrospinal fluid after epidural administration to the pregnant ewe. *Basic Clin. Pharmacol. Toxicol.* **2019**, *125*, 430–438. [CrossRef]

17. RSPCA. What Are Some of the Painful Procedures Experienced by Sheep on Farm? 2022. Available online: https://kb.rspca.org.au/knowledge-base/what-are-some-of-the-painful-procedures-experienced-by-sheep-on-farm/ (accessed on 1 September 2022).
18. Australia, A.H. *Australian Animal Welfare Standards and Guidelines for Sheep*; Animal Health Australia: Canberra, Australia, 2016.
19. Sa, L. *Pain Relief for Invasive Procedures*; Meat and Livestock Australia: Sydney, Australia, 2022.
20. Passler, T. Regulatory and Legal Considerations of Anesthetics and Analgesics Used in Food-producing Animals. In *Farm Animal Anesthesia*; Wiley Blackwell: Hoboken, NJ, USA, 2022; pp. 263–284. [CrossRef]
21. AWI. *Anaesthetics and Analgesics Widely Adopted by Woolgrowers*; Australian Wool Innovation: Sydney, Australia, 2019.
22. APVMA. Public Chemical Registration Information Search (PubCRIS). 2022. Available online: https://portal.apvma.gov.au/pubcris (accessed on 31 August 2022).
23. Small, A.H.; Marini, D.; le Floch, M.; Paull, D.; Lee, C. A pen study evaluation of buccal meloxicam and topical anaesthetic at improving welfare of lambs undergoing surgical mulesing and hot knife tail docking. *Res. Vet. Sci.* **2018**, *118*, 270–277. [CrossRef]
24. Small, A.H.; Marini, D.; Dyall, T.; Paull, D.; Lee, C. A randomised field study evaluating the effectiveness of buccal meloxicam and topical local anaesthetic formulations administered singly or in combination at improving welfare of female Merino lambs undergoing surgical mulesing and hot knife tail docking. *Res. Vet. Sci.* **2018**, *118*, 305–311. [CrossRef]
25. Small, A.H.; Belson, S.; Holm, M.; Colditz, I.G. Efficacy of a buccal meloxicam formulation for pain relief in Merino lambs undergoing knife castration and tail docking in a randomised field trial. *Aust. Vet. J.* **2014**, *92*, 381–388. [CrossRef] [PubMed]
26. Small, A.H.; Belson, S.; Brewer, H.; Schmoelzl, S.M. Marking to weaning production aspects of lambs provided with NSAID analgesia compared with lambs receiving no analgesia at the time of elastrator ring marking. *Aust. Vet. J.* **2021**, *99*, 40–43. [CrossRef] [PubMed]
27. Colditz, I.G.; Paull, D.R.; Hervault, G.; Aubriot, D.; Lee, C. Development of a lameness model in sheep for assessing efficacy of analgesics. *Aust. Vet. J.* **2011**, *89*, 297–304. [CrossRef] [PubMed]
28. Colditz, I.G.; Paull, D.R.; Lloyd, J.B.; Johnston, L.; Small, A.H. Efficacy of meloxicam in a pain model in sheep. *Aust. Vet. J.* **2019**, *97*, 23–32. [CrossRef]
29. Copeland, A.; Speckels, A.; Merkatoris, P.; Breuer, R.M.; Schleining, J.A.; Smith, J. Laser ablation and management of a retropharyngeal abscess caused by *Corynebacterium pseudotuberculosis* in a ram. *Vet. Rec. Case Rep.* **2020**, *8*, e001010. [CrossRef]
30. Depenbrock, S.; Urbano, T.; Ziegler, J.; Wetzlich, S.; Clapham, M.O.; Tell, L.A. Pharmacokinetic Parameters and Tissue Withdrawal Intervals for Sheep Administered Multiple Oral Doses of Meloxicam. *Animals* **2021**, *11*, 2797. [CrossRef] [PubMed]
31. Dunbar, M.L.; Walkowiak, K.J.; Faustich, J.S.; Rendahl, A.K.; Graham, M.L. Preliminary Evaluation of Sustained-release Compared with Conventional Formulations of Meloxicam in Sheep (*Ovis aries*). *J. Am. Assoc. Lab. Anim. Sci.* **2019**, *58*, 339–345. [CrossRef] [PubMed]
32. Durmic, Z.; McGrath, P.; Wilmot, M.; Adams, N.; Tan, T.; Callahan, L.; Mayberry, C. Surgical and postoperative events during permanent fistulation of sheep rumen by the Schalk and Amadon method. *Aust. Vet. J.* **2015**, *93*, 234–239. [CrossRef]
33. Inglis, L.; Hancock, S.; Laurence, M.; Thompson, A. Behavioural measures reflect pain-mitigating effects of meloxicam in combination with Tri-Solfen® in mulesed Merino lambs. *Animal* **2019**, *13*, 2586–2593. [CrossRef]
34. Kells, N.J.; Beausoleil, N.J.; Godfrey, A.J.R.; Littlewood, K.E.; Ward, R.N.; Johnson, C.B. Effect of analgesic strategies on pain behaviour associated with combined ring castration and hot iron tail docking in Merino lambs. *Appl. Anim. Behav. Sci.* **2020**, *222*, 104914. [CrossRef]
35. Olagaray, K.E.; Bradford, B.J.; Sordillo, L.M.; Gandy, J.C.; Mamedova, L.K.; Swartz, T.H.; Jackson, T.D.; Persoon, E.K.; Shugart, C.S.; Youngs, C.R. Postpartum meloxicam administration alters plasma haptoglobin, polyunsaturated fatty acid, and oxylipid concentrations in postpartum ewes. *J. Anim. Sci. Biotechnol.* **2020**, *11*, 68. [CrossRef] [PubMed]
36. Paull, D.R.; Small, A.H.; Lee, C.; Palladin, P.; Colditz, I.G. Evaluating a novel analgesic strategy for ring castration of ram lambs. *Vet. Anaesth. Analg.* **2012**, *39*, 539–549. [CrossRef] [PubMed]
37. Silva, N.; Trindade, P.H.E.; Oliveira, A.R.; Taffarel, M.O.; Moreira, M.A.P.; Denadai, R.; Rocha, P.B.; Luna, S.P.L. Validation of the Unesp-Botucatu composite scale to assess acute postoperative abdominal pain in sheep (USAPS). *PLoS ONE* **2020**, *15*, e0239622. [CrossRef] [PubMed]
38. Harris, C.; White, P.J.; Hall, E.; Van der Saag, D.; Lomax, S. Evaluation of Electroencephalography, Behaviour and Eye Temperature in Response to Surgical Castration in Sheep. *Animals* **2021**, *11*, 637. [CrossRef] [PubMed]
39. Grant, E.P.; Wickham, S.L.; Anderson, F.; Barnes, A.L.; Fleming, P.A.; Miller, D.W. Preliminary Findings on a Novel Behavioural Approach for the Assessment of Pain and Analgesia in Lambs Subject to Routine Husbandry Procedures. *Animals* **2020**, *10*, 1148. [CrossRef] [PubMed]
40. Kaler, J.; Daniels, S.L.; Wright, J.L.; Green, L.E. Randomized clinical trial of long-acting oxytetracycline, foot trimming, and flunixine meglumine on time to recovery in sheep with footrot. *J. Vet. Intern. Med.* **2010**, *24*, 420–425. [CrossRef] [PubMed]
41. Straticò, P.; Varasano, V.; Suriano, R.; Mariscoli, M.; Robbe, D.; Giammarco, M.; Vignola, G.; Petrizzi, L. Analgesic effects of intravenous flunixin and intrafunicular lidocaine or their combination for castration of lambs. *Vet. Rec. Open* **2018**, *5*, e000266. [CrossRef]
42. Marini, D.; Pippia, J.; Colditz, I.G.; Hinch, G.; Petherick, J.C.; Lee, C. Randomised trial of the bioavailability and efficacy of orally administered flunixin, carprofen and ketoprofen in a pain model in sheep. *Aust. Vet. J.* **2015**, *93*, 265–270. [CrossRef]
43. Marini, D.; Pippia, J.; Colditz, I.G.; Hinch, G.N.; Petherick, C.J.; Lee, C. Palatability and pharmacokinetics of flunixin when administered to sheep through feed. *PeerJ* **2016**, *4*, e1800. [CrossRef]

44. Dória, R.G.S.; Ferraz, G.R.L.; Filippo, P.A.D.; Lacerenza, M.D.; Fernandes, L.M.; Oleskovicz, N.; Valadão, C.A.A. Subarachnoid ketamine and ketamine s (+) associated with lidocaine in sheep and goats anesthesia. *Vet. Anim. Sci.* **2021**, *11*, 100148. [CrossRef]
45. Durand, D.; Faure, M.; de la Foye, A.; de Boyer des Roches, A. Benefits of a multimodal analgesia compared to local anesthesia alone to alleviate pain following castration in sheep: A multiparametric approach. *Animal* **2019**, *13*, 2034–2043. [CrossRef]
46. Gondaliya, S.R.; Bhavsar, S.K.; Singh, R.D.; Patel, J.H.; Thaker, A.M. Pharmacokinetics and intramuscular bioavailability of ketoprofen in Patanwadi sheep. *J. Vet. Pharmacol. Toxicol.* **2015**, *14*, 53–55.
47. Schoiswohl, J.; Spergser, J.; Kofler, J. Polyarthritis caused by Erysipelothrix rhusiopathiae in three Austrian sheep flocks—Diagnosis, treatment and management measures. *Schweiz. Arch. Tierheilkd.* **2020**, *162*, 771–780. [CrossRef]
48. Wagner, A.E.; Mama, K.R.; Ruehlman, D.L.; Pelkey, S.; Turner, A.S. Evaluation of effects of sciatic and femoral nerve blocks in sheep undergoing stifle surgery. *Lab. Anim.* **2011**, *40*, 114–118. [CrossRef] [PubMed]
49. Bortolami, E.; Della Rocca, G.; Di Salvo, A.; Giorgi, M.; Kim, T.W.; Isola, M.; De Benedictis, G.M. Pharmacokinetics and antinociceptive effects of tramadol and its metabolite O-desmethyltramadol following intravenous administration in sheep. *Vet. J.* **2015**, *205*, 404–409. [CrossRef] [PubMed]
50. Habibian, S.; Bigham, A.S.; Aali, E. Comparison of lidocaine, tramadol, and lidocaine-tramadol for epidural analgesia in lambs. *Res. Vet. Sci.* **2011**, *91*, 434–438. [CrossRef]
51. Bedendo, L.H.; Gasparotto, J.C.; Vaccarin, C.V.; Segat, H.J.; Favaretto, B.P.; Soares, A.V. Postoperative analgesic comparison of tramadol or epidural morphine in sheep submitted to cesarean section. *PUBVET* **2019**, *13*, 20203244774. [CrossRef]
52. Ajadi, R.A.; Sobanke, O.A.; Adeniyi, A.A.; Adeusi, A.A.; Adebiyi, A.; Akinloye, A.K. Influence of Tramadol on Anaesthetic Indices and Physiological Parameters of Epidural Lignocaine in West African Dwarf Sheep Undergoing Laparo-Ovariectomy. *Niger. J. Physiol. Sci.* **2017**, *32*, 165–170. [PubMed]
53. Hakomäki, H.; Kokki, H.; Lehtonen, M.; Ranta, V.P.; Räsänen, J.; Voipio, H.M.; Kokki, M. Pharmacokinetics of buprenorphine in pregnant sheep after intravenous injection. *Pharmacol. Res. Perspect.* **2021**, *9*, e00726. [CrossRef] [PubMed]
54. Padgett, A.L.; Lepiz, M.L.; Mackay, E.E.; Driskill, A.J.; Ivanov, I.V.; Fajt, V.R.; Konarik, M.M.; Mays, T.P.; Washburn, S.E. Comparison of analgesic efficacy and fetal effects between transdermal administration of fentanyl and intramuscular administration of buprenorphine in pregnant sheep. *Am. J. Vet. Res.* **2020**, *81*, 581–593. [CrossRef]
55. Walkowiak, K.J.; Graham, M.L. Pharmacokinetics and Antinociceptive Activity of Sustained-Release Buprenorphine in Sheep. *J. Am. Assoc. Lab. Anim. Sci.* **2015**, *54*, 763–768.
56. Baljinnyam, T.; Radnaa, E.; Niimi, Y.; Fukuda, S.; Prough, D.S.; Enkhbaatar, P. Cutaneous burn diminishes beneficial effect of intravenously administered mesenchymal stem cells on acute lung injury induced by smoke inhalation in sheep. *Burns* **2020**, *46*, 1914–1923. [CrossRef]
57. Bellini, L.; Benedictis, G.M.d. Effect of three opioid-based analgesic protocols on the perioperative autonomic-mediated cardiovascular response in sheep. *Lab. Anim.* **2019**, *53*, 491–499. [CrossRef] [PubMed]
58. DeRossi, R.; Pagliosa, R.; Módolo, T.C.; Maciel, F.B.; Macedo, G.G. Thoracic epidural analgesia via the lumbosacral approach using multiport catheters with a low concentration of bupivacaine and morphine in sheep. *Vet. Anaesth. Analg.* **2012**, *39*, 306–314. [CrossRef]
59. Murdoch, F.R.; Maker, G.L.; Nitsos, I.; Polglase, G.R.; Musk, G.C. Intraperitoneal medetomidine: A novel analgesic strategy for postoperative pain management in pregnant sheep. *Lab. Anim.* **2013**, *47*, 66–70. [CrossRef]
60. DeRossi, R.; Jardim, P.H.; Hermeto, L.C.; Pagliosa, R.C. Comparison of analgesic and systemic effects of bupivacaine, methadone, or bupivacaine/methadone administered epidurally in conscious sheep. *Aust. Vet. J.* **2015**, *93*, 164–169. [CrossRef]
61. Buchholz, T.; Hildebrand, M.; Heider, A.; Stenger, V.; Arens, D.; Spadavecchia, C.; Zeiter, S. Transdermal Fentanyl Uptake at Two Different Patch Locations in Swiss White Alpine Sheep. *Animals* **2020**, *10*, 1675. [CrossRef]
62. Christou, C.; Oliver, R.A.; Rawlinson, J.; Walsh, W.R. Transdermal fentanyl and its use in ovine surgery. *Res. Vet. Sci.* **2015**, *100*, 252–256. [CrossRef]
63. Musk, G.C.; Murdoch, F.R.; Tuke, J.; Kemp, M.W.; Dixon, M.J.; Taylor, P.M. Thermal and mechanical nociceptive threshold testing in pregnant sheep. *Vet. Anaesth. Analg.* **2014**, *41*, 305–311. [CrossRef] [PubMed]
64. Small, A.H.; Jongman, E.C.; Niemeyer, D.; Lee, C.; Colditz, I.G. Efficacy of precisely injected single local bolus of lignocaine for alleviation of behavioural responses to pain during tail docking and castration of lambs with rubber rings. *Res. Vet. Sci.* **2020**, *133*, 210–218. [CrossRef] [PubMed]
65. Small, A.; Marini, D.; Colditz, I. Local Anesthetic Delivered with a Dual Action Ring and Injection Applicator Reduces the Acute Pain Response of Lambs during Tail Docking. *Animals* **2021**, *11*, 2242. [CrossRef]
66. Small, A.; Fetiveau, M.; Smith, R.; Colditz, I. Three Studies Evaluating the Potential for Lidocaine, Bupivacaine or Procaine to Reduce Pain-Related Behaviors following Ring Castration and/or Tail Docking in Lambs. *Animals* **2021**, *11*, 3583. [CrossRef]
67. Simpson, K.M.; Van Metre, D.C.; Applegate, T.J.; Taylor, J.D.; Johnson, J.; Brooks, K.S.; Mama, K.R. Evaluation of the 4-point regional nerve block using 2% lidocaine in sheep. *Can. Vet. J.* **2022**, *63*, 269–274.
68. Rostami, M.; Vesal, N. Comparison of lidocaine, lidocaine/epinephrine or bupivacaine for thoracolumbar paravertebral anaesthesia in fat-tailed sheep. *Vet. Anaesth. Analg.* **2011**, *38*, 598–602. [CrossRef] [PubMed]
69. Mattos-Junior, E.; Flaherty, D.; Nishimura, L.T.; Carregaro, A.B.; de Carvalho, L.L. Clinical effects of epidurally administered dexmedetomidine with or without lidocaine in sheep. *Vet. Rec.* **2020**, *186*, 534. [CrossRef]

70. Lizarraga, I.; Janovyak, E.; Beths, T. Comparing lidocaine, bupivacaine and a lidocaine-bupivacaine mixture as a metacarpal block in sheep. *Vet. J.* **2013**, *197*, 515–518. [CrossRef]
71. Harris, C.; White, P.J.; Mohler, V.L.; Lomax, S. Electroencephalography Can Distinguish between Pain and Anaesthetic Intervention in Conscious Lambs Undergoing Castration. *Animals* **2020**, *10*, 428. [CrossRef]
72. Ghadirian, S.; Vesal, N. Brachial plexus block using lidocaine/epinephrine or lidocaine/xylazine in fat-tailed sheep. *Vet. Res. Forum* **2013**, *4*, 161–167.
73. Ghadirian, S.; Vesal, N.; Maghsoudi, B.; Akhlagh, S.H. Comparison of lidocaine, lidocaine-morphine, lidocaine-tramadol or bupivacaine for neural blockade of the brachial plexus in fat-tailed lambs. *Vet. Anaesth. Analg.* **2016**, *43*, 109–116. [CrossRef]
74. Rostami, M.; Vesal, N. The effects of adding epinephrine or xylazine to lidocaine solution for lumbosacral epidural analgesia in fat-tailed sheep. *J. S. Afr. Vet. Assoc.* **2012**, *83*, 1. [CrossRef]
75. DeRossi, R.; Silva-Neto, A.B.; Pompermeyer, C.T.; Frazílio, F.O.; Jardim, P.H.; de Barros, A.C. The efficacy and safety of levobupivacaine administered by lumbosacral epidural route in conscious sheep. *Res. Vet. Sci.* **2012**, *92*, 278–282. [CrossRef]
76. Molony, V.; Kent, J.E.; Viñuela-Fernández, I.; Anderson, C.; Dwyer, C.M. Pain in lambs castrated at 2 days using novel smaller and tighter rubber rings without and with local anaesthetic. *Vet. J.* **2012**, *193*, 81–86. [CrossRef] [PubMed]
77. Lacasta, D.; Reina, R.; Ruiz de Arcaute, M.; Ferrer, L.M.; Benito, A.A.; Tejedor, M.T.; Echeverria, I.; Ruiz, H.; Martinez Cardenas, S.; Windsor, P.A. Effect of a Topical Formulation on Infective Viral Load in Lambs Naturally Infected with Orf Virus. *Vet. Med.* **2021**, *12*, 149–158. [CrossRef] [PubMed]
78. Lomax, S.; Sheil, M.; Windsor, P.A. Duration of action of a topical anaesthetic formulation for pain management of mulesing in sheep. *Aust. Vet. J.* **2013**, *91*, 160–167. [CrossRef] [PubMed]
79. Centonze, P.; Lacitignola, L.; Pumilia, P.; Luca, P.D.; Guarracino, A.; Esposito, C.; Pesce, A.; Crovace, A.M.; Crovace, A.; Staffieri, F. *Analgesic, Sedative and Cardiovascular Effects of Clonidine as an Adjuvant for Spinal Anesthesia in Sheep Un-Dergoing Orthopedic Surgery*; Società Italiana delle Scienze Veterinarie (SISVet): Perugia, Italy, 2015; p. 237.
80. Genccelep, M.; Karasu, A. Evaluation of analgesic and sedative effects of repeated and increasing doses of xylazine HCl in sheep. *Med. Weter.* **2017**, *73*, 468–472. [CrossRef]
81. Moolchand, M.; Kachiwal, A.B.; Soomro, S.A.; Bhutto, Z.A. Comparison of sedative and analgesic effects of xylazine, detomidine, and medetomidine in sheep. *Egypt. J. Sheep Goat Sci.* **2014**, *9*, 43–48.
82. Hyndman, T.H.; Musk, G.C.; Murdoch, F.R.; Maker, G.L.; Whittem, T. The bioavailability of medetomidine in eight sheep following oesophageal administration. *Res. Vet. Sci.* **2015**, *103*, 137–142. [CrossRef]
83. Uemura, K.; Shimazutsu, K.; McClaine, R.J.; McClaine, D.J.; Manson, R.J.; White, W.D.; Benni, P.B.; Reynolds, J.D. Maternal and preterm fetal sheep responses to dexmedetomidine. *Int. J. Obs. Anesth.* **2012**, *21*, 339–347. [CrossRef]
84. DeRossi, R.; Pompermeyer, C.T.; Silva-Neto, A.B.; Barros, A.L.; Jardim, P.H.; Frazílio, F.O. Lumbosacral epidural magnesium prolongs ketamine analgesia in conscious sheep. *Acta Cir. Bras.* **2012**, *27*, 137–143. [CrossRef]
85. Kania, B.F.; Wrońska, D.; Bracha, U. Pain, pathophysiological mechanisms, and new therapeutic options for alternative analgesic agents in sheep: A review and investigation. *Animals* **2021**, *11*, 909. [CrossRef]
86. Mathurkar, S.; Singh, P.; Kongara, K.; Chambers, P. Pharmacokinetics of Salicylic Acid Following Intravenous and Oral Administration of Sodium Salicylate in Sheep. *Animals* **2018**, *8*, 122. [CrossRef]
87. Munn, R.; Woodward, A.; Beths, T.; Whittem, T. Observations on the use of a pain numbing device for repetitive percutaneous sampling in sheep. *Aust. Vet. J.* **2021**, *99*, 445–448. [CrossRef]
88. Ratajczak-Enselme, M.; Grégoire, N.; Estebe, J.P.; Dollo, G.; Chevanne, F.; Bec, D.; Ecoffey, C.; Couet, W.; Le Corre, P. Population Pharmacokinetics of Amitriptyline After Intrathecal, Epidural, and Intravenous Administration in Sheep. *Reg. Anesth. Pain. Med.* **2015**, *40*, 681–686. [CrossRef] [PubMed]
89. Manual, M. Analgesia Pharmacology. 2016. Available online: https://www.msdvetmanual.com/management-and-nutrition/pain-assessment-and-management/analgesic-pharmacology?query=opioids (accessed on 23 August 2022).
90. Australian Pesticides and Veterinary Medicines Authority. *Veterinary Anti-Inflammatories*; Australian Pesticides and veterinary Medicines authorit: Sydney, Australi, 2018.
91. Australian Pesticides and Veterinary Medicine Authority—Veterinary Anti-Inflammatories. Available online: https://apvma.gov.au/node/922 (accessed on 6 June 2022).
92. Products, D.V. Tri-Solfen. 2021. Available online: https://trisolfen.com.au/ (accessed on 24 August 2022).
93. Gómez, A.P. *Postoperative Pain Management in Companion Animals: An Update*; Veterinary Business Development Ltd.: Peterborough, UK, 2017.
94. VIN. VIN Veterinary Drug Handbook. In *Lidocaine*; Veterinary Information Network: Davis, CA, USA, 2017.
95. Stillman, M.W.; Whittaker, A.L. Use and Efficacy of Analgesic Agents in Sheep (*Ovis aries*) Used in Biomedical Research. *J. Am. Assoc. Lab. Anim. Sci.* **2019**, *58*, 755–766. [CrossRef] [PubMed]
96. Kästner, S.B.R. A_2-agonists in sheep: A review. *Vet. Anaesth. Analg.* **2006**, *33*, 79–96. [CrossRef]
97. AVA. *Guidelines for Prescribing, Authorising and Dispensing Veterinary Medicines*; Australian Veterinary Association: Sydney, Australia, 2013.
98. Fajt, V.R.; Pugh, D. Commonly used drugs in sheep and goats: Suggested dosages. In *Sheep and Goat Medicine*; Elsevier: Amsterdam, The Netherlands, 2012; pp. 579–595.
99. ACLAM. *ACLAM e-Formulary*; ACLAM: Ann Arbor, MI, USA, 2021.

100. Lin, H. Pain Management for Farm Animals. In *Farm Animal Anesthesia*; Wiley Blackwell: Hoboken, NJ, USA, 2022; pp. 207–246. [CrossRef]
101. Brosnan, R.J.; Cenani, A.; Costa, L.R.; Condy, P.; Snell, C. Analgesic effect of the mint terpenoid L-carvone in sheep. *Vet. Anaesth. Analg.* **2023**, *50*, 459–465. [CrossRef] [PubMed]
102. MLA. Sheep Projections. 2022. Available online: https://www.mla.com.au/prices-markets/Trends-analysis/sheep-projections/#:~:text=The%20national%20flock%20is%20predicted,MLA)%20latest%20Sheep%20Industry%20Projections (accessed on 1 July 2023).
103. Kelly, J. Pain Management in Sheep and Cattle. 2022. Available online: https://www.mla.com.au/contentassets/a16e282ff6e941a99183eaeac02e5fed/jillian-kelly_gm.pdf (accessed on 1 July 2023).
104. Percie du Sert, N.; Hurst, V.; Ahluwalia, A.; Alam, S.; Avey, M.T.; Baker, M.; Browne, W.J.; Clark, A.; Cuthill, I.C.; Dirnagl, U. The ARRIVE guidelines 2.0: Updated guidelines for reporting animal research. *J. Cereb. Blood Flow Metab.* **2020**, *40*, 1769–1777. [CrossRef] [PubMed]
105. Carbone, L.; Austin, J. Pain and laboratory animals: Publication practices for better data reproducibility and better animal welfare. *PLoS ONE* **2016**, *11*, e0155001. [CrossRef] [PubMed]
106. Prank, M.R.; Paul, S.K.; Hoque, M.A.; Al Faruk, M.S. Clinical management of paracetamol poisoning in cat. *Bangladesh J. Vet. Anim. Sci.* **2022**, *10*, 71–74.
107. KuKanich, K.; George, C.; Roush, J.K.; Sharp, S.; Farace, G.; Yerramilli, M.; Peterson, S.; Grauer, G.F. Effects of low-dose meloxicam in cats with chronic kidney disease. *J. Feline Med. Surg.* **2021**, *23*, 138–148. [CrossRef] [PubMed]
108. Antiorio, A.T.F.B.; Aleman-Laporte, J.; de Freitas, A.P.P.; Yamamoto, P.K.; Cintra, L.; Mori, C.M.C. Administration of meloxicam to improve the welfare of mice in research: A systematic review (2000–2020). *Vet. Res. Commun.* **2022**, *46*, 1–8. [CrossRef]
109. Conlee, K.; Stephens, M.; Rowan, A.N. Addressing Distress and Pain in Animal Research: The Veterinary, Research, Societal, Regulatory and Ethical Contexts for Moving Forward. 2009. Available online: https://www.wellbeingintlstudiesrepository.org/cgi/viewcontent.cgi?article=1060&context=acwp_arte (accessed on 1 July 2023).

animals

MDPI

Review

Pain Assessment in Goat Kids: Focus on Disbudding

Kavitha Kongara [1,*], Preet Singh [1], Dinakaran Venkatachalam [2] and John Paul Chambers [1]

[1] School of Veterinary Science, Massey University, Tennent Drive, Palmerston North 4442, New Zealand; p.m.singh@massey.ac.nz (P.S.); j.p.chambers@massey.ac.nz (J.P.C.)
[2] Ministry for Primary Industries, Wellington 6011, New Zealand; dinakaran.venkatachalam@mpi.govt.nz
* Correspondence: k.kongara@massey.ac.nz

Simple Summary: Disbudding is a routine husbandry procedure performed in goat kids in their first few weeks of life. Behavioural and physiological changes following the procedure suggest significant pain and distress to the goat kid, which is a welfare concern. Pain assessment is fundamental to implementation of effective pain treatment and/or management protocols. This review provides details on pain assessment methods in goat kids following different methods of disbudding. Commonly used pain assessment methods in other young farm animals were also included.

Abstract: Farm animals are routinely subjected to painful husbandry procedures for various purposes. Goat kids are disbudded to improve goat welfare and to ensure safety of other livestock, farm personnel, attending veterinarians and for various other production and managemental procedures. Disbudding is commonly performed on dairy goat farms, in kids under 3 weeks of age. Many scientific studies reported physiological and behavioural changes indicating pain and distress following disbudding, and this can be a significant cause of welfare compromise in goat kids. Recognition and measurement of pain is important to treat and/or manage pain and distress following painful procedures. This review focuses on pain assessment in goat kids following disbudding, using both physiological and behavioural measures. As only a limited information is available on the topic of interest, relevant studies in other young farm animals have also been discussed to compare the status quo in goat kids.

Keywords: goat kids; disbudding pain assessment; physiology behaviour

1. Introduction

Farm animals are routinely subjected to painful husbandry procedures for various purposes. Goat kids are disbudded to reduce injury to other goats in the flock, farm personnel, and attending veterinarians. Hornless goats are also less likely to get their heads entangled in equipment such as milking machines. Disbudding is commonly performed on dairy goat farms, in kids under 3 weeks of age. Many scientific studies reported physiological and behavioural changes indicating pain and distress following disbudding, and this can be a significant cause of welfare compromise in goat kids. Accurate recognition and assessment of pain is important to implement correct treatments at appropriate doses and intervals to ensure animal wellbeing [1]. The ability to report one's own experience of pain and its intensity helps in developing effective management strategies in humans. Pain assessment in pre-verbal infants, people with cognitive impairment and animals is often challenging. In most instances, pain is assessed based on observation of animal's behaviour [1]. Although this approach is useful in clinical situations, it can be difficult to assess subtle signs of pain in prey animals, such as sheep and goat, that do not manifest overt signs of pain [2].

Physiological and behavioural measures that are likely to indicate pain and/or nociception and related distress have been developed and used for assessment of pain and the efficacy of analgesic strategies in animal studies. The main aim of this review is to describe the methods of pain assessment in goat kids following husbandry procedures such as disbudding. Commonly used pain assessment methods in other young farm animals was also included, as little information is available in goat kids compared to calves, lambs, and piglets.

2. Physiological Measures

2.1. Plasma Cortisol, Glucose and Lactate

The physiological variable that has been historically studied to assess the pain related distress following routine husbandry procedures is the total plasma cortisol [3]. Cortisol is a hormone, which is released in high concentrations into circulation after activation of hypothalamo-pituitary- adrenal (HPA) axis in response to stressful conditions including pain [4]. It thus gives an indication of distress, rather than pain directly. Increases in plasma cortisol concentrations following disbudding and other painful husbandry procedures such as castration and tail docking, and its alleviation by analgesic administration has commonly been used to assess pain in calves, lambs, and goat kids [5–7]. The degree of plasma cortisol response was also used to compare the pain intensity between different methods of goat kid disbudding [8]. Disbudding of kids with liquid nitrogen or caustic paste has been found to cause more severe changes in cortisol, indicating intense acute distress, and possibly pain, compared to cautery disbudding [8].

To measure plasma cortisol, blood samples are commonly collected from the jugular veins and cortisol concentrations are measured using different assay techniques, including the enzyme-linked immunosorbent assay and radioimmunoassay [5]. Although many studies have used plasma cortisol for the assessment of pain induced distress, there are some caveats to this method. Plasma cortisol levels were also influenced by diurnal rhythms [9] and stressors that are not associated with pain such as animal handling [10], feeding [11], and sexual excitement [12]. Trends in plasma cortisol concentrations measured from blood samples collected before and at multiple time points after a specific procedure would give a more reliable estimate of the stress associated with pain following a noxious procedure in production animals (although blood sampling will increase plasma cortisol). In analgesia studies, comparing the cortisol levels between sham-handled and treatment groups, and correlating cortisol response with other pain assessment measures such as pain behaviours, would make plasma cortisol evaluations a more valid way to estimate the efficacy of analgesic drugs [13]. It should be noted that some drugs which are used as analgesics in ruminants, such as alpha 2 adrenergic agonists, probably have a direct depressant effect on the adrenal glands in goats, as they do in pigs [14].

Plasma metabolites, such as glucose and lactate, were measured in conjunction with plasma cortisol following disbudding of goat kids [8,15]. These blood constituents were measured based on the hypothesis that increased secretion of cortisol in response to pain induced distress stimulates mobilization of glycogen, which results in increased production of glucose and lactate [16]. None of the studies found significant changes in plasma lactate and glucose despite a significant elevation of plasma cortisol after disbudding. The researchers opined that these two metabolites are not as useful as plasma cortisol responses to acute pain associated with disbudding.

2.2. Autonomic Responses

Other physiological measures resulting from the activation of sympathetic nervous outflow in response to noxious stimulation, such as heart rate, respiratory rate, blood pressure and body temperature can be monitored to assess pain following disbudding in goat kids [6]. Although changes in these variables are influenced by the type of noxious stimulus based on the method of husbandry procedure; in general, animals in pain often have an increase in these variables [17,18]. A major limitation to the use of the sympathetically driven variables, similar to cortisol, is that changes are not necessarily specific to pain [19], and are usually limited to acute pain responses and hence less useful for the assessment of longer lasting pain. Also, changes in these autonomic variables have poor correlation with pain perception as these are modulated by centres of the brain stem below the level of the higher centres involved in cognitive pain perception [1].

2.3. Electroencephalography (EEG)

The EEG is a record of the spontaneous electrical activity of the cerebral cortex. It is an objective tool for assessment of neural (sensory) processing of noxious stimuli, which is known as nociception. EEG provides a direct and quick profile of cortical neuronal activity in response to noxious stimulation in animals. The characteristic configuration of neurones (particularly pyramidal type) in the layers of the cerebral cortex leads to formation of electrical vectors that facilitates transmission of electrical currents. The EEG is the far-field potential (electrical field created by inherent neuronal activity and recorded away from them) of the vector currents recorded using electrodes on the surface of the head [20]. Fast Fourier transformation, a mathematical procedure, converts the raw EEG signal into its component sine waves of different frequency, characterized by corresponding amplitude. Thus, the power spectrum generated is simply a distribution and derivation of spectral EEG variables from the corresponding frequencies and amplitudes [20]. There are also a number of other proprietary methods of analysing raw EEG data, which are used in instruments designed for people.

Changes in the EEG power spectrum have been used for the quantification of nociception in minimally anaesthetised red-deer, lambs, and calves following noxious procedures such as velvet antler removal, castration, and slaughter through a ventral neck-cut, respectively [21–23]. Changes in frequency spectra of EEG in lightly anaesthetized animals (with halothane, which is minimally analgesic) reflect changes in activity of cerebral cortex in response to pain perception [22,23]. This concept is further supported by studies in goats that did not find a difference in EEG spectral frequencies between conscious animals and those that were lightly anaesthetised during slaughter [24]. Advantages of the minimal anaesthesia model developed for recording EEG for quantification of pain in animals include the minimization of the impact on the EEG of extraneous electrical activity (such as from muscles) and loss of conscious pain perception by study animals, thereby reducing suffering from pain during noxious stimulation [20].

In goats, EEG power spectra have been used to delineate loss of awareness from cognitive perception of pain following the application of various techniques for euthanasia and pre-slaughter stunning [25]. Also, there have been studies that described the EEG frequency band analysis to find the effect of pre-slaughter stress, experimental pain and ontological changes in brain activity in goats [26–28]. So far, no studies are available on the use of EEG spectral frequency changes to quantify pain associated with husbandry procedures such as disbudding/dehorning in goats.

Although an electroencephalogram can non-invasively capture the sensory neuronal processing of noxious stimuli and indirectly reflect cortical pain perception, it may not represent the motivational states such as aversiveness, fear and anxiety associated with pain perception, all of which are subjective experiences. Another limitation to the use of the EEG is that it can only be used for acute pain assessment, as an animal can be kept minimally anaesthetized for only a limited amount of time. A combined approach such as concurrent evaluation of cognitive pain perception manifested by behavioural changes will be more reliable for pain assessment using EEG in farm animals.

2.4. Mechanical Nociceptive Threshold Testing

Mechanical stimuli, such as pressure, are usually applied to the skin, and can be used to produce quantifiable nociception or pain in farm animals, including goats. The degree of mechanical pressure that an animal can tolerate before showing an avoidance (behavioural) response or withdrawal reflex is defined as a mechanical nociceptive (pressure) threshold of that animal [29,30]. Mechanical stimuli are used to induce experimental pain to test or compare analgesics in normal animals. In animals subjected to noxious procedures, these stimuli are used to test the pain sensitivity at the site of tissue damage (wound site) and/or areas surrounding the wound. A variety of commercial, hand-held algometers are available (e.g., Force one FDIX 50, Wagner Instruments, Riverside, CT, USA; ProdPlus algometer, TopCat Metrology Ltd., Ely, UK), which are validated for use in various farm animal species, including goats [18,31,32]. Mechanical stimulation of the wound area (injured tissue at the disbudded area) can also be performed to test the tactile sensitivity, using von Frey filaments [31,33,34]. Wounds remained sensitive to tactile stimuli for at least 75 h after disbudding in calves [33].

Mechanical nociceptive thresholds are usually measured in Newtons (N) or Kilogram Force (Kgf). Force is applied perpendicular to the skin surface at a constant rate and in gradual increments to avoid a sudden, jerky increase in readings without corresponding attainment of threshold manifested by signs of discomfort specific to the species being tested. Head withdrawal with a specific ear flick has been noted as the end response by goat kids being tested at the disbudded site [30,34]. To attain a good correlation between threshold values and the amount of pressure applied, selecting an algometer probe that matches the size of the animal is also important. A round rubber tip probe of 1 cm^2 has been used on a Force One Wagner algometer [30], and a 2 mm diameter metal probe has been used on a ProdPlus algometer in disbudded goat kids [32,34]. Also, other factors such as age and body weight of the animal should be considered while testing nociceptive thresholds for a study in piglets; age and body weight were demonstrated to affect the pressure threshold testing responses [35]. In goat kids, pressure algometry has been used to compare the pain sensitivity between three different methods of disbudding, and cautery disbudding was found to cause less acute pain hypersensitivity than the caustic paste and cryosurgery methods [30]. In other studies, it was used to find the welfare benefits of alternative methods of disbudding [34] and the duration of wound hypersensitivity after hot-iron disbudding in goat kids [32]. In the study by Frahm et al. [34] development of acute mechanical hypersensitivity around the horn buds following the injection of clove oil or isoeugenol has been reported [34]. Tissue irritation by the injected substances was suggested as the potential cause of heightened pain sensitivity and, therefore, the welfare benefits of the method over hot-iron disbudding are questionable. Persistent mechanical hypersensitivity throughout the wound healing period (average 7-weeks) has been reported by Alvarez et al. [32].

Animal handling prior to the application of the stimulus has the potential to influence the threshold readings. A good human–animal relationship, prior habituation of the animal to the device, and testing in home pens close to dam/other pen mates can minimize the confounding effect of handling stress on threshold measurements [34].

Other types of nociceptive stimuli that have been used to test the effect of local anaesthetic (LA) administration prior to goat kid disbudding were pin pricks [36]. The observer gently pricked the skin around the horn buds with a needle point at approximately 30 s to 1 min interval until no response is observed to confirm analgesia after infiltration with LA. This qualitative analgesia testing method is rapid and practical to use in field conditions, and reliable to assess the onset of analgesia prior to disbudding.

In summary, all the devices and methods developed to test the mechanical nociceptive thresholds of farm animals let us assess the level of pain hypersensitivity, efficacy of analgesics and/or analgesic approaches, and development of 'plasticity' in the nervous system, following noxious husbandry procedures in field conditions.

2.5. Infra-Red Thermography

Infra-red thermography (IRT) is a non-invasive method of recording infrared radiation emitted by bodies, i.e., heat [37]. It can measure changes in surface temperature due to activation of autonomic nervous system in response to stress. A stress-induced secretion of catecholamines causes an increase in internal body temperature called stress induced hyperthermia, and also causes a reduction in blood flow in the skin around eyes. A drop in eye temperature, assessed using IRT, has been reported to be associated with the onset of acute pain in calves following hot-iron dehorning [38,39]. Sympathetically mediated vasoconstriction in response to acute pain has been proposed to be the cause of the drop in eye temperature. The IRT recording can be taken at a distance from the animal without a need to contact/restrain the animal. This overcomes the effect of animal handling stress on actual data, which is the drawback associated with the measurement of other physiological (and autonomic) variables in response to stress and pain [38] Care must be taken that the hair coat must be free of dirt, grease and foreign material while obtaining IRT images of the area of interest [39].

A study in disbudded goat kids used IRT to take the images of the horn bud (~3 cm diameter around horn bud), and surface skin temperature was calculated from a ~2 cm diameter area around the horn buds [8]. Disbudded goat kids were found to have a higher skin surface temperature around the horn buds than sham kids. Another study has used IRT to assess the wound inflammation and surface temperature during healing after cautery disbudding of goat kids [32], and reported that the necrotic tissue formed during the inflammatory phase was hotter than epithelium. Studies are required to find the use of the IRT to evaluate the effect of analgesic agents and strategies after disbudding in goat kids.

2.6. Biomarkers

Immunological, inflammatory and pain biomarkers were measured in tissues and body fluids of young farm animals following husbandry procedures [31,40–42]. Acute phase inflammatory proteins such as haptoglobins were measured in goat kids disbudded using different methods [8]. Clove oil injection under the horn bud caused a significant increase in serum haptoglobin concentrations 24 h after treatment, which indicates marked inflammation associated with the method of disbudding [8]. Measurement of immunoreactive proteins such as β-endorphin concentration in the blood plasma has been shown to be a useful method of acute pain assessment following hot-iron disbudding of goat kids [43]. With the advent of new molecular biological techniques in animals, it has been possible to demonstrate the efficacy of non-steroidal anti-inflammatory drugs in suppressing the expression of mRNA (in peripheral leucocytes of a blood sample) for inflammatory cytokine and nociceptive marker genes in calves disbudded by thermocautery [42]. More advanced techniques such as plasma proteome analysis allowed identification of candidate protein biomarkers that are associated with nociceptive and inflammatory processes following surgical dehorning of calves [44]. Biomarkers such as these may be useful in the future to assess various methods of pain mitigation for husbandry procedures in farm animals including goats.

3. Behavioural Measures

Changes in normal behaviour are commonly observed in animals in response to painful stimuli. Hence, monitoring normal behavioural patterns and deviations from these is a significant component of pain assessment in animals [45]. Species-specific behavioural changes that can be identified and quantified have been used to evaluate pain and analgesia in goat kids, lambs, piglets, and calves following routine husbandry procedures [4,6,46]. A person assessing pain behaviours, either directly from an animal or indirectly from video recordings, should be familiar with the animal's normal behavioural patterns, have adequate training and experience in recognition of changes in normal behaviours in response to noxious procedures [47,48].

A multitude of factors including severity of insult to the tissues and neuronal pathways influence the manifestation of pain behaviours by animals. Age, previous pain experience, social hierarchy in the herd, human presence, environmental conditions are among the others [49]. In young farm animals such as goat kids, lambs and calves, a significant alteration in their normal behaviour, body posture, locomotor activity, orientation toward dam, and response to manipulations have been used as a basic template for the behavioural assessment of pain and analgesia following husbandry procedures, including disbudding.

There appears to be two phases to goat kids' behaviour in response to hot iron disbudding—acute avoidance type behaviours in responses to application of the iron and abnormal behaviour in the subsequent hours. Behavioural indicators of acute stress and pain during cautery disbudding in goat kids have been first reported by Alvarez et al. [6] Struggles (kicks), such as a vigorous movement of the legs and attempts to escape, and vocalisations in the form of bleats with an open or closed mouth were reported. A change in the frequency of behaviours along with plasma cortisol levels was used to find the effect of analgesic treatment on disbudding pain [7].

In a sham controlled trial, Hempstead et al. [50] recorded behavioural changes associated with cautery disbudding of female Saanen dairy goat kids. Video recordings of pre- and post-treatment behaviours were analysed by a trained and an experienced observer to build an ethogram of behaviour patterns. Although the frequency of 11 behaviours in the ethogram was chosen to study, individual behaviours such as head shaking, rubbing, scratching, and body shaking were significantly different between treated and control kids. Due to a large inter-individual variation, self-grooming behaviour did not reach statistical significance despite an apparent increase in its frequency in the disbudded group. Individual variation in a behavioural response, within the same species, to a noxious procedure is one important confounding factor in pain assessment studies. This could be due to differences in pain perception and expression between individual animals [51]. The following table (Table 1) provides an overview of behavioural changes in goat kids following different methods of disbudding.

Table 1. Individual behaviours and ethograms described in various studies investigating pain and pain mitigation strategies in disbudded goat kids.

Title of the Study	Behavioural Indicators Used	Other Variables Measured
Physiological and behavioural alterations in disbudded goat kids with and without local anaesthesia [6].	Struggles and vocalisations	Plasma cortisol, Heart rate and respiratory rate pre- and post (up to 4 h) disbudding
Evaluation of alternatives to cautery disbudding of dairy goat kids using behavioural measures of post-treatment pain [52].	Head and body shaking, head scratching, feeding and self-grooming noted from video recordings 24 h pre- and post-disbudding	Accelerometery measures such as lying bouts and duration recorded 24 h pre- and post-disbudding
Effect of isoflurane alone or in combination with meloxicam on the behavior and physiology of goat kids following cautery disbudding [53].	------Do----	Plasma cortisol, glucose and lactate measured pre- and post (up to 120 min) disbudding
Acute cortisol and behavior of dairy goat kids administered local anesthesia, topical anesthesia or systemic analgesia prior to cautery disbudding [54].	Rump movements, tail shakes and vocalisations recorded during disbudding	Plasma cortisol levels measured pre- and post-disbudding
Can Isoflurane and Meloxicam Mitigate Pain Associated with Cautery Disbudding of 3-Week-Old Goat Kids? [15]	Head and body shaking, head scratching, feeding and self-grooming noted from video recordings 1 h pre- and post-disbudding	Plasma cortisol, glucose and lactate measured pre- and post (up to 120 min) disbudding
Evaluation of Pain Mitigation Strategies in Goat Kids after Cautery Disbudding [55]	State events (head scratching, self-grooming, allogrooming, feeding/drinking, exploration, standing, lying, social play, etc.) and point events (vocalisation, shaking, tail movements, stretching, etc.) in an ethogram built using BORIS * software (https://www.boris.unito.it/ accessed on 1 May 2023) from a 3 h video recording after disbudding.	--------------

* BORIS; Behavioural Observation Research Interactive Software.

In addition to the assessment of frequency of specific behaviours, some studies used a visual analogue scale (from 1–10) to score pain behaviours to evaluate the efficacy of anaesthetic and analgesic treatments administered to goat kids [56,57]. A summary of the pain behaviours was used to assign a score from the scale. Also, a vocalisation score, which is the total number of vocalizations during the disbudding procedure for both horn buds, was used to compare the alternative methods to cautery disbudding [58].

From these studies in goat kids, it is evident that disbudding induces specific changes in behaviour, which include increased frequency of struggles, vocalisations, head shaking, scratching, and rubbing, body and tail shaking, lying bouts and duration. Changes in these behaviours have been used to evaluate relative noxiousness of different (alternative) methods of disbudding and efficacy of anaesthetic and analgesic protocols in various studies.

Facial Grimace Scale

Due to the limitations of behavioural methods (such as laborious and time-consuming video and audio recording and analyses and the need for specialised equipment), and physiological measures (such as expensive laboratory analytical procedures), Lou et al. (2020) developed a goat kid grimace scale for pain assessment following thermal disbudding [59]. It is a pain assessment method that enables instantaneous identification and assessment of changes in facial expression after a noxious procedure [60]. The basic template for the scale has been drawn from sheep and lamb grimace scales [61,62] and adapted for a goat face. Orbital and lip tightening, nostril dilatation and change in ear position have been scored on a 3-point scale (0 = Not Present, 1 = Moderately Present, and 2 = Obviously Present), from photographic images, to find the effect of analgesic administration and thermal disbudding in goat kids. Orbital tightening and ear position scores were found to be more reliable indicators of treatment effects than lip tightening and nostril dilation. More research needs to be conducted to further explore the validity of this facial grimace scale in goat kids.

4. Conclusions

Both the behavioural and physiological systems are involved in the response to stress and pain, and in majority of research studies, a combination of both methods is used in pain assessment. Individual behaviours or an ethogram in conjunction with plasma cortisol measurement has been the basic pain assessment paradigm in a large number of goat disbudding studies. Some studies used other physiological methods such as mechanical nociceptive threshold testing, infrared thermography and plasma inflammatory marker analyses, heart rate and body temperature changes, and body weight gains, etc. Only one study used a facial grimace scale in combination with plasma cortisol evaluations in disbudded goat kids. Using a combination of pain assessment methods allows to offset some of the disadvantages associated with each technique and to gain as much accurate and comprehensive information as possible from an experimental study [63]. In clinical practice, a combination of pin prick testing for analgesia before disbudding and behavioural assessment afterwards is probably most practical.

Author Contributions: Conceptualisation, K.K.; writing—original draft preparation, K.K.; writing—review and editing, K.K., P.S., D.V. and J.P.C. All authors have read and agreed to the published version of the manuscript.

Funding: This research received no external funding.

Institutional Review Board Statement: Not applicable.

Informed Consent Statement: Not applicable.

Data Availability Statement: Data is contained within the article.

Conflicts of Interest: The authors declare no conflict of interest.

References

1. Holton, L.L.; Scott, E.M.; Nolan, A.M.; Reid, J.; Welsh, E. Relationship between physiological factors and clinical pain in dogs scored using a numerical rating scale. *J. Small Anim. Pract.* **1998**, *39*, 469–474. [CrossRef]
2. Matthews, J.; Dustan, B. Disbudding of goat kids. *Practice* **2019**, *41*, 433–444. [CrossRef]
3. Mellor, D.J.; Stafford, K.J. Acute castration and tailing distress and its alleviation in lambs. *N. Z. Vet. J.* **2000**, *48*, 33–43. [CrossRef] [PubMed]
4. Molony, V.J.; Kent, E. Assessment of acute pain in farm animals using behavioral and physiological measurements. *J. Anim. Sci.* **1997**, *75*, 266–272. [CrossRef] [PubMed]
5. Sutherland, M.A.; Tucker, C. The long and short of it; A review of tail docking in farm animals. *Appl. Anim. Behav. Sci.* **2011**, *135*, 179–191. [CrossRef]
6. Alvarez, L.; Nava, R.A.; Ramirez, A.; Ramirez, E.; Gutierrez, J. Physiological and behavioural alterations in disbudded goat kids with and without local anaesthesia. *Appl. Anim. Behav. Sci.* **2009**, *117*, 190–196. [CrossRef]
7. Alvarez, L.; Gutierrez, J. A first description of the physiological and behavioural responses to disbudding in goat kids. *Anim. Welf.* **2010**, *19*, 55–59. [CrossRef]
8. Hempstead, M.N.; Waas, J.R.; Stewart, M.; Cave, V.M.; Sutherland, M.A. Evaluation of alternatives to cautery disbudding of dairy goat kids using physiological measures of immediate and longer-term pain. *J. Dairy Sci.* **2018**, *101*, 5374–5387. [CrossRef] [PubMed]
9. Andersson, H.; Lillipers, K.; Rydhmer, L.; Forsberg, M. Influence of light environment and photoperiod on plasma melatonin and cortisol profiles in young domestic boars, comparing two commercial melatonin assays. *Domest. Anim. Endocrinol.* **2000**, *19*, 261–274. [CrossRef]
10. Weary, D.M.; Niel, L.; Flower, F.C.; Fraser, D. Identifying and preventing pain in animals. *Appl. Anim. Behav. Sci.* **2006**, *100*, 64–76. [CrossRef]
11. Eriksson, L.; Teräväinen, T.L. Circadian rhythm of plasma cortisol and blood glucose in goats. *Aust. J. Appl. Sci.* **1989**, *2*, 202–203. [CrossRef]
12. Colborn, D.R.; Thompson, D.L.; Roth, T.L.; Capehart, J.S.; White, K.L. Responses of cortisol and prolactin to sexual excitement and stress in stallions and geldings. *J. Anim. Sci.* **1991**, *69*, 2556–2562. [CrossRef] [PubMed]
13. Tschoner, T. Methods for Pain Assessment in Calves and Their Use for the Evaluation of Pain during Different Procedures—A Review. *Animals* **2021**, *11*, 1235. [CrossRef] [PubMed]
14. Jager, L.P.; De Graaf, G.J.; Widjaja-Greefkes, H. Effects of atipamezole, detomidine and medetomidine on release of steroid hormones by porcine adrenocortical cells in vitro. *Eur. J. Pharmacol.* **1998**, *346*, 71–76. [CrossRef] [PubMed]
15. Hempstead, M.N.; Waas, J.R.; Stewart, M.; Cave, V.M.; Sutherland, M.A. Can Isoflurane and Meloxicam Mitigate Pain Associated with Cautery Disbudding of 3-Week-Old Goat Kids? *Animals* **2020**, *10*, 878. [CrossRef]
16. Prunier, A.; Mounier, A.M.; Hay, M. Effects of castration, tooth resection, or tail docking on plasma metabolites and stress hormones in young pigs. *J. Anim. Sci.* **2005**, *83*, 216–222. [CrossRef] [PubMed]
17. Stubsjoen, S.M.; Flo, A.S.; Moe, R.O.; Janczak, A.M.; Skjerve, E.; Valle, P.S.; Zanella, A.J. Exploring non-invasive methods to assess pain in sheep. *Physiol. Behav.* **2009**, *98*, 640–648. [CrossRef] [PubMed]
18. Stubsjoen, S.M.; Bohlin, J.; Skjerve, E.; Valle, P.S.; Zanella, A.J. Applying fractal analysis to heart rate time series of sheep experiencing pain. *Physiol. Behav.* **2010**, *101*, 74–80. [CrossRef]
19. Peers, A.; Mellor, D.J.; Wintour, E.M.; Dodic, M. Blood pressure, heart rate, hormonal and other acute responses to rubber-ring castration and tail docking of lambs. *N. Z. Vet. J.* **2002**, *50*, 56–62. [CrossRef]
20. Murrell, J.; Johnson, C.B. Neurophysiological techniques to assess pain in animals. *J. Vet. Pharm. Ther.* **2006**, *295*, 325–335. [CrossRef]
21. Gibson, T.J.; Johnson, C.B.; Murrell, J.C.; Hulls, C.M.; Mitchinson, S.L.; Stafford, K.J.; Johnstone, A.C.; Mellor, D.J. Electroencephalographic responses of halothane-anaesthetised calves to slaughter by ventral-neck incision without prior stunning. *N. Z. Vet. J.* **2009**, *57*, 77–83. [CrossRef] [PubMed]
22. Johnson, C.B.; Woodbury, W.M.; Caulkett, N.; Wilson, P. Comparison of lidocaine and antler pedicle compression for analgesia during antler removal in red deer (*Cervus elaphus*) anaesthetised by halothane in oxygen: EEG effects. *Vet. Anaesth. Analg.* **2005**, *32*, 16–71. [CrossRef] [PubMed]
23. Johnson, C.B.; Stafford, K.J.; Sylvester, S.P.; Ward, R.N.; Mitchinson, S.; Mellor, D.J. Effects of age on the electroencephalographic response to castration in lambs anaesthetised using halothane in oxygen. *N. Z. Vet. J.* **2005**, *53*, 433–437. [CrossRef]
24. Sabow, A.B.; Goh, Y.M.; Zulkifli, I.; Sazili, A.Q.; Kadir, M.Z.A.A.; Kaka, U.; Khadijah, N.; Adeyemi, K.D.; Ebrahimi, M. Electroencephalographic responses to neck cut and exsanguination in minimally anaesthetized goats. *S. Afr. J. Anim. Sci.* **2017**, *47*, 34–40. [CrossRef]
25. Sutherland, M.A.; Watson, T.J.; Johnson, C.B.; Millman, S.T. Evaluation of the efficacy of a non-penetrating captive bolt to euthanase neonatal goats up to 48 hours of age. *Anim. Welf.* **2016**, *25*, 471–479. [CrossRef]
26. Raghazli, R.; Othman, A.H.; Kaka, U.; Abubakar, A.A.; Imlan, J.C.; Hamzah, H.; Sazili, A.Q.; Goh, Y.M. Physiological and electroencephalogram responses in goats subjected to pre-and during slaughter stress. *Saudi J. Biol. Sci.* **2021**, *28*, 6396–6407. [CrossRef] [PubMed]

27. Huozha, R.; Rastogi, S.K.; Korde, J.P.; Madan, A.K. Electroencephalographic changes during experimental pain induction in goats. *Vet. Arhiv.* **2011**, *81*, 359–368.
28. Bergamasco, L.; Macchi, E.; Facello, C.; Badino, P.; Odore, R.; Re, G.; Osella, M.C. Electroencephalographic power spectral analysis of growing goat kids (*Capra hircus*). *Small Rumin. Res.* **2006**, *66*, 265–272. [CrossRef]
29. MN Hempstead, M.N.; Stewart, M.; Waasb, J.R.; Sutherland, M.A. Effect of disbudding on pain sensitivity and weight-gain of dairy goat kids. *Proc. N. Z. Soc. Anim. Prod.* **2017**, *77*, 159–160.
30. Hempstead, M.N.; Waas, J.R.; Stewart, M.; Zobel, G.; Cave, V.M.; Julian, A.F.; Sutherland, M.A. Pain sensitivity and injury associated with three methods of disbudding goat kids: Cautery, cryosurgical and caustic paste. *Vet. J.* **2018**, *239*, 42–47. [CrossRef]
31. Mirra, A.; Spadavecchia, C.; Bruckmaier, R.; Gutzwiller, A.; Casoni, D. Acute pain and peripheral sensitization following cautery disbudding in 1- and 4-week-old calves. *Physiol. Behav.* **2018**, *184*, 248–260. [CrossRef] [PubMed]
32. Alvarez, L.; Adcock, S.J.J.; Tucker, C.B. Sensitivity and wound healing after hot-iron disbudding in goat kids. *J. Dairy Sci.* **2019**, *102*, 10152–10162. [CrossRef] [PubMed]
33. Mintline, E.M.; Stewart, M.; Rogers, A.R.; Cox, N.R.; Verkerk, G.A.; Stookey, J.M.; Webster, J.R.; Tucker, C.B. Play behavior as an indicator of animal welfare: Disbudding in dairy calves. *Appl. Anim. Behav. Sci.* **2013**, *144*, 22–30. [CrossRef]
34. Frahm, S.; Di Giminiani, P.; Stanitznig, A.; Schoiswohl, J. Nociceptive Threshold of Calves and Goat Kids Undergoing Injection of Clove Oil or Isoeugenol for Disbudding. *Animals* **2020**, *10*, 1228. [CrossRef] [PubMed]
35. Janczak, A.M.; Ranheim, B.; Fosse, T.K.; Hild, S.; Nordgreen, J.; Moe, R.O.; Zanella, A.J. Factors affecting mechanical (nociceptive) thresholds in piglets. *Vet. Anaesth. Analg.* **2012**, *39*, 628–635. [CrossRef] [PubMed]
36. Venkatachalam, D.; Chambers, J.P.; Kongara, K.; Ward, N.; Jacob, A.; Singh, P. Pharmacokinetics, efficacy, and convulsive dose of articaine hydrochloride in goat kids. *Vet. Anaesth. Analg.* **2021**, *48*, 264–271. [CrossRef]
37. Travain, T.; Valsecchi, P. Infrared Thermography in the Study of Animals' Emotional Responses: A Critical Review. *Animals* **2021**, *11*, 2510. [CrossRef]
38. Stewart, M.; Webster, J.R.; Schaefer, A.L.; Cook, N.J.; Scott, S.L. Infrared thermography as a non-invasive tool to study Animal Welfare. *Anim. Welf.* **2005**, *14*, 319–325. [CrossRef]
39. Stewart, M.; Stafford, K.; Dowling, S.; Schaefer, A.; Webster, J. Eye temperature and heart rate variability of calves disbudded with or without local anaesthetic. *Physiol. Behav.* **2008**, *93*, 789–797. [CrossRef]
40. Pang, W.Y.; Earley, B.; Murray, M.; Sweeney, T.; Gath, V.; Crowe, M.A. Banding or Burdizzo castration and carprofen administration on peripheral leukocyte inflammatory cytokine transcripts. *Res. Vet. Sci.* **2011**, *90*, 127–132. [CrossRef]
41. Sandercock, D.A.; Barnett, M.W.; Coe, J.E.; Downing, A.C.; Nirmal, A.J.; Di Giminiani, P.; Edwards, S.A.; Freeman, T.C. Transcriptomics Analysis of Porcine Caudal Dorsal Root Ganglia in Tail Amputated Pigs Shows Long-Term Effects on Many Pain-Associated Genes. *Front. Vet. Sci. Sec. Anim. Behav. Welf.* **2019**, *6*, 314. [CrossRef]
42. Kongara, K.; Dukkipati, V.S.R.; Tai, H.M.; Heiser, A.; Murray, A.; Webster, J.; Johnson, C.B. Differential Transcription of Selected Cytokine and Neuroactive Ligand-receptor Genes in Peripheral Leukocytes from Calves in Response to Cautery Disbudding. *Animals* **2020**, *10*, 1187. [CrossRef]
43. Greenwood, P.L.; Shutt, D.A. Effects of management practices on cortisol, b-endorphin and behavior in young goats. *Aust. Soc. Anim. Prod. Adel. Aust.* **1990**, *18*, 224–227.
44. Ghodasara, P.; Satake, N.; Sadowski, P.; Kopp, S.; Mills, P.C. Investigation of cattle plasma proteome in response to pain and inflammation using next generation proteomics technique, SWATH-MS. *Mol. Omics* **2022**, *18*, 133–142. [CrossRef] [PubMed]
45. Morton, D.B.; Griffiths, P.H. Guidelines on the recognition of pain, distress and discomfort in experimental animals and hypothesis for assessment. *Vet. Rec.* **1985**, *116*, 431–436. [CrossRef] [PubMed]
46. Leslie, E.; Hernández-Jover, M.; Newman, R.; Holyoake, P. Assessment of acute pain experienced by piglets from ear tagging, ear notching and intraperitoneal injectable transponders. *Appl. Anim. Behav. Sci.* **2010**, *127*, 86–95. [CrossRef]
47. Sanford, J.; Ewbank, R.; Molony, V.; Tavenor, W.D.; Uvarov, O. Guidelines for the recognition and assessment of pain in animals. *Vet Rec.* **1986**, *118*, 334–338.
48. Bufalari, A.; Adami, C.; Angeli, G.; Short, C.E. Pain Assessment in Animals. *Vet. Res. Commun.* **2007**, *31* (Suppl. S1), 55–58. [CrossRef] [PubMed]
49. Fitzpatrick, J.; Scott, M.; Nolan, A. Assessment of pain and welfare in sheep. *Small Rumin. Res* **2006**, *62*, 55–61. [CrossRef]
50. Hempstead, M.N.; Waas, J.R.; Stewart, M.; Cave, V.M.; Sutherland, M.A. Behavioural response of dairy goat kids to cautery disbudding. *Appl. Anim. Behav. Sci.* **2017**, *194*, 42–47. [CrossRef]
51. Small, A.; Fisher, A.; Lee, C.; Colditz, I. Gap Evaluation of Pain Alleviation Research. Final Report to AWI. 2020. Available online: Wool.com/globalassets/wool/sheep/research-publications/welfare/improved-pain-relief/project-final-report-on-gap-evaluation-of-pain-alleviation.pdf (accessed on 1 May 2023).
52. Hempstead, M.N.; Waas, J.R.; Stewart, M.; Cave, V.M.; Sutherland, M.A. Evaluation of alternatives to cautery disbudding of dairy goat kids using behavioural measures of post-treatment pain. *Appl. Anim. Behav. Sci.* **2018**, *206*, 32–38. [CrossRef]
53. Hempstead, M.N.; Waas, J.R.; Stewart, M.; Dowling, S.K.; Cave, V.M.; Lowe, G.L.; Sutherland, M.A. Effect of isoflurane alone or in combination with meloxicam on the behavior and physiology of goat kids following cautery disbudding. *J. Dairy Sci.* **2018**, *101*, 3193–3204. [CrossRef]

54. Hempstead, M.N.; Lindquist, T.M.; Shearer, J.K.; Shearer, L.C.; Sutherland, M.A.; Plummer, P.J. Acute cortisol and behavior of dairy goat kids administered local anesthesia, topical anesthesia or systemic analgesia prior to cautery disbudding. *Physiol. Behav.* **2020**, *222*, 112942. [CrossRef] [PubMed]
55. Ajuda, I.; Battini, M.; Mattiello, S.; Arcuri, C.; Stilwell, G. Evaluation of Pain Mitigation Strategies in Goat Kids after Cautery Disbudding. *Animals* **2020**, *10*, 277. [CrossRef]
56. Ingvast-Larsson, C.; Hogberg, M.; Mengistu, U.; Olsen, L.; Bondesson, U.; Olsson, K. Pharmacokinetics of meloxicam in adult goats and its analgesic effect in disbudded kids. *J. Vet. Pharmacol. Ther.* **2011**, *34*, 64–69. [CrossRef] [PubMed]
57. Wagmann, N.; Spadavecchia, C.; Morath-Huss, U.; Schüpbach-Regula, G.; Zanolari, P. Evaluation of anaesthesia and analgesia quality during disbudding of goat kids by certified Swiss farmers. *BMC Vet. Res.* **2018**, *14*, 220. [CrossRef] [PubMed]
58. Still Brooks, K.M.; Hempstead, M.N.; Anderson, J.L.; Parsons, R.L.; Sutherland, M.A.; Plummer, P.J.; Millman, T.S. Characterization of Efficacy and Animal Safety across Four Caprine Disbudding Methodologies. *Animals* **2021**, *11*, 430. [CrossRef]
59. Lou, M.E. The Development of a Goat Kid Grimace scale Following Thermal Disbudding. Master's Thesis Chapter, University of Minnesota Twin Cities, Minneapolis, MN, USA, 2020.
60. Langford, D.J.; Bailey, A.L.; Chanda, M.L.; Clarke, S.E.; Drummond, T.E. Coding of facial expressions of pain in the laboratory mouse. *Nat. Methods* **2010**, *7*, 447–449. [CrossRef]
61. Guesgen, M.J.; Beausoleil, N.J.; Leach, M.; Minot, E.O.; Stewart, M.; Stafford, K.J. Coding and quantification of a facial expression for pain in lambs. *Behav. Process.* **2016**, *132*, 49–56. [CrossRef]
62. Häger, C.; Biernot, S.; Buettner, M.; Glage, S.; Keubler, L.M.; Held, N.; Bleich, E.M.; Otto, C.; Müller, C.W.; Decker, S.; et al. The Sheep Grimace Scale as an indicator of post-operative distress and pain in laboratory sheep. *PLoS ONE* **2017**, *12*, 0175839. [CrossRef]
63. Johnson, C.B. Research Tools for the Measurement of Pain and Nociception. *Animals* **2016**, *6*, 71. [CrossRef] [PubMed]

animals

Review

Pain Mitigation Strategies for Disbudding in Goat Kids

Preet Singh *, Dinakaran Venkatachalam, Kavitha Kongara and Paul Chambers

Tāwharau Ora School of Veterinary Science, Massey University, Palmerston North 4410, New Zealand; k.kongara@massey.ac.nz (K.K.); j.p.chambers@massey.ac.nz (P.C.)
* Correspondence: p.m.singh@massey.ac.nz

Simple Summary: The process of removing horn buds (disbudding) is painful for young animals. People are increasingly concerned about the well-being of animals, so finding ways to reduce the pain and stress caused by disbudding is important. This review discusses various methods to ease pain during disbudding in goat kids, including using drugs to sedate and relieve pain, blocking nerves, and giving anti-inflammatory medications afterward. It also mentions the potential harm of certain drugs. The recommended approach is to use a combination of sedation, nerve blocking, and anti-inflammatory drugs for the best results in reducing pain. This review ends by suggesting directions for more research to further improve the well-being of young goats during the disbudding process.

Abstract: Pain mitigation strategies for disbudding in goat kids have gained significant attention in recent years because of growing concerns for animal welfare. Disbudding, the removal of horn buds in young goats, is a common practice to enhance safety and manage herd dynamics. However, the procedure will cause pain and distress if not managed effectively. This review covers the array of pain mitigation techniques currently available for disbudding, including the efficacy of these strategies in reducing pain and stress during the disbudding process, with specific attention to the potential toxicity associated with local anesthetics. The current best practice for disbudding on the farm suggests sedation/analgesia with an alpha-2 agonist, the placement of a two-point cornual nerve block, and then an NSAID for postoperative pain. In conclusion, this review offers recommendations for future research directions aimed at enhancing the welfare of young goats subjected to the disbudding procedure. These suggestions hold the promise of fostering significant improvements in the overall well-being of these animals.

Keywords: disbudding; goat kids; analgesia

1. Introduction

Goat kids, particularly females from milking herds, are commonly subjected to disbudding to reduce potential injury to animals and humans from horns or entanglement in equipment in the milking shed. Goat kids are typically disbudded within the first week of life, most commonly via thermal cauterization using a hot iron. Hot iron disbudding is a painful procedure that requires pain relief to be provided to the animals undergoing this procedure to avoid a negative impact on their welfare. Hot iron disbudding is considered a significant surgical process and should be performed by a veterinarian or under their supervision. This is particularly important in goat kids because there is a significant risk of thermal brain damage and death.

The need for disbudding can be circumvented by selectively breeding for polledness. Although this approach is effective in cattle, it tends to trigger severe reproductive complications in certain dairy goats with European lineage, such as the Saanen, Alpine, and Toggenburg breeds [1]. In these breeds, the existence of horns is governed by a recessive gene that leads to infertility. Consequently, female goats with a homozygous polled genotype will mature as infertile intersex individuals, while male goats with the same genotype face a heightened likelihood of developing sperm granulomas. Therefore, selective breeding for polledness is not beneficial for these goat breeds, and disbudding remains the only tool to eliminate the horns [1].

The pain from hot iron disbudding has at least two phases, which probably require different approaches to alleviation. The application of a hot iron causes intense acute pain, and then, the release of inflammatory mediators from the burnt tissue will cause a lower-grade but longer-lasting pain. Both phases should be treated to ensure animal welfare. There are a variety of ways of accomplishing this: general anesthetics and analgesics with or without local anesthetics and then non-steroidal anti-inflammatory drugs (NSAIDs) for postoperative pain. The efficacy of local anesthetics and general anesthetics for disbudding pain in goat kids has not been extensively investigated or reported in the literature compared with calves. Most of the drugs recommended or referenced in this review are used off-label, as they are not registered for use in food-producing animals. Additionally, the registration status and availability of these drugs differ across countries and regions.

2. General Anesthesia and Systemic Analgesia

Xylazine hydrochloride is probably the most commonly used alpha-2 adrenergic agonist in ruminant veterinary practice despite its association with significant complications like hypoxemia and pulmonary edema. These complications are linked to the activation of pulmonary intravascular macrophages [2] and may involve the release of TNF-Alpha. TNF suppression agents, such as choline chloride given before xylazine, have a small beneficial effect [3]. Other inflammatory mediators are almost certainly involved as well. In adult sheep, the degree of hypoxemia is similar to xylazine, romifidine, detomidine, and medetomidine [4]. Goats exhibit higher sensitivity to xylazine compared with other ruminants [5]. As a result, careful dose calculation and monitoring of the effects of xylazine are particularly important in goats, especially in goat kids, when compared with other ruminants.

Sedatives such as xylazine have proven effective in reducing stress during restraint and offer partial analgesic effects, yet they do not provide comprehensive pain relief during disbudding. Wagmann et al. (2018) investigated the efficacy of a mixture of xylazine (0.05 mg/kg) and ketamine (20 mg/kg) administered prior to disbudding by certified Swiss farmers [6]. The authors concluded that this mixture did not provide adequate anesthesia and analgesia in goat kids and suggested that refinement to this protocol is required. Dexmedetomidine administered intramuscularly may be more effective for analgesia as compared with the administration of lidocaine around the horn bud and intramuscular injections of meloxicam [7]. In the UK, where disbudding is carried out under general anesthesia in the first week of life, xylazine overdose is considered the most common cause of death [8]. Alphaxalone has been recommended for disbudding performed on the farm at a dose of 6 mg/kg administered intravenously [9]. As it provides negligible analgesia [10], a pre- and postsurgical analgesic protocol should be followed.

Alpha-2 agonists are easy to administer while performing this procedure on the farm; however, they produce adverse effects such as hypothermia and cardiovascular and respiratory depression and often require a reversal agent such as yohimbine or atipamezole postoperatively (which will also reverse analgesia).

Nonsteroidal anti-inflammatory drugs have demonstrated their ability to alleviate post-disbudding pain; however, they have fallen short in preventing the acute pain induced during the disbudding process itself [11]. Another disadvantage is that, as goats are disbudded shortly after birth, drugs requiring liver metabolism may have a variable, but probably long, duration of action. Goat kids, like other neonates, almost certainly have a restricted capacity to metabolize and eliminate drugs. The administration of sedatives such as xylazine may reduce stress and make it easier to administer local anesthetics.

Inhalant general anesthetics, especially isoflurane, have demonstrated effectiveness in mitigating pain during disbudding. However, their practicality in commercial farming scenarios is questionable given the need for anesthetic equipment and the potential for increased disbudding costs [12]. Using inhalation anesthetics in oxygen delivered by an anesthetic machine via a face mask used to be considered the best practice [9] but can be dangerous in the presence of a hot iron, as oxygen supports combustion.

The United Kingdom and several European countries mandate that disbudding in goats is exclusively carried out by a veterinarian, utilizing appropriate anesthetics and analgesics [6]. Consequently, in New Zealand and Australia, a significant shift has occurred, and pain relief measures have become mandatory for disbudding goat kids [13]. This represents a noteworthy advancement in animal welfare practices.

3. Local Anesthesia

The administration of a local anesthetic to produce a nerve block is a common strategy for alleviating pain during the disbudding process [1]. Usually, two nerves are blocked on each side: the cornual branches of the infratrochlear nerve (a branch of the ophthalmic division of the trigeminal nerve CNV_1) and the zygomaticotemporal nerve (also referred to as the cornual branch of the lacrimal nerve, a branch of the maxillary division of the trigeminal nerve CNV_2) (Figure 1).

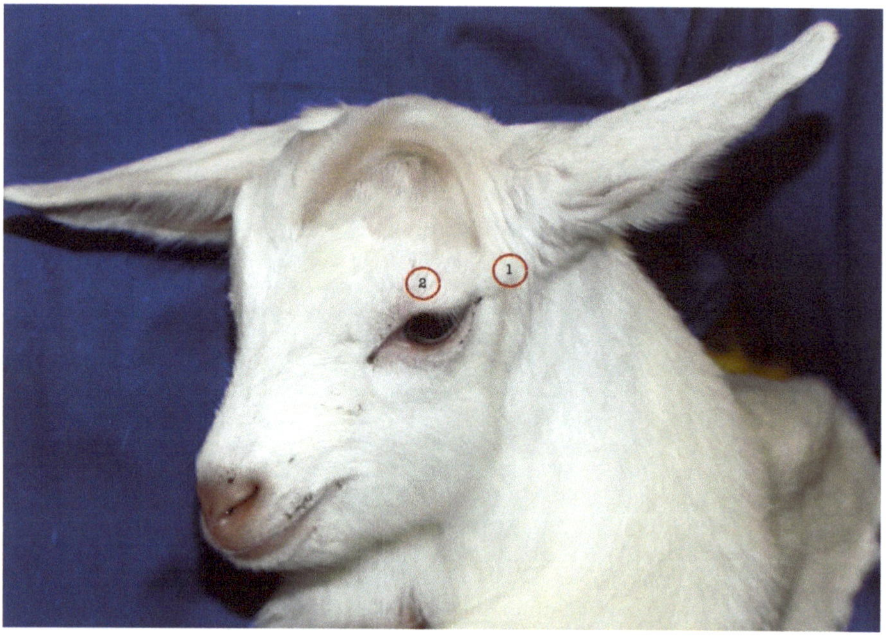

Figure 1. Each horn bud in the goat kids is innervated by two cornual nerves, one from the zygomaticotemporal nerve and the other from the infratrochlear nerve. Therefore, two injection sites (as shown in the image) on each side are recommended to effectively alleviate pain during the process of disbudding.

Usually, 0.5 mL of a local anesthetic is injected at each site, but because there may be several branches in the infratrochlear nerve [14], sometimes larger volumes are used in order to achieve a greater spread.

All local anesthetics have a potential for toxicity [15]. Goat kids, given their small size, are susceptible to overdosing. Overdose can cause cardiac effects (signs of reduced cardiac output), sedation, convulsions, and death. The need to block four nerves for both horn buds (in contrast to a single nerve per side in calves) increases the amount of local anesthetic required. Additionally, the vascularity at the site of the nerve block [16] exacerbates the risk.

Goat kids are not small calves [17], as they undergo disbudding at a younger age, with thinner skulls and much lower body weights [17]. For instance, goat breeds like pygmy and Nigerian Dwarf, which can weigh under 2 kg, might encounter issues when 2 mL of 2% lidocaine (20 mg/mL), equivalent to 20 mg/kg, is injected for nerve blocks. This dosage could easily lead to plasma concentrations sufficient to cause convulsions [18].

It is crucial to emphasize that the techniques used for disbudding calves should not be directly applied to goat kids. Thinner skulls in goat kids increase the risk of brain damage due to thermal injuries, resulting in convulsions and death. Pathological findings from goat kids disbudded with hot irons have revealed central areas of cavitation in the brain both in gray and white matter. Histological lesions included extensive hemorrhages and coagulation necrosis [19].

The thermal lesions caused by cautery disbudding can be infected by bacteria, leading to a potential risk of bacterial invasion and the development of meningoencephalitis [20]. This can be treated with a course of broad-spectrum antibiotics [19]. Surviving goat kids display signs of incoordination, paraplegia, and convulsions even up to 3 weeks. The application time is also crucial, as the placement of a cautery iron for 15 to 20 s has been shown to cause severe brain injuries [21].

4. Lidocaine

Lidocaine is the most commonly used local anesthetic in veterinary practice [5] in most places, apart from the EU, and is thus used in goat medicine. It is a cheap, effective, usually safe, and readily available drug. Only one study has reported convulsions in a goat kid following the intramuscular injection of lidocaine at approximately 10 mg/kg [22]. In a dose-ranging investigation, a dosage of 7 mg/kg of body weight administered intravenously over a 60 s interval yielded no observable signs of toxicity [18]. Consequently, this dosage is presumed safe for cornual nerve blocks in goat kids, apart from possibly accidental intravenous administration. This dose may also be deemed safe for other localized and regional nerve blocks in goat kids, although more extensive safety studies involving more animals are necessary.

The minimum dose necessary to induce seizures in young goats (12.42 mg/kg) [18] is lower than the comparable dose observed in newborn lambs (18.40 mg/kg) [22], and the average plasma level associated with convulsions in young goats (13.59 $\pm$ 2.34 μg/mL) [18] is lower than the level observed in newborn lambs (16.6 $\pm$ 1.2 μg/mL) [22]. Comparatively, in dogs, the lidocaine concentration leading to toxicity averages 8.21 + 1.69 μg/mL, which is notably below the level recorded in young goats (13.59 $\pm$ 2.34 μg/mL) [23]. Meyer et al. (2001) found that horses exhibited intoxication at a serum concentration of 3.24 $\pm$ 0.74 μg/mL [24]. These variations might stem from the diverse criteria used to assess toxicity across different species. In dogs, the toxic indicator is the tonic extension phase, and horses are evaluated based on skeletal muscle fasciculation [23,24], while for goat kids, the endpoint is the occurrence of convulsions [18].

An additional factor contributing to these observed disparities might be the variation in the rate of drug administration across different species or the fact that young animals display reduced sensitivity to lidocaine toxicity when compared with adult animals, a phenomenon likely attributable to the higher volume of distribution, a characteristic of younger individuals [22].

The subcutaneous administration of 0.5 mL of 1% lidocaine (such as around a superficial nerve) exhibits rapid absorption in goat kids, with an average T_{max} of 0.33 $\pm$ 0.11 h [18]. Rapid absorption has also been reported following cornual nerve blocks in goat kids [25]. The C_{max} (0.58 $\pm$ 0.17 μg mL^{-1}) was around four times less than the concentration (2.55 $\pm$ 0.41 μg mL^{-1}) observed at 1 min following intravenous administration of 8 mg kg^{-1} for over 60 s, the maximum dose that did not show any observable toxicity signs. The elimination rate of both lidocaine and its main metabolite in most species, monoethylglycinexylidide (MEGX), is moderate, indicated by mean $t_{1/2\lambda z}$ values of 2.28 h and 3.20 h, respectively. After subcutaneous administration, the mean peak plasma concentration of lidocaine (2.12 ± 0.81 μg/mL) is roughly 6.5 times lower than the mean plasma concentration associated with convulsions (13.59 ± 2.34 μg/mL) [11]. Decreasing peak plasma concentrations lowers the likelihood of encountering toxicity [26]. Given that the C_{max} resulting from a 0.5 mL/site injection of 1% lidocaine hydrochloride is significantly below the toxic plasma concentration, this dosage is likely to be safe for cornual nerve blocking in goat kids. No pain-related behavioral signs have been observed during disbudding, indicating that it is also effective. Nevertheless, most of the goat kids begin exhibiting behaviors like head scratching and head shaking around 20 min after nerve blocks, indicating that the anesthetic effect lasts only for about 20 min.

Another major concern of using lidocaine in food-producing animals is its metabolism to dimethylaniline, (DMA, 2,6 xylidine). Amide-type local anesthetics like lidocaine undergo hepatic biotransformation through specific cytochrome P450 isoforms (CYP3A4) [27]. In the liver, lidocaine is transformed into various metabolites, including MEGX and glycinexylidide (GX), via oxidative N-dealkylation and DMA via hydrolysis [28]. MEGX further undergoes biotransformation into DMA, although MEGX does not appear to be produced in significant quantities in adult cattle [29]. DMA is oxidized in the liver into N-(2,6-dimethylphenyl) hydroxylamine (DMHA) and 4-amino-3,5-dimethylphenol (DMAP). DMHA then goes through phase II biotransformation, specifically acetylation, forming reactive esters that ultimately convert into a reactive nitrenium ion through phase 2 metabolism [27]. Both DMHA and the nitrenium ion can bind covalently to DNA, potentially leading to the development of tumors. DMHA can also react with hemoglobin, resulting in the formation of hemoglobin adducts through covalent binding with cysteine residues [27]. The other metabolite, DMAP, undergoes oxidation into an iminoquinone, a highly reactive electrophile with genotoxic properties [30]. DMAP can also be generated from the nitrenium ion or DMHA. DMA is also rapidly converted into 4-hydroxyDMA, which accounts for most urinary excretions in adult cattle [29]. DMA, based on toxicology studies in rats, has been categorized as a possible carcinogen (Group B) by the International Agency for Research on Cancer [31]. Their study revealed that the chronic oral administration of DMA at a dosage of 3000 mg/kg resulted in the development of various cancers in rats, including nasal papilloma and carcinoma, rhabdomyosarcoma, subcutaneous fibromas, and fibrosarcomas [32]. Furthermore, the mortality rates were higher in animals that received 1000 and 3000 mg/kg of DMA compared with a control group. These findings led to the conclusion that DMA acts as a carcinogen in rats. In dogs, chronic oral administration of DMA causes weight reduction, hyperbilirubinemia, hypoproteinemia, and significant fatty degeneration. In vitro studies have further confirmed the genotoxic and mutagenic properties of DMA [27]. Several human studies have also indicated a potential link between DMA and an increased risk of bladder cancer [33,34]. Collectively, these reports strongly suggest that DMA may indeed be a carcinogen (at high doses), making the use of lidocaine in animals undesirable.

5. Articaine

Articaine is also an amino-amide class of local anesthetic and has a distinct molecular structure with a thiophene ring instead of a benzene ring, as well as an ester group. The presence of the thiophene ring enhances its lipid solubility, enabling articaine to penetrate nerve fibers more efficiently and rapidly [35]. Moreover, its unique molecular composition allows for improved penetration through both bone and soft tissues compared with other local anesthetics [36,37]. The ester group present in articaine makes it noteworthy among amino-amide local anesthetics, as it undergoes rapid hydrolysis via esterases in tissues and plasma [35–37] into inactive articainic acid. Its structure means that DMA is not formed during metabolism. Therefore, articaine hydrochloride could be a safer and better option than lidocaine in goat kids.

An investigation involving goat kids was conducted to assess the toxicity and pharmacokinetics of articaine [25]. This study involved the intravenous administration of articaine hydrochloride IV over a 60 s interval, and cornual nerve blocks were administered using 1.5% articaine hydrochloride at a dosage of 0.5 mL per site. The goat kids exhibited no indications of toxicity associated with nerve blocks throughout these procedures. The average IV dosage required to produce convulsions was 16.24 ± 1.79 mg kg^{-1}, and the average plasma concentrations of articaine and articainic acid at convulsions were 9.90 ± 2.38 μg mL^{-1} and 1.52 ± 0.91 μg mL^{-1}, respectively.

A cornual nerve block (0.5 mL/site) using 1.5% articaine hydrochloride took about 4 min to anesthetize the horn buds based on the absence of a withdrawal response and vocalization during disbudding. This confirms the effective analgesic properties of articaine. However, post-procedure observations of the animals indicated the emergence of pain-related behaviors, including head scratching and head shaking, approximately 25 min after administration [25].

The absorption of articaine after cornual nerve block is rapid with a mean C_{max} of 586.58 ± 175.10 ng mL^{-1} at 0.22 ± 0.09 h (T_{max}) [25]. The rapid absorption of articaine has also been reported in red deer [38] and humans [39]. This could be because of vasodilatation, similar to most other local anesthetics [40,41]. The short elimination half-lives ($t_{\frac{1}{2}\lambda z}$) of articaine following intravenous administration (0.66 ± 0.14 h) and subcutaneous administration (1.26 ± 0.34 h) indicate that articaine is rapidly eliminated following systemic absorption [25]. Rapid elimination has also been reported in deer [38] and people [39]. The rapid elimination of articaine may be due to rapid hydrolysis via plasma esterases into articainic acid [37]. Articainic acid is an inactive metabolite, whereas several primary metabolites of lidocaine are active and can increase the risk of toxicity during accidental intravenous administration or overdosage [35,42]. The plasma clearance of articaine was rapid in goat kids with a mean CLss of 5.33 ± 0.66 L kg^{-1} in [25].

There have been no documented cases of toxicity associated with articaine in humans, except, rarely, paresthesia. Paresthesia is characterized by persistent anesthesia or altered sensations in the form of neuropathy [43]. The underlying causes of paresthesia are still not fully understood, although it is believed to be related to the concentration of local anesthetics used. It has been observed that paresthesia occurs more frequently following the administration of 4% local anesthetic formulations, such as articaine and prilocaine, compared with the use of 2% local anesthetic formulations [43] (Puccini et al., 2015).

While articaine is generally regarded as a safe local anesthetic, it can potentially cause systemic toxicity (including central nervous system and cardiovascular toxicity) in the same way as other local anesthetics when toxic concentrations are achieved through inadvertent intravenous administration or overdose. However, the risk of systemic toxicity from overdosing is relatively low compared with lidocaine and other amide-type local anesthetics, as articaine undergoes rapid hydrolysis following systemic absorption [35]. The clearance of articaine was found to be 10 times greater than that of lidocaine in humans [44]. Rapid hydrolysis into an inactive metabolite and rapid elimination indicate that articaine may be safer than lidocaine for cornual nerve blocks in goat kids. Because of its wider margin of safety, articaine is clinically used as a 4% solution, whereas lidocaine is used as a 2% solution [37].

6. Bupivacaine

Both lidocaine and articaine are effective for less than 30 min. The most commonly used local anesthetic in people, bupivacaine, has a much longer duration in most species. It was used for nerve blocks in adult goats in [45] for the surgical examination of the stifle joint conducted under general anesthesia and sciatic–femoral nerve blocks. In that study, 0.5% bupivacaine provided effective analgesia with minimal adverse effects. A group receiving a higher dose of bupivacaine exhibited unilateral motor blockade. There are no reports of its use in goat kids. Extended-release formulations of bupivacaine have been used for disbudding in calves [46]. However, long-acting local anesthetics such as bupivacaine may not be a good choice for goat kids, as bupivacaine possesses a greater risk of cardiac toxicity than articaine and lidocaine. This is an area that needs further research.

7. Postoperative Analgesia

Since many of the currently available local anesthetics for farm animals are short-acting, additional pain relief, such as systemic NSAIDs, may be required to control post-disbudding pain. Meloxicam has been reported to reduce post-disbudding pain in goat kids [12]. The subcutaneous administration of meloxicam one hour prior has been shown to minimize the expression of inflammatory cytokines in calves [47]. The administration of both local anesthesia and systemic NSAIDs appears to have the potential to provide better analgesia (compared with their single use) for disbudding pain. However, this did not provide complete postoperative pain relief in goat kids [11,48]. A mixture of meloxicam, lidocaine, and xylazine demonstrated greater effectiveness in the initial hour compared with using only lidocaine [48]. Therefore, a multimodal analgesic approach is likely to prove more beneficial than the use of a single class of drugs.

8. Alternative Methods of Disbudding

Various alternative methods to thermal cautery disbudding have been explored in goat kids [49]. These include cryosurgical and chemical disbudding techniques. Cryosurgery involves the use of liquid nitrogen, while the chemical method employs caustic paste (usually sodium, calcium, or potassium hydroxide paste) to destroy horn buds. However, both cryosurgical and caustic paste methods have been shown to induce more pain than cautery disbudding, as indicated by physiological and behavioral changes [49]. Caustic paste can also pose the risk of damaging the eyes of goat kids and the udders of does.

There is a need for a simple, safe, cost-effective technique to prevent the growth of horn buds in goat kids. An approach involving clove oil injection into the buds has been explored for horn bud destruction both in goat kids [50] and in calves [51], but the degree of distress experienced during or after the procedure has not been extensively reported. Histopathological changes in horn buds injected with clove oil have revealed coagulative necrosis of the epidermis and the infiltration of neutrophils. In a pilot study involving 12 calves injected with clove oil and isoeugenol at different volumes, it was found that the injection volume plays a crucial role in successful disbudding in calves [52]. Although clove oil or isoeugenol injections cause less tissue damage compared with hot iron disbudding, the success rate is lower, as scur formation was observed six months post-procedure [53,54]. Therefore, clove oil injection may not be entirely effective in preventing horn bud growth, as evidenced by the emergence of scurs in a substantial number of goat kids.

Cloves have been used as an anesthetic for fish [55] and a product (Aqui-S, Aqui-S New Zealand Ltd., Lower Hutt, New Zealand) is currently registered to enable handling. It has traditionally been used as a local anesthetic agent in human dentistry; however, at higher doses, it exhibits cytotoxic effects [56]. As cloves and their oil have been used since antiquity as food flavorings, there are fewer concerns about residues in food-producing animals. Consequently, these characteristics render them a viable alternative for disbudding.

Concerns regarding adverse effects in goat kids, including swelling around the horn buds, have been raised because of hypersensitivity reactions caused by eugenol, the main component of clove oil [49]. The pain experienced during the administration of clove oil for disbudding in calves has been found to be significantly less than the cautery disbudding method. However, calves injected with clove oil still exhibited signs of discomfort in [57].

A similar study with calves subjected to clove oil injection showed that, while it might delay horn bud growth, complete elimination is not always achieved. Over a 16-month observation period, calves developed horns or scurs, suggesting that clove oil injection might not offer full prevention of horn bud growth [58]. These findings emphasize the complexities and limitations associated with alternative disbudding methods in terms of both effectiveness and potential adverse reactions.

While clove oil appears promising as an alternative to cautery disbudding, it falls short of the effectiveness demonstrated by traditional methods. Further efforts should be directed toward refining the formulation of clove oil or eugenol to enhance its distribution in the horn bud. Research is essential to understanding the pharmacokinetics and systemic absorption of eugenol/clove oil and determining if a slow-release formulation is necessary. Such advancements could potentially reduce the volume used and mitigate associated side effects.

9. Conclusions

The ideal analgesia for goat kid disbudding is probably a combination of general anesthetic with postoperative analgesia, but this is not practical or economical on a farm. The current best practice for disbudding on the farm suggests sedation and analgesia with an alpha-2 agonist, the placement of a two-point nerve block, and then an NSAID for postoperative pain. Even this is unlikely to be completely effective in all cases, and more research is needed in this area.

Local anesthetics are an essential part of the analgesic protocol, but goat kids are susceptible to overdose. Articaine has been shown to be a safe and effective local anesthetic for cornual nerve block in goat kids with some potential advantages over lidocaine. However, comprehensive future studies, involving different doses and concentrations of articaine hydrochloride within a larger population, are essential to definitively establishing its safety and efficacy for disbudding. Articaine is not registered for use in livestock and lacks defined minimum residual limits (MRLs). This absence of MRLs is a significant obstacle hindering its use in livestock. Residue studies to allow an MRL to be set for articaine would require substantial financial investment and a collaborative effort from the animal industry.

In addition to local anesthetic nerve blockades, sedation with an alpha-2 agonist will minimize stress and pain during local anesthetic injections. As recovery may be prolonged in neonatal animals, an antagonist should be available. Postoperative pain management using NSAIDs is essential, although long-acting local anesthetics may be available in the future. Evaluating the safety and efficacy of this protocol and its administration via alternative routes, such as transdermal or oral, for disbudding in goat kids should be a focus of future studies.

Looking ahead, further research into drugs that will prevent the growth of horn buds safely and effectively is needed. Addressing the issue of pain associated with disbudding in goat kids necessitates a collaborative effort across multiple disciplines, including clinical veterinarians, pharmacologists, chemists, and animal welfare scientists. By pooling expertise from these diverse fields, a more comprehensive and effective solution may be developed to enhance the well-being of goat kids during the disbudding process. This multidisciplinary approach is crucial for the successful resolution of the problem and the development of humane practices in goat farming.

Funding: This research received no external funding.

Institutional Review Board Statement: Not applicable.

Informed Consent Statement: Not applicable.

Data Availability Statement: Not applicable.

Conflicts of Interest: The authors declare no conflict of interest.

References

1. Smith, M.C.; Sherman, D.M. *Goat Medicine*; John Wiley & Sons: Hoboken, NJ, USA, 2023; ISBN 9780813818825.
2. Celly, C.S.; Atwal, O.S.; McDonell, W.N.; Black, W.D. Histopathologic Alterations Induced in the Lungs of Sheep by Use of Alpha2-Adrenergic Receptor Agonists. *Am. J. Vet. Res.* **1999**, *60*, 154–161. [CrossRef] [PubMed]
3. Singh, P.M.; Reid, K.; Gaddam, R.; Bhatia, M.; Smith, S.; Jacob, A.; Chambers, P. Effect of Choline Chloride Premedication on Xylazine-Induced Hypoxaemia in Sheep. *Vet. Anaesth. Analg.* **2017**, *44*, 1149–1155. [CrossRef]
4. Celly, C.S.; McDonell, W.N.; Young, S.S.; Black, W.D. The Comparative Hypoxaemic Effect of Four Alpha 2 Adrenoceptor Agonists (Xylazine, Romifidine, Detomidine and Medetomidine) in Sheep. *J. Vet. Pharmacol. Ther.* **1997**, *20*, 464–471. [CrossRef] [PubMed]
5. Taylor, P.M. Anaesthesia in Sheep and Goats. *Practice* **1991**, *13*, 31–36. [CrossRef]
6. Wagmann, N.; Spadavecchia, C.; Morath-Huss, U.; Schüpbach-Regula, G.; Zanolari, P. Evaluation of Anaesthesia and Analgesia Quality during Disbudding of Goat Kids by Certified Swiss Farmers. *BMC Vet. Res.* **2018**, *14*, 220. [CrossRef] [PubMed]

7. Nfor, O.N.; Chan, J.P.-W.; Kere, M.; Peh, H.-C. Disbudding Pain: The Benefits of Disbudding Goat Kids with Dexmedetomidine Hydrochloride. *Small Rumin. Res.* **2016**, *139*, 60–66. [CrossRef]
8. Matthews, J.; Dustan, B. Disbudding of Goat Kids. *Practice* **2019**, *41*, 433. [CrossRef]
9. Buttle, H.; Mowlem, A.; Mews, A. Disbudding and Dehorning of Goats. *Practice* **1986**, *8*, 63. [CrossRef]
10. Varga, M. *Textbook of Rabbit Medicine*, 2nd ed.; Part II: Rabbit Med.; Elsevier: Amsterdam, The Netherlands, 2014; pp. 178–202. [CrossRef]
11. Ingvast-Larsson, C.; Högberg, M.; Mengistu, U.; Olsén, L.; Bondesson, U.; Olsson, K. Pharmacokinetics of Meloxicam in Adult Goats and Its Analgesic Effect in Disbudded Kids. *J. Vet. Pharmacol. Ther.* **2011**, *34*, 64–69. [CrossRef]
12. Hempstead, M.N.; Waas, J.R.; Stewart, M.; Dowling, S.K.; Cave, V.M.; Lowe, G.L.; Sutherland, M.A. Effect of Isoflurane Alone or in Combination with Meloxicam on the Behavior and Physiology of Goat Kids Following Cautery Disbudding. *J. Dairy Sci.* **2018**, *101*, 3193–3204. [CrossRef]
13. *Code of Welfare: Painful Husbandry Procedures*; Ministry of Primary Industries: Wellington, New Zealand, 2018.
14. Vitums, A. Nerve and Arterial Blood Supply to the Horns of the Goat with Reference to the Sites of Anesthesia for Dehorning. *J. Am. Vet. Méd. Assoc.* **1954**, *125*, 284–286. [PubMed]
15. Lemke, K.A.; Dawson, S.D. Local and Regional Anesthesia. *Vet. Clin. N. Am. Small Anim. Pract.* **2000**, *30*, 839–857. [CrossRef]
16. Harwood, D. Disbudding Goats. *Vet. Rec.* **2012**, *170*, 343. [CrossRef] [PubMed]
17. Hempstead, M.; Waas, J.; Stewart, M.; Sutherland, M. Goat Kids Are Not Small Calves: Species Comparisons in Relation to Disbudding. *Anim. Welf.* **2020**, *29*, 293–312. [CrossRef]
18. Venkatachalam, D.; Chambers, P.; Kongara, K.; Singh, P. Toxicity and Pharmacokinetic Studies of Lidocaine and Its Active Metabolite, Monoethylglycinexylidide, in Goat Kids. *Animals* **2018**, *8*, 142. [CrossRef] [PubMed]
19. Thompson, K.; Bateman, R.; Morris, P. Cerebral Infarction and Meningoencephalitis Following Hot-Iron Disbudding of Goat Kids. *N. Z. Vet. J.* **2005**, *53*, 368–370. [CrossRef] [PubMed]
20. Sanford, S.E. Ontario. Meningoencephalitis Caused by Thermal Disbudding in Goat Kids. *Can. Vet. J. Rev. Vet. Can.* **1989**, *30*, 832.
21. Hempstead, M.N.; Shearer, J.K.; Sutherland, M.A.; Fowler, J.L.; Smith, J.S.; Smith, J.D.; Lindquist, T.M.; Plummer, P.J. Cautery Disbudding Iron Application Time and Brain Injury in Goat Kids: A Pilot Study. *Front. Vet. Sci.* **2021**, *7*, 568750. [CrossRef]
22. Morishima, H.O.; Pedersen, H.; Finster, M.; Sakuma, K.; Bruce, S.L.; Gutsche, B.B.; Stark, R.I.; Covino, B.G. Toxicity of Lidocaine in Adult, Newborn, and Fetal Sheep. *Anesthesiology* **1981**, *55*, 57–61. [CrossRef]
23. Wilcke, J.R.; Davis, L.E.; Neff-Davis, C.A. Determination of Lidocaine Concentrations Producing Therapeutic and Toxic Effects in Dogs. *J. Vet. Pharmacol. Ther.* **1983**, *6*, 105–111. [CrossRef]
24. Meyer, G.A.; Lin, H.C.; Hanson, R.R.; Hayes, T.L. Effects of Intravenous Lidocaine Overdose on Cardiac Electrical Activity and Blood Pressure in the Horse. *Equine Vet. J.* **2001**, *33*, 434–437. [CrossRef] [PubMed]
25. Venkatachalam, D.; Chambers, J.P.; Kongara, K.; Ward, N.; Jacob, A.; Singh, P.M. Pharmacokinetics, Efficacy and Convulsive Dose of Articaine Hydrochloride in Goat Kids. *Vet. Anaesth. Analg.* **2021**, *48*, 264–271. [CrossRef] [PubMed]
26. Vnuk, D.; Lemo, N.; Radisic, B.; Nesek-Adam, V.; Musulin, A.; Kos, J. Serum Lidocaine Concentration after Epidural Administration in Dogs. *Vet. Med.* **2006**, *51*, 432–436. [CrossRef]
27. Tydén, E.; Tjälve, H.; Larsson, P. Metabolic Activation of 2,6-Xylidine in the Nasal Olfactory Mucosa and the Mucosa of the Upper Alimentary and Respiratory Tracts in Rats. *Toxicol. Sci.* **2004**, *81*, 263–272. [CrossRef] [PubMed]
28. Bill, T.J.; Clayman, M.A.; Morgan, R.F.; Gampper, T.J. Lidocaine Metabolism Pathophysiology, Drug Interactions, and Surgical Implications. *Aesthetic Surg. J.* **2004**, *24*, 307–311. [CrossRef] [PubMed]
29. Hoogenboom, R.L.A.P.; Zuidema, T.; Essers, M.; van Vuuren, A.M.; van Wikselaar, P.G.; van Eijkeren, J.C.H.; Mengelers, M.J.B.; Zeilmaker, M.J.; Bulder, A.S. Concentrations of Dimethylaniline and Other Metabolites in Milk and Tissues of Dairy Cows Treated with Lidocaine. *Food Addit. Contam. Part A* **2015**, *32*, 1256–1264. [CrossRef] [PubMed]
30. Gan, J.; Skipper, P.L.; Tannenbaum, S.R. Oxidation of 2,6-Dimethylaniline by Recombinant Human Cytochrome P450s and Human Liver Microsomes. *Chem. Res. Toxicol.* **2001**, *14*, 672–677. [CrossRef]
31. IARC Monographs on the Evaluation of Carcinogenic Risks to Humans, Vol 57, Occupational Exposures of Hairdressers and Barbers and Personal Use of Hair Colourants; Some Hair Dyes, Cosmetic Colourants, Industrial Dyestuffs and Aromatic Amines. *Anal. Chim. Acta* **1995**, *300*, 340. [CrossRef]
32. Program, N.T. NTP Toxicology and Carcinogenesis Studies of 2,6-Xylidine (2,6-Dimethylaniline) (CAS No. 87-62-7) in Charles River CD Rats (Feed Studies). *Natl. Toxicol. Program Tech. Rep. Ser.* **1990**, *278*, 1–138.
33. Tao, L.; Day, B.W.; Hu, B.; Xiang, Y.-B.; Wang, R.; Stern, M.C.; Gago-Dominguez, M.; Cortessis, V.K.; Conti, D.V.; Berg, D.V.D.; et al. Elevated 4-Aminobiphenyl and 2,6-Dimethylaniline Hemoglobin Adducts and Increased Risk of Bladder Cancer among Lifelong Nonsmokers—The Shanghai Bladder Cancer Study. *Cancer Epidemiol. Biomark. Prev.* **2013**, *22*, 937–945. [CrossRef]
34. Duan, J.-D.; Jeffrey, A.M.; Williams, G.M. Assessment of the Medicines Lidocaine, Prilocaine, and Their Metabolites, 2,6-Dimethylaniline and 2-Methylaniline, for DNA Adduct Formation in Rat Tissues. *Drug Metab. Dispos.* **2008**, *36*, 1470–1475. [CrossRef]
35. Yapp, K.E.; Hopcraft, M.S.; Parashos, P. Articaine: A Review of the Literature. *Br. Dent. J.* **2011**, *210*, 323–329. [CrossRef] [PubMed]
36. Skjevik, Å.A.; Haug, B.E.; Lygre, H.; Teigen, K. Intramolecular Hydrogen Bonding in Articaine Can Be Related to Superior Bone Tissue Penetration: A Molecular Dynamics Study. *Biophys. Chem.* **2011**, *154*, 18–25. [CrossRef] [PubMed]
37. Oertel, R.; Rahn, R.; Kirch, W. Clinical Pharmacokinetics of Articaine. *Clin. Pharmacokinet.* **1997**, *33*, 417–425. [CrossRef] [PubMed]

38. Venkatachalam, D.; Chambers, J.P.; Kongara, K.; Singh, P. Pharmacokinetics of Articaine Hydrochloride and Its Metabolite Articainic Acid after Subcutaneous Administration in Red Deer (Cervus Elaphus). *N. Z. Vet. J.* **2017**, *66*, 16–20. [CrossRef] [PubMed]
39. Snoeck, M. Articaine: A Review of Its Use for Local and Regional Anesthesia. *Local Reg. Anesth.* **2012**, *5*, 23–33. [CrossRef] [PubMed]
40. Becker, D.E.; Reed, K.L. Local Anesthetics: Review of Pharmacological Considerations. *Anesth. Prog.* **2012**, *59*, 90–101; quiz 102–103. [CrossRef]
41. Becker, D.E.; Reed, K.L. Essentials of Local Anesthetic Pharmacology. *Anesth. Prog.* **2006**, *53*, 98–108; quiz 109–110. [CrossRef]
42. Strong, J.M.; Mayfield, D.E.; Atkinson, A.J.; Burris, B.C.; Raymon, F.; Webster, L.T. Pharmacological Activity, Metabolism, and Pharmacokinetics of Glycinexylidide. *Clin. Pharmacol. Ther.* **1975**, *17*, 184–194. [CrossRef]
43. Piccinni, C.; Gissi, D.B.; Gabusi, A.; Montebugnoli, L.; Poluzzi, E. Paraesthesia after Local Anaesthetics: An Analysis of Reports to the FDA Adverse Event Reporting System. *Basic Clin. Pharmacol. Toxicol.* **2015**, *117*, 52–56. [CrossRef]
44. Simon, M.A.M.; Vree, T.B.; Gielen, M.J.M.; Booij, L.H.D.J. Comparison of the Effects and Disposition Kinetics of Articaine and Lidocaine in 20 Patients Undergoing Intravenous Regional Anaesthesia during Day Case Surgery. *Pharm. World Sci.* **1998**, *20*, 88–92. [CrossRef] [PubMed]
45. Adami, C.; Bergadano, A.; Bruckmaier, R.M.; Stoffel, M.H.; Doherr, M.G.; Spadavecchia, C. Sciatic-Femoral Nerve Block with Bupivacaine in Goats Undergoing Elective Stifle Arthrotomy. *Vet. J.* **2011**, *188*, 53–57. [CrossRef] [PubMed]
46. Venkatachalam, D.; Kells, N.; Chambers, P.; Jacob, A.; Ward, N.; Singh, P. Pharmacokinetics and Efficacy of a Novel Long-Acting Bupivacaine Formulation for Cornual Nerve Block in Calves. *Front. Vet. Sci.* **2022**, *9*, 1060951. [CrossRef] [PubMed]
47. Kongara, K.; Dukkipati, V.S.R.; Tai, H.M.; Heiser, A.; Murray, A.; Webster, J.; Johnson, C.B. Differential Transcription of Selected Cytokine and Neuroactive Ligand-Receptor Genes in Peripheral Leukocytes from Calves in Response to Cautery Disbudding. *Animals* **2020**, *10*, 1187. [CrossRef] [PubMed]
48. Knauer, W.A.; Barrell, E.A.; Guedes, A.G.P.; Ventura, B.A. Effects of Multimodal Pain Management Strategies on Acute Physiological and Behavioral Response to Cautery Disbudding in Neonatal Goat Kids. *J. Dairy Sci.* **2023**, *106*, 2830–2845. [CrossRef] [PubMed]
49. Hempstead, M.N.; Waas, J.R.; Stewart, M.; Cave, V.M.; Sutherland, M.A. Evaluation of Alternatives to Cautery Disbudding of Dairy Goat Kids Using Physiological Measures of Immediate and Longer-Term Pain. *J. Dairy Sci.* **2018**, *101*, 5374–5387. [CrossRef] [PubMed]
50. Molaei, M.M.; Mostafavi, A.; Kheirandish, R.; Azari, O.; Shaddel, M. Study of Disbudding Goat Kids Following Injection of Clove Oil Essence in Horn Bud Region. *Vet. Res. Forum Int. Q. J.* **2015**, *6*, 17–22.
51. Molaei, M.M.; Azari, O.; Esmaeilzadeh, S. Study of Calves Disbudding Following Injection of Clove Oil under Horn Bud. *J. Vet. Res.* **2014**, *4*, 363–369.
52. Schoiswohl, J.; Stanitznig, A.; Waiblinger, S.; Frahm, S.; Krametter-Froetscher, R.; Wittek, T. Suppression of Horn Growth in Cattle by Clove Oil and Isoeugenol. *J. Vet. Behav.* **2020**, *36*, 1–3. [CrossRef]
53. Schoiswohl, J.; Stanitznig, A.; Sigmund, M.; Kneissl, S.; Thaller, D.; Frahm, S.; Waiblinger, S.; Palme, R.; Tichy, A.; Wittek, T.; et al. Comparison of Alternative Disbudding Methods with Hot-Iron Dehorning of Goat Kids. *J. Vet. Behav.* **2021**, *46*, 31–39. [CrossRef]
54. Hempstead, M.N.; Waas, J.R.; Stewart, M.; Cave, V.M.; Turner, A.R.; Sutherland, M.A. The Effectiveness of Clove Oil and Two Different Cautery Disbudding Methods on Preventing Horn Growth in Dairy Goat Kids. *PLoS ONE* **2018**, *13*, e0198229. [CrossRef]
55. Sladky, K.K.; Swanson, C.R.; Stoskopf, M.K.; Loomis, M.R.; Lewbart, G.A. Comparative Efficacy of Tricaine Methanesulfonate and Clove Oil for Use as Anesthetics in Red Pacu (Piaractus Brachypomus). *Am. J. Vet. Res.* **2001**, *62*, 337–342. [CrossRef]
56. Markowitz, K.; Moynihan, M.; Liu, M.; Kim, S. Biologic Properties of Eugenol and Zinc Oxide-Eugenol A Clinically Oriented Review. *Oral Surg. Oral Med. Oral Pathol.* **1992**, *73*, 729–737. [CrossRef]
57. Sutherland, M.A.; Larive, J.; Cave, V.; Zobel, G. Behavioural and Physiological Responses to Clove Oil Injected under the Horn Bud of Calves. *Appl. Anim. Behav. Sci.* **2018**, *204*, 29–36. [CrossRef]
58. Sutherland, M.; Julian, A.; Huddart, F. Clove Oil Delays Rather Than Prevents Scur/Horn Growth in Dairy Cattle. *Vet. Sci.* **2019**, *6*, 102. [CrossRef]

animals

MDPI

Review

Pig Sedation and Anesthesia for Medical Research

Ruxandra Costea *, Ioana Ene and Ruxandra Pavel

Faculty of Veterinary Medicine, University of Agronomic Sciences and Veterinary Medicine, 011464 Bucharest, Romania

* Correspondence: ruxandra.costea@fmvb.usamv.ro

Simple Summary: Anesthesia plays a crucial role in ensuring the ethical treatment of research animals and obtaining reliable and accurate data. Pig anesthesia is a significant aspect of clinical veterinary practice, especially when performing surgical procedures, diagnostic imaging, various medical interventions, and scientific research procedures. Proper anesthesia protocols ensure that the animals are kept unconscious and do not experience pain or distress, which is not only ethically responsible but also needed by regulatory bodies and animal welfare standards. This article is a narrative review that presents considerations for sedation and anesthesia of pigs, highlighting species particularities and reviewing the agents and protocols commonly used for medical and scientific research.

Abstract: In clinical veterinary practice, proper training and expertise in anesthesia administration and monitoring are essential. Pigs are suitable experimental animals for many surgical techniques because they are similar in size to humans and have a short reproductive cycle. This makes them ideal for research concerning organ transplantation, cardiovascular surgery, and other procedures that require a large animal model. Sedation and premedication should be administered at the lowest dose to be effective with predictable results and reduced adverse effects, to ensure the safety of both the animal and the team involved in the procedure, with a fast onset and optimizing the induction and maintenance of anesthesia. The goal of induction is to achieve a safe and effective level of anesthesia that ensures patient safety and facilitates research. Most of the time, inhalation anesthesia with endotracheal intubation is the ideal choice for maintenance of anesthesia. The difficulties related to endotracheal intubation of pigs can be overcome by knowing the anatomical peculiarities. Effective analgesia tailored to the specific procedure, the pig's condition, and individual responses to medications should complete the maintenance and recovery protocols, reducing perioperative complications.

Keywords: sedation; anesthesia; pig; research models; protocols

1. Physical Examination

Anesthesia ensures the welfare of the animal, enables safe and effective procedures, and allows accurate data collection [1]. Pigs are commonly used in medical and scientific research as models for studying various aspects of human health, physiology, and disease due to their physiological and anatomical similarities to humans [2–5]. Pigs are known to be highly sensitive to stress; consequently, they should be conditioned at the research facility for approximately 7–14 days before anesthesia, in order to have time to adapt to the experimental environment, to avoid stress-induced respiratory disease or diarrhea [6,7].

Physical preanesthetic examination must be performed in a low-stress environment with a focus on evaluation of respiratory and cardiovascular system function. Age and maturity criteria should be considered when choosing a model. The majority of pigs utilized in research projects weigh 15 to 30 kg and are 8 to 12 weeks old [6]. The decision to withhold food and water preoperatively in pigs should involve consideration of the animals' age, growth rate, breed, pregnancy status, clinical status, and the procedure to be performed. Food and water withdrawal regimens have a wide variation of 2–12 h, with particularly aggressive fasting regimes for gastrointestinal or abdominal surgery [8]. Although fasting may reduce the risk of regurgitation, fasting is recommended, as aspiration of regurgitated material can occur and may cause airway obstruction, irritation, and ultimately aspiration pneumonia. Aspiration of acidic stomach fluid may cause immediate reflexive airway closure and destruction of type II alveolar cells and pulmonary capillary lining cells. Consequently, pulmonary edema and hemorrhage may develop along with bronchospasm, dyspnea, hypoxemia, and cyanosis. Recovery from aspiration pneumonia, which may take a few days to develop, depends on the pH of the material aspirated. Swine tend to have very acidic stomach fluid with a pH as low as 1.5–2.5 [9,10]. Alfalfa and other types of hay can delay gastric emptying time, which means that vomiting and aspiration may still occur even after a 12-h fasting period. To avoid this, alfalfa or other forms of hay should be eliminated from the regular diet 2–3 days before general anesthesia [11]. Piglets, who are prone to hypoglycemia, should be denied suckling for only 1–2 h before anesthetic induction [9].

Following the preanesthetic physical examination, pigs can be included in a corresponding anesthetic risk classification system according to the American Society of Anesthesiologists (ASA) physical status classification system modified for veterinary medicine, which is a valuable prognostic tool, recommended to identify an increased risk of anesthetic complications and mortality [12,13].

2. Recommendations for Injectable Administration

Injections should be performed slowly, if possible, to minimize pain associated with injection and tissue damage [13]. The dimensions of the needle must be selected with consideration of the size of the animal and the liquid consistency of the injectate (aqueous or oily). For subcutaneous (SC) or intramuscular injections (IM), an extension line can be used to connect the syringe and cannula to reduce the risk associated with any evasive movements of the pig [14]. As the skin of swine can only be tented to a minor degree, only small-volume SC injections can be delivered [15]. Two locations are suitable for SC injections and are recommended: the knee fold (body weight under 20 kg) or caudal to the ear base for larger pigs [14,16]. The muscles of the caudal thigh region, semimembranosus and semitendinosus, and the gluteal muscles of the cranial thigh are generally selected as suitable sites for large-volume intramuscular injections (IM), while for small volumes to be injected, it is preferred to access the dorsolateral neck region. The injection can be performed in a less stressful way for the pig if it is possible to feed it simultaneously [16,17]. Intravenous access (IV) can be challenging because pigs resist restraint and they have very few superficial veins accessible for IV injection or catheterization [9,17]. The auricular veins, jugular vein, and femoral vein are all commonly used for drawing blood or administering fluids in pigs. The auricular veins located on the lateral and medial dorsal ear margins offer the easiest access for intravenous injection [18,19]. Topical application of a eutectic mixture of lidocaine 2.5% and prilocaine 2.5% for anesthesia has been used for various procedures in human medicine and although studies in animals are limited, it appears to facilitate various procedures in veterinary medicine, including venipuncture [20]. Puncture of the ear vein requires physical restraint of the swine or heavy sedation. After occluding blood flow at the base of the ear, the vessels are easy to identify (Figure 1). Catheterization of the jugular or femoral veins can be challenging and should only be performed by experienced personnel [14,15].

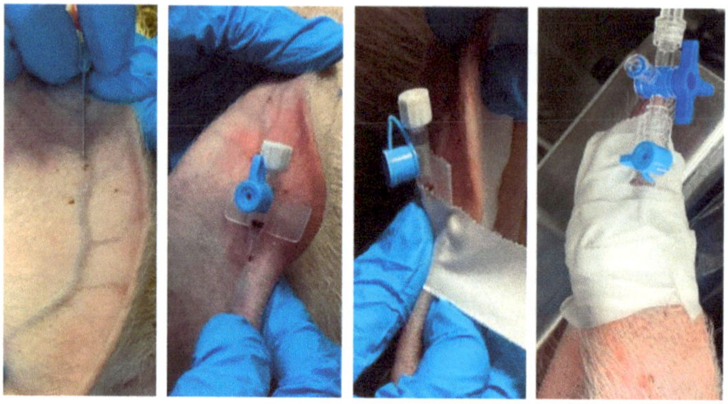

Figure 1. Mounting and fixing a peripheral catheter in the auricular vein.

3. Sedation and Premedication

Sedation is often suitable for minor procedures, such as physical examination and diagnostic imaging. or it represents premedication for anesthesia. The choice of an appropriate sedative protocol should be based on the procedure's type, animal health status, age, and size. Other factors, such as the desired level of sedation and the duration of the procedure, influence the selection of medication. Sedatives should be administered at the lowest effective dose to minimize the risk of adverse effects and calculated based on the pig's weight. After sedation, pigs may still need some level of physical restraint to ensure the safety of both the animal and the people involved in the procedure [12,15].

Stress during handling and restraint can lead to increased vocalizations, making the process of injection of sedative drugs challenging. Small pigs (<10 kg) may be more easily restrained compared to larger ones and less prone to stress-related vocalizations [1].

Multiple classes of agents may be considered for sedation in pigs. A detailed chart of dosage, route, and other considerations is listed in Table 1. Short, minimally invasive procedures may require lighter sedation with a focus on anxiolysis, achieved through benzodiazepines and alpha-2 adrenoreceptor agonists. Some examples of sedation protocols include azaperone, acepromazine, diazepam, midazolam, xylazine, and medetomidine, used alone or in combination. For more invasive surgeries, a combination of sedatives, analgesics, and anesthetics may be employed to ensure deep sedation, pain control, and a stable anesthetic plane. When deeper sedation is necessary, ketamine can be added to the combinations. The combination of tiletamine and zolazepam produces heavy sedation and immobilization with a relatively small volume of injection, making it particularly suitable for larger animals [1]. If pain is present or anticipated for the procedure, the protocols may also include opioids such as buprenorphine, morphine, or methadone.

Premedication refers to the administration of medications prior to the induction of anesthesia, minimizing stress and anxiety, providing pre-emptive analgesia, and optimizing the induction and maintenance of anesthesia. The ideal premedication agent must be effective with predictable results and fast onset, easy to administer, reversible, and offer analgesia and muscle relaxation with minimum cardiovascular and respiratory depression. Medication and protocols will be decided based on the preanesthetic evaluation (ASA status, temperament, procedure, level of pain expected), anesthetist's level of experience, and equipment available [13].

Protocols for premedication usually include multiple agents, to achieve the maximum effect with minimum secondary effects. The use of the anticholinergics glycopyrrolate and atropine has the potential to reduce salivation and bronchial secretions, but should be performed with caution considering their cardiovascular effects [15].

Table 1. Premedication and sedative drugs used in pigs.

Agent	Dose	Route	Considerations, References
Azaperone	1–8 mg/kg (2–5 mg/kg mean)	IM	20 min to effect, sedative [21]
Acepromazine	0.03–1.1 mg/kg	IM, IV	tranquilizer [21,22]
Alfaxalone	5 mg/kg	IM	sedation [23,24]
Diazepam	0.2–1 mg/kg	IV	mild sedative [21,24]
Midazolam	0.1–0.5 mg/kg	IM, IV	sedation [21,24]
Xylazine	1–2 mg/kg	IM, IV	pigs are the least sensitive to xylazine [11]
Medetomidine	0.03–0.08 mg/kg	IM, IV	sedation and muscle relaxation [11,22]
Ketamine	2–30 mg/kg	IM, IV	poor muscle relaxation and analgesia [21,24,25]
Buprenorphine	0.01–0.05 mg/kg q 8–12 h.	IM, SC	significant respiratory depression [14,26]
Butorphanol	0.1–0.3 mg/kg q 4–6 h.	IM, IV	analgesia, short duration [21,25]
Tiletamine/Zolazepam Telazol®	2–8.8 mg/kg	IM, IV	sedation or anesthesia for minor surgery, 20–30 min, reversed with flumazenil 0.08 mg/kg [23]
Naloxone	0.5–2 mg/kg	IV	[21]
Glycopyrrolate	0.005–0.01 mg/kg	IM, IV	correct bradycardia, decrease salivation [9,15]
Atropine	0.02–0.04 mg/kg	IM, IV	correct bradycardia, decrease salivation [9,26]
Combinations			
Azaperone Midazolam	4 mg/kg azaperone 1 mg/kg midazolam	IM	[27,28]
Azaperone Xylazine	2 mg/kg azaperone 2 mg/kg xylazine	IM	[27,28]
Azaperone Butorphanol Ketamine	5 mg azaperone, 0.2 mg butorphanol 15 mg ketamine	IM	[28,29]
Azaperone Xylazine Ketamine	6 mg/kg azaperone 2 mg/kg xylazine 15 mg/kg ketamine	IM	[28,30]
Azaperone Midazolam Ketamine	2 mg/kg azaperone 0.3 mg/kg midazolam 15 mg/kg ketamine	IM	[21,28]
Acepromazine Ketamine	1.1 mg/kg acepromazine 33 mg/kg ketamine	IM	[21]
Alfaxalone Butorphanol Medetomidine	4 mg/kg alfaxalone 0.4 mg/kg butorphanol 40 µg/kg medetomidine	IM	[31]
Dexmedetomidine Ketamine Methadone	10 µg/kg dexmedetomidine 10 mg/kg ketamine 0.25–0.4 mg/kg methadone	IM	Premedication, facilitate intubation [32]
Xylazine Ketamine	1–2 mg/kg xylazine 10–20 mg/kg ketamine	IM	Premedication, short-term anesthesia [12,33]
Medetomidine Ketamine	0.04–08 mg/kg medetomidine 10 mg/kg ketamine 10 mg/kg ketamine	IV, IM	Premedication, short-term anesthesia [34]

In the authors' practice, the most common combination for sedative drugs used for pigs includes IM administration of ketamine (10–20 mg/kg), xylazine (1–2 mg/kg), and midazolam (0.1–0.2 mg/kg), with alternative combinations that include medetomidine or dexmedetomidine [12]. The lower doses are usually used for sedation and the higher are intended for anesthetic premedication.

Detailed considerations regarding the dosage, route of administration, and relevant data for sedation and premedication are shown in Table 1.

3.1. Butyrophenones

Azaperone is a neuroleptic sedative medication that belongs to the class of butyrophenone derivatives. It is widely used in pigs to provide sedation, reduce anxiety, calm animals, and combat aggression and stress in pigs [35,36]. Azaperone works at central adrenergic dopamine D2 receptors located in the reticular activating system, leading to its sedative and anti-anxiety effects [37]. Vasodilation, hypotension, and hypothermia may occur following the administration of azaperone so it should not be used in debilitated, hypovolemic, or hypotensive pigs. It can also be used for maiden sows after their first litter to reduce the rejection of piglets [22]. Azaperone given alone by the intramuscular route has a rapid onset of action (5–20 min) with a duration of action of 2–6 h (maximal effects within 30 min), while intravenous injection often results in excitation [9]. Oral or intranasal administration of azaperone at a dose of 4 mg/kg induces sedation in piglets that is clinically comparable to an intramuscular administration of 2 mg/kg [38,39]. Deeper sedation with fewer adverse effects can be achieved by combining azaperone with ketamine and butorphanol [22,29] or azaperone with ketamine and an alpha-2 adrenoreceptor agonist [30,40]. Susceptible Pietrain pigs were protected against halothane-induced malignant hyperthermia with azaperone at doses of 0.5–2 mg/kg IM [11,41].

3.2. Phenothiazines

Acepromazine (0.11–1.1 mg/kg IM, IV, SC) is commonly used alone for tranquilization [21]. This drug decreases spontaneous motor activity and may cause hypotension and hypothermia [9]. The recommended dose of 0.1–0.4 mg/kg IV or IM may be used in combination with other drugs to improve the quality of premedication [40]. The combination of acepromazine with ketamine or tiletamine/zolazepam produces reliable sedation and muscle relaxation [36]. Acepromazine 1.1–1.65 mg/kg IM has been reported to reduce the incidence of malignant hyperthermia related to anesthesia [41,42].

3.3. Benzodiazepines

Benzodiazepines are a class of sedative and anxiolytic drugs that are commonly used in both human and veterinary medicine. They work by enhancing the effects of a neurotransmitter called gamma-aminobutyric acid (GABA), which leads to sedative, anxiolytic (anti-anxiety), muscle relaxant, and anticonvulsant effects [43]. Midazolam, when compared with diazepam, is water-soluble, is absorbed rapidly, has a higher affinity for receptors, stronger potency, and quicker onset with a shorter duration of effect [9]. Diazepam and midazolam can be used in combination with ketamine, alpha-2 adrenoreceptor agonists, and opioids. When used in combination with ketamine, muscle relaxation will be improved during anesthesia [36], and when used in combination with alfaxalone (5 mg/kg IM), muscle relaxation and sedation levels increase [9]. Intranasal administration of midazolam (0.2 mg/kg) provides reliable sedation (effect in 3–4 min) [44]. Less commonly used benzodiazepines include flurazepam 2 mg/kg IV [43] and lorazepam 0.1 mg/kg [15]. Flumazenil 0.02–0.08 mg/kg is a selective benzodiazepine antagonist reversal agent that can be used to counteract the effects of benzodiazepines in cases of overdose or adverse reactions, or to facilitate recovery from sedation or anesthesia [45].

3.4. Alpha-2 Adrenoreceptor Agonists

Alpha-2 adrenoreceptor agonists are a class of medications that activate specific receptors in the body. These medications have various effects, including sedation, analgesia, muscle relaxation, and vasoconstriction. Alpha-2 adrenoreceptor agonists are often used for sedation, preanesthetic medication, and pain management in pigs, alone or as part of a balanced anesthesia protocol in combination with other medication, such as anesthetics and analgesics [46]. Pigs are more resistant to alpha-2 adrenoceptor agonists than ruminants and other domestic animals and require a higher dosage for mild to moderate sedation [46,47]. While alpha-2 agonists have beneficial effects, they can also cause side effects such as bradycardia, decreased respiratory rate, hypotension, decreased gastrointestinal motility, and hypothermia. Reversal agents (e.g., atipamezole, yohimbine, tolazoline, vatinoxan) are available to antagonize the effects of the alpha-2 adrenoceptor agonists [36].

Intramuscular administration of medetomidine at doses ranging from 0.04 to 0.08 mg/kg induced sedation and muscle relaxation, with an increasing effect observed at higher doses [34]. However, increasing the dose above 0.1 mg/kg did not further intensify sedation or muscle relaxation, but instead prolonged the duration of these effects. Medetomidine (0.04 mg/kg IV or 0.08 mg/kg IM) in combination with ketamine has been utilized in pigs for short-term anesthesia [34]. Medetomidine, when combined with butorphanol (0.2 mg/kg IM) and ketamine (10 mg/kg IM), produced prolonged anesthesia in pigs compared to a combination of xylazine (2 mg/kg IM), butorphanol (0.2 mg/kg IM), and ketamine (10 mg/kg IM). The achieved muscle relaxation was adequate for tracheal intubation, but moderate cardiovascular depression was observed after using the combination of medetomidine, butorphanol, and ketamine for anesthesia [48]. In a specific study involving young pigs, the administration of a combination of 0.08 mg/kg medetomidine and 0.2 mg/kg butorphanol did not provide adequate sedation to facilitate blood sampling in all animals [49].

3.5. Dissociative Anesthetics

Ketamine is an NMDA (N-methyl D aspartate) receptor antagonist drug that can be used for sedation in pigs. It works by antagonizing the effects of the neurotransmitter glutamate, resulting in sedation, analgesia, and dissociation from the environment. Ketamine is often used in combination with other medications to achieve the desired level of sedation or anesthesia. Ketamine can cause side effects such as increased muscle tone, muscle fasciculations, poor muscle relaxation, and analgesia when used alone [36]. Occasionally, pigs may experience a period of disorientation and ataxia during recovery from ketamine sedation and might need a comfortable environment to prevent injury during this phase. These effects can be managed and minimized through appropriate dosing and the use of ketamine combined with other medications [50]. In healthy animals, ketamine has a good analgesic effect and only slightly modifies heart rate. When ketamine is administered alone, the ability of the swallowing reflex is unaffected, but excitation and excessive salivation can develop during anesthesia and recovery [22]. Tiletamine is a dissociative anesthetic used in veterinary medicine in combination with zolazepam (Telazol® tiletamine/zolazepam) to induce sedation or anesthesia in pigs. Tiletamine is approximately twice as potent as ketamine and has a longer duration of action [51]. Telazol® (tiletamine/zolazepam, 4.4 mg/kg) and xylazine (2.2 mg/kg) IM provide rapid sedation and can be used for sedation and induction [47]. Pigs often experience prolonged and rough recovery characterized by swimming motions, with repeated attempts to right themselves when recovering from Telazol anesthesia, similar to that observed when ketamine is used alone [41,52]. Studies have shown that tiletamine and zolazepam are both eliminated more slowly in pigs than in other species and that tiletamine has a longer effect than zolazepam in pigs [52]. Flumazenil can be used to antagonize zolazepam, but care should be granted to avoid residual effects of tiletamine leading to excitation, muscular tone, and fasciculations [23,45].

3.6. Opioids

Opioids are a class of medication commonly used for pain management and sedation in pigs, acting by binding to specific receptors in the nervous system (opioid receptors), which results in pain relief, sedation, and other effects [12,37]. Opioids can be used for sedation in pigs, particularly for pain management and calming effects. Opioids can be used in combination with other sedatives, anesthetics, or analgesics to achieve the desired level of sedation and pain control; pure μ agonists result in a strong analgesic effect, and partial μ agonists can be used in protocols for moderate pain along with μ- antagonists/K-agonists. Opioids can cause side effects such as vocalization, excitations, respiratory depression, decreased heart rate, and constipation. Butorphanol, administered at 0.2 mg/kg intramuscularly, resulted in important behavioral changes in piglets, resembling panic attacks, which have not been described in this species before [53]. The administration of buprenorphine did not decrease piglet vocalizations during the castration procedure but proved to be highly successful in mitigating pain behaviors [54]. In the post-surgery recovery, buprenorphine alleviated pain related to different surgical procedures, but had reduced effectiveness in addressing pain symptoms associated with inflammation, organ failure, or systemic disease when compared to pain associated with surgical incisions, orthopedic, dental, or ophthalmic procedures [55]. Buprenorphine has a relatively long duration of effect and low rate of side effects, but doses higher than 0.01 mg/kg must be used bearing in mind a possible respiratory depression [14]. Fentanyl, a short-acting opioid, can be used in pigs as a constant intravenous infusion at rates varying from 10 to 100 μg/kg/h without major side effects [46]. Boluses of morphine and fentanyl infusions will decrease the minimum alveolar concentration (MAC) levels of isoflurane [41]. Fentanyl and buprenorphine can also be used as transdermal patches, providing long-term analgesia, with a reduced incidence of side effects [56]. An example of ensuring preoperative and postoperative analgesia is represented by the protocol consisting of epidural morphine (0.1 mg/kg) prior to abdominal surgery, and a transdermal fentanyl patch (50 mg/h) postoperatively, which contributes to almost immediate restoration of normal activity levels and weight gain after recovery from general anesthesia [57]. Reversal agents available (e.g., antagonist naloxone 0.5–2 mg/kg IV [21]) can counteract negative side effects of opioids and can be used in unexpected reactions or overdose. In these cases, analgesic effects will also be reversed.

3.7. Alfaxalone

Alfaxalone is a neurosteroid anesthetic agent used for sedation, induction, and maintenance of anesthesia, with a rapid onset and relatively short duration of action. Alfaxalone can be administered both IV and IM in pigs [58,59]. Alfaxalone can cause side effects such as respiratory depression, decreased heart rate, and a decrease in blood pressure. Alfaxalone has been used in pigs to induce and maintain anesthesia with minimal cardiovascular effects [31,53]. A combination of alfaxalone and dexmedetomidine can be used to maintain long-duration total intravenous anesthesia in pigs [32,60].

3.8. Local Anesthetics

Lidocaine and bupivacaine are local anesthetic medications commonly used for various purposes in pigs, including local anesthesia for surgical procedures, postoperative pain management, and nerve blocks [24]. While local anesthetics are generally well-tolerated, some pigs may experience hypersensitivity or allergic reactions to the medications [37]. Careful observation of adverse reactions is important [61]. Lidocaine is widely used intravenously in different species to provide analgesia and as an adjunct to general anesthesia. In one experimental model of lung transplantation, intravenous lidocaine was associated with an attenuation of the histological markers of lung damage in the early stages of reperfusion [62]. Administration of lidocaine may help to prevent lung injury during surgery with one lung ventilation, reducing the expression of proinflammatory cytokines and lung apoptosis [63].

3.9. Neurokinin-1 (NK-1) Receptor Antagonists—Maropitant

Maropitant is a potent, selective neurokinin (NK-1) receptor antagonist primarily administered before anesthetic premedication (1 mg/kg q 24 h, IM) as an antiemetic medication [64]. The MAC of sevoflurane is decreased by maropitant, indicating a potential role as an adjunct visceral analgesic, as demonstrated in other animals [65]. Thus, there is a potential for future applications for swine.

3.10. Non-Depolarizing Neuromuscular Blocking Agents (NMBs)

In biomedical research, the use of non-depolarizing neuromuscular blocking agents (NMBs) involves profound muscle relaxation and prevents accidental awareness in conditions of inadequate anesthesia or analgesia. NMBs are widely recommended for tracheal intubation, which is relatively difficult in swine. Studies are quite controversial regarding the achievement of these objectives [66]. When using NMBs, pigs must be unconscious and controlled ventilation must be used. NMBs are not recommended for routine use or without advanced monitoring, which includes measuring arterial blood pressure and neuromuscular blockade assessment with a peripheral nerve stimulator. The NMBs can be administrated as boluses or continuous-rate infusions. Reversal of the neuromuscular blockade involves administration of an acetylcholinesterase inhibitor (neostigmine, edrophonium), which can also generate side effects such as bradycardia and gastrointestinal stimulation. To reduce parasympathetic stimulation, it is recommended to administer an anticholinergic (atropine, glycopyrrolate) before the antagonization of the NMBs. The most common NMBs used are pancuronium, vecuronium, atracurium, and rocuronium. Although monitoring of neuromuscular blockade is possible in pigs, neuromuscular blockade is rarely objectively monitored and is often administered based on clinical signs such as the return of spontaneous ventilation [67].

4. Induction of Anesthesia

The induction of anesthesia is the process of administering medication to initiate general anesthesia. Preoxygenation with supplemental oxygen via a mask or a flow-by technique can increase the oxygen concentration in the lungs and bloodstream, reducing the risks for hypoxia during induction. The goal of induction is to achieve an adequate depth of anesthesia to prevent any perception or response to the procedures being performed. Induction agents are administered by the inhaled or intravenous route, or a combination (Table 2), depending on the patient and surgical setting. Inhalational induction is not preferred as a method for the induction of anesthesia in pigs, due to the lack of predictable effects, the high volume of volatile agents necessary, and increased risks for the personnel. Ketamine, thiopental, propofol, and alfaxalone are the drugs most commonly used for inducing anesthesia in pigs, due to their fast-acting effects and short recovery time. Thiopental is a thiobarbiturate used for maintenance of anesthesia with tracheal intubation and positive pressure ventilation, as apnea may occur. Ketamine administration alone is not recommended but it can be combined with propofol for endotracheal intubation [68].

Table 2. Induction agents in pigs.

Agent	Dose	Route	Considerations, References
Propofol	2–5 mg/kg	IV	[37,68]
Propofol Fentanyl	2 mg/kg 5 μg/kg	IV	allows intubation [14,46]
Dexmedetomidine Propofol	20–40 μg/kg dexmedetomidine 2–4 mg/kg propofol		[46]
Propofol Ketamine	1–1.5 mg/kg propofol 0.5–1 mg/kg ketamine	IV	sedation, induction, no respiratory depression, good recovery [68,69]
Alfaxalone	0.6–1.1 mg/kg	IV, IM	[46]
Etomidate	2–4 mg/kg	IV	provides cardiovascular stability [46,69]
Thiopental	10–20 mg/kg	IV	apnea, prolonged recovery [9]

5. Endotracheal Intubation

Endotracheal intubation is necessary to protect the airway, preventing aspiration and maintaining positive pressure ventilation during anesthesia [70]. Swine intubation is challenging, technically difficult, and requires experience due to anatomical features: the shape of the head, thick, muscular, long tongue, long and narrow oropharyngeal space, small larynx, and an undersized trachea compared to many other animals. The elongated soft palate can hide the epiglottis and partially obstruct the airway, making breathing more difficult, especially in brachycephalic breeds of pigs [18]. The pharyngeal diverticulum is an anatomical structure found in pigs that protrudes from the wall of the pharynx, above the esophagus. The presence and length of the pharyngeal diverticulum (3–4 cm in adults, 1 cm in piglets), can vary among individuals and affect the ease of intubation [19]. The porcine larynx is tubular and lies caudal to the intermandibular space. The structural elements are divided into the thyroid cartilage, the cricoid cartilage, and some primitive arytenoid cartilage. This organ creates a characteristic obtuse angle with the trachea [19]. This anatomical characteristic, along with the existence of the lateral laryngeal ventricles, or ventricles of Morgagni, has been cited as the cause of the difficulty that may be encountered when intubation is performed [18,19]. The vocal cords are positioned caudoventrally [18] and can be easily traumatized if too much pressure is applied during tracheal intubation [19].

Both dorsal and ventral recumbency are described as positions for endotracheal intubation, but ventral recumbency is crucial in facilitating safe and fast intubation and reduces the risk of airway obstruction determined by overextension of the head [19,22,41]. Ventral recumbency can be advantageous if compared to dorsal, especially for operators lacking experience in anesthetizing animals [71]. To decrease the risk of laryngeal spasm, the arytenoids can be sprayed with 2–4% lidocaine a minute before intubation is attempted [46,72].

A laryngoscope with a long, straight blade and a plastic guide wire (bougie) can be used to facilitate introduction of the endotracheal tube (ETT) [73]. Some techniques are described using a urinary catheter, a rigid stylet through the tube [41], or a rigid semiflexible intubating stylet adapted manually [12]. The laryngoscope should be introduced until the base of the epiglottis, pressing the tongue followed by lifting the soft palate with the tip of the tube. The ETT is advanced under direct visualization into the trachea (Figure 2). If the ETT cannot be advanced, it should be gently rotated around its longitudinal axis. Straight tubes made of soft material may be advantageous in diminishing the risk of laryngeal trauma. To avoid any aspiration, it is recommended to use cuffed endotracheal tubes and to have available equipment for suction if regurgitation appears. Due to the anatomical particularities in many situations, a flexible connector can be added between the endotracheal tube and the circuit. Ideally, successful and smooth intubation should be performed on the first attempt. If resistance is encountered during intubation at the level of the arytenoid cartilages, a smaller ETT should be used. Repeated attempts during a standard intubation procedure can determine laryngospasm and laryngeal trauma [72,74]. Extubating is performed gently to avoid any traumatization of the tissues; their edema can cause obstructions of the airways during the awakening period. Each time the patient's position changes, the endotracheal tube must first be disconnected from the respiratory circuit. As an alternative to ETT, a laryngeal mask can be used. The mask is designed to be positioned over the larynx and enable positive pressure ventilation if required [1]. In neonatal piglets, ETT can be very difficult, so the use of a bougie to guide a laryngeal mask during placement can reduce the potential of airway obstruction [75].

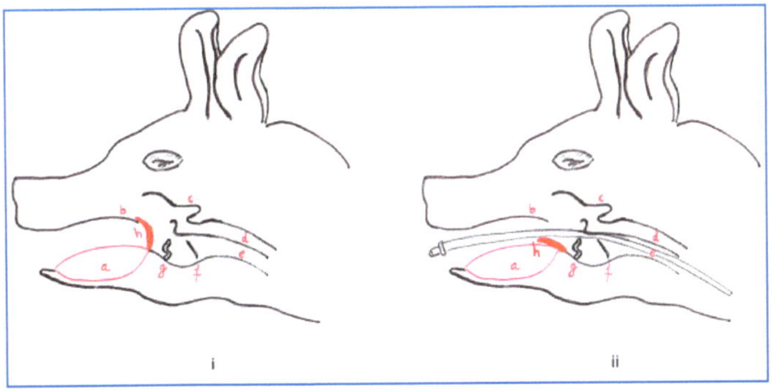

Figure 2. Anatomical features of the oropharyngeal region in pigs (**i**), advancement of the endotracheal tube (**ii**)—a, tongue; b, soft palate; c, pharyngeal diverticulum; d, esophagus; e, trachea; f, lateral ventricle; g, vocal cord; h, epiglottis.

6. Maintenance of Anesthesia

Maintenance (Table 3) of anesthesia can be performed by administering intravenous anesthetics (total intravenous anesthesia—TIVA), volatile/inhaled anesthetics, or mixed (partial intravenous anesthesia—PIVA) [76,77]. A hypermetabolic response to potent volatile anesthetic gases such as halothane, sevoflurane, desflurane, and isoflurane can trigger malignant hyperthermia, a pharmacogenetic disorder of skeletal muscle [78]. Maintenance of anesthesia can be complimented with a multimodal approach by the use of local anesthesia. Lumbosacral epidural anesthesia is the most commonly used form of regional analgesia in swine [9]. For maintenance of anesthesia, in the authors' practice [77], the most common PIVA protocol used isoflurane (1–1.5%) in combination with IV infusion of ketamine (1–3 mg/kg/h) and lidocaine (3–6 mg/kg/h).

Effective analgesia, in a pre-emptive approach tailored to the specific procedure, can prevent the onset of pain and minimize the sensitization of pain pathways, reducing the overall pain experience. Using a combination of different classes of analgesic drugs can provide more comprehensive pain relief [14,79]. Multimodal analgesia involves using opioids, nonsteroidal anti-inflammatory drugs (NSAIDs), local anesthetics, and other pain-relieving medications. NSAIDs are commonly used to reduce inflammation and inhibit pain signaling pathways. NSAIDs alone might not provide sufficient pain control for more invasive procedures, so they should be used in combination with other analgesic medications or techniques [80,81]. The specific choice of NSAIDs and its dosing regimen should be determined by the individual pig's health status, the procedure being performed, and other relevant factors, to ensure the safety and well-being of the animals. NSAIDs such as meloxicam or flunixin meglumine can help reduce inflammation and provide analgesia [80]. They are particularly useful for managing postoperative pain and are often used in combination with opioids. Local anesthetics such as lidocaine or bupivacaine can be administered via various nerve blocks or wound infiltration to provide targeted pain relief to specific areas and to reduce the need for systemic analgesics and in some cases, continuous infusion of analgesic medications can maintain a consistent level of pain relief throughout the procedure and into the recovery period [82]. Effective pain management should continue into the recovery period and protocols should be adjusted based on the pig's response and pain level. Crystalloid fluids during anesthesia are used to maintain homeostasis, to cover losses, to restore blood volume, and for stabilization, usually given at a rate of 5–10 mL/kg/h IV. For patients younger than 12 weeks, glucose 5% can be given to prevent hypoglycemia [83].

Table 3. Maintenance agents in pigs.

Agent	Dose	Route	Considerations, References
Isoflurane	1.6–1.9% MAC	ETT	[84]
Sevoflurane	2.4–2.66% MAC	ETT	[85]
Propofol	2–3 mg/kg, followed by 0.1–0.2 mg/kg/min	IV	[24]
Alfaxalone	4.8 mg/kg/h	IV	[31]
Fentanyl	50 μg/kg, followed by CRI 30–100 μg/kg/h.	IV	[23,46]
Alfaxalone Dexmedetomidine	5.3 mg/kg/h alfaxalone 3.0 μg/kg/h dexmedetomidine	IV	[32]
Alfaxalone Dexmedetomidine Ketamine	5 mg/kg/h alfaxalone 4 μg/kg/h dexmedetomidine 5 mg/kg/h ketamine	IV	[60]
Medetomidine Butorphanol Ketamine	0.03–0.08 mg/kg medetomidine 0.2 mg/kg butorphanol 10 mg/kg ketamine	IM	Longer sedation than Xylazine-Butorphanol-Ketamine [48]
Xylazine Ketamine Midazolam	2 mg/kg xylazine 0.25 mg/kg midazolam 10–20 mg/kg ketamine	IM	Immobilization in 2 min, effect for 50–90 min [8]
Tiletamine/Zolazepam Telazol® Xylazine	4.4–6 mg/kg tiletamine/zolazepam 2–2.2 mg/kg xylazine	IM	Provides rapid sedation and can be used for sedation and induction [45,47]
Tiletamine/Zolazepam Telazol® Medetomidine	5 mg/kg tiletamine/zolazepam 0.005 mg/kg medetomidine	IM	Provides rapid sedation and can be used for sedation and induction [45,47,56]
Guaifenesin Ketamine Xylazine "Triple drip"	50 mg Guaifenesin 2 mg Ketamine 1 mg Xylazine CRI 2.2 mL/kg/h	IV	Recovery in 30–45 min, Guaifenesin- centrally acting muscle relaxant [23,47]
Flunixin Meglumine	1–4 mg/kg q 24 h.	IV	managing postoperative pain [23]
Meloxicam	0.4 mg/kg	IM	managing postoperative pain [8,22]
Carprofen	1–4 mg/kg q 12 h. 2 mg/kg q 24 h.	IM, IV	managing postoperative pain [8]

7. Perianesthetic Monitoring and Complications

Safely managing anesthesia requires a thorough understanding of the indicators linked to the depth of anesthesia and the continuous surveillance of both the patient and the anesthetic apparatus. Monitoring during anesthesia enables evaluation of the depth of anesthesia, adjustment depending on patient particularities, and lastly, the monitoring of body functions during the procedure and in the recovery. Assessing anesthesia depth should be performed every 5–10 min, by evaluation of muscle relaxation, of the jaw tone, absence of movements, and absence of palpebral during anesthesia. If ketamine is included in the anesthetic protocol, ocular reflexes are not reliable [14].

For short surgeries, basic monitoring is recommended, while for surgeries that last more than 60 min or with patients who belong to the risk group ASA III-V, additional monitoring is recommended. Basic monitoring should include heart rate, pulse rate and quality, respiratory rate, mucous membrane color, capillary refill time, oxygen saturation, and temperature [40]. The pulse can be detected by feeling the auricular artery, the brachial artery, the saphenous artery, or the sublingual artery on the ventral surface of the tongue. Pulse oximetry measures both pulse rate and the percentage of oxygenated hemoglobin.

The probe can be best placed on the pig's tongue, lip, or ear, but also on the eyelid [40,41], tip of its tail, or in the interdigital space for unpigmented animals. Direct auscultation of the heart should also be performed. In swine, the normal heart rate typically falls within the range of 60 to 90 beats per minute. During anesthesia, drugs such as ketamine and alpha-2 adrenoreceptor agonists can have a significant effect on the heart rate, causing tachycardia or bradycardia, respectively. Rate, rhythm, and pattern of respiration should be assessed during anesthesia. Temperature should be periodically assessed, and appropriate warming methods should be applied during anesthesia in order to prevent hypothermia.

Additional monitoring involves capnography, arterial blood pressure measurement, electrocardiography (ECG), assessment of urinary output, and blood glucose concentration [9]. Capnography analyzes the CO_2 concentration in the gases expired by the patient and evaluates the adequacy of ventilation, equipment integrity, and the cardiovascular system. ECG monitoring for detecting dysrhythmias can be easily performed in pigs, especially using patch electrodes. Pigs have a prolonged Q-T interval compared to other species [6]. Non-invasive blood pressure measurement is relatively easy in pigs, with either oscillometric or Doppler flow monitors, and a cuff that should be between 40% and 60% of the circumference of the limb [40].

If non-depolarizing neuromuscular blocking agents are used in the protocols, monitoring of the neuromuscular blockade is mandatory, including measuring arterial blood pressure and neuromuscular blockade assessment. Possible complications include incomplete recovery from non-depolarizing neuromuscular blocking agents (postoperative residual curarization) and upper airway obstruction. Mechanomyography and acceleromyography techniques are the most used methods for neuromuscular blockade monitoring. Acceleromyography, due to its ease of use for research purposes, was presented in several studies involving pigs [86,87].

During the recovery phase from inhalation anesthesia, diligent and frequent monitoring is imperative, as life-threatening complications can arise [41]. Hypotension with mean arterial pressures less than 65 mmHg or systolic arterial pressures less than or equal to 85 mmHg is common in miniature pigs and may need intervention with dopamine or dobutamine (1–10 mg/kg/min continuous rate IV infusion for either), colloids, or fluid support [88].

When sedating pigs, respiratory obstruction can be a major concern. Oxygen can be supplied via the anesthesia machine or an oxygen demand valve, ideally with the pig placed in a sternal position [22]. Dorsal soft palate displacement, leading to airway obstruction, can develop in nonintubated pigs during anesthesia or after extubation [88,89]. One study on the majority of anesthesia-related complications during experimental invasive surgical procedures on pigs showed that, within the group of individuals at high anesthetic risk for invasive surgical operation, complications occurred in 20.31% of cases [12]. The majority of anesthetic difficulties involved intubation (14.06%), which led to the adjustment of the anesthetic approach by performing an emergency tracheotomy (6.25%) and keeping the anesthesia through an endotracheal tube attached to this level [12]. These types of complications need immediate attention and medical stabilization, as they can become life-threatening. In a liver injury model in pigs, vasopressin, as opposed to fluid resuscitation or saline placebo, resulted in prolonged survival and complete recovery from uncontrolled and otherwise fatal hemorrhagic shock [90]. Some complications may appear in correlation with the conditions in which the pigs are housed. Consequently, care should be used for any possible material to be ingested that can determine gastrointestinal foreign body blockages [88]. Limiting the number of pigs in stalls is important because bite wounds are common complications and can be a source of infection for experiments that involve surgical management [91].

Malignant hyperthermia (MH) is a disorder of skeletal muscle that starts as a hypermetabolic response that can be triggered in susceptible pigs by stress, a warm environment, volatile anesthetic gases, and the muscle relaxant succinylcholine [92]. Porcine stress syndrome and malignant hyperthermia can develop in genetically susceptible pigs when they interact with stressors, such as exertion, heat, or social interaction, or when they are exposed to certain medications or anesthetics that stimulate skeletal muscle [93]. MH affects humans, horses, dogs, and certain pig breeds and can be clinically manifested by hyperthermia, tachycardia, tachypnea, increased carbon dioxide production, increased oxygen consumption, acidosis, hyperkalemia, muscle rigidity, and rhabdomyolysis [63]. Halothane is traditionally considered the most likely volatile inhalant to trigger MH, but delayed onset of MH can also occur with exposure to isoflurane and desflurane [41]. Rhabdomyolysis is not a classic symptom of MH, but it can occur as a late complication during MH when muscle tissue breaks down and releases potassium and myoglobin into the bloodstream [92]. The effectiveness of injecting azumolene into pigs susceptible to MH is not fully understood but, as an analog of dantrolene (which is currently the only drug used to treat MH), azumolene is effective in reversing MH crisis in pigs in some studies [94,95]. A nanocrystalline dantrolene sodium suspension is also described as effective in the treatment of malignant hyperthermia and comparable to that of standard dantrolene sodium in pigs [96], but more research is needed to confirm its efficacy and safety.

8. Recovery

Proper post-anesthesia care, in a calm environment with the pig positioned in a sternal recumbency as soon as possible, is essential during recovery to ensure that the pig wakes up safely and without complications. It is advisable to retain the endotracheal tube until the pig begins moving its head spontaneously or can no longer tolerate the tube. Ideally, the pig should be placed with the head elevated and the neck extended to help maintain a patent airway [41]. Continuous monitoring of vital signs, which include heart rate, respiratory rate, body temperature, and oxygen saturation, is crucial during the recovery period and should be assessed for all major procedures at least every 15 min during recovery as it regains consciousness [91]. It is advisable to be ready to take action in the event of complications or any adverse reactions to anesthesia. Maintaining a warm and controlled environment to prevent the pig from getting too cold is essential, as pigs are susceptible to hypothermia during anesthesia and recovery. Mild hypothermia improved survival in a clinically relevant pig model of hemorrhagic shock and trauma [97]. Pain should be assessed and managed appropriately during recovery. The recovery area should be kept quiet and free from unnecessary disturbances, allowing a gradual and safe recovery.

9. Conclusions

Pigs share many anatomical and physiological similarities with humans, allowing extensive surgical procedures and monitoring, making them suitable for complex experiments. Proper anesthesia management is essential when conducting experiments involving animals and researchers must acquire a thorough knowledge of the techniques and protocols to be conducted [98]. Anesthesia is essential to minimize pain and distress in research animals. Pig anesthesia safeguards animal welfare, enables accurate data collection, facilitates standardized experiments, ensures the safety of both animals and researchers, and supports the development and validation of medical interventions. Pigs offer a level of consistency and reproducibility in experiments that may be more challenging to achieve with smaller animals. Researchers must adhere to strict ethical guidelines and obtain appropriate approvals. Continuing education and research procedures in terms of the Three Rs (replacement; reduction; refinement) are needed to ensure minimal use of pigs in research, along with a maximized welfare [99].

Author Contributions: Conceptualization, R.C. and R.P.; methodology, R.C. and R.P.; writing—original draft preparation, R.C., I.E. and R.P. All authors have read and agreed to the published version of the manuscript.

Funding: This research received no external funding.

Institutional Review Board Statement: Not applicable.

Informed Consent Statement: Not applicable.

Data Availability Statement: Not applicable.

Conflicts of Interest: The authors declare no conflict of interest.

References

1. Flecknell, P. *Laboratory Animal Anaesthesia*; Academic Press: Cambridge, MA, USA, 2015; pp. 238–239.
2. Sullivan, T.P.; Eaglstein, W.H.; Davis, S.C.; Mertz, P. The Pig as a Model for Human Wound Healing. *Wound Repair Regen.* **2001**, *9*, 66–76. [CrossRef]
3. Kuzmuk, K.N.; Schook, L.B. Pigs as a Model for Biomedical Sciences. In *The Genetics of the Pig*; CABI: Wallingford, UK, 2011; pp. 426–444.
4. Lunney, J.K.; Van Goor, A.; Walker, K.E.; Hailstock, T.; Franklin, J.; Dai, C. Importance of the Pig as a Human Biomedical Model. *Sci. Transl. Med.* **2021**, *13*, eabd5758. [CrossRef] [PubMed]
5. Clark, S.C.; Sudarshan, C.D.; Khanna, R.; Roughan, J.V.; Flecknell, P.A.; Dark, J.H. A New Porcine Model of Reperfusion Injury after Lung Transplantation. *Lab. Anim.* **1999**, *33*, 135–142. [CrossRef] [PubMed]
6. Smith, A.C.; Swindle, M.M. Preparation of swine for the laboratory. *ILAR J.* **2006**, *47*, 358–363. [CrossRef]
7. Grandin, T. Minimizing Stress in Pig Handling in the Research Lab. *Lab Anim.* **1986**, *15*, 15–20.
8. Bradbury, A.G.; Clutton, R.E. Review of practices reported for preoperative food and water restriction of laboratory pigs (*Sus scrofa*). *J. Am. Assoc. Lab. Anim. Sci.* **2016**, *55*, 35–40. [PubMed]
9. Anderson, D.E.; Mulon, P.Y. Anesthesia and Surgical Procedures in Swine. In *Diseases of Swine*; John Wiley & Sons, Inc.: Hoboken, NJ, USA, 2019; pp. 171–196.
10. DeRouchey, J.; Goodband, B.; Tokach, M.; Dritz, S.; Nelssen, J. Digestive System of the Pig: Anatomy and Function. *N. Am. Vet. Commun. Conf.* **2009**, *23*, 375–376.
11. Lin, H. Perioperative Monitoring and Management of Complications. In *Farm Animal Anesthesia: Cattle, Small Ruminants, Camelids, and Pigs*; John Wiley & Sons, Inc.: Hoboken, NJ, USA, 2022; pp. 135–158.
12. Costea, R.; Tudor, R.; Degan, A.; Girdan, G. Anesthesia Complications Related to Swine Experimental Invasive Surgical Procedures. *Sci. Works. Ser. C Vet. Med.* **2019**, *65*, 2065-1295.
13. Portier, K.; Ida, K.K. The ASA Physical Status Classification: What Is the Evidence for Recommending Its Use in Veterinary Anesthesia?—A Systematic Review. *Front. Vet. Sci.* **2018**, *5*, 204. [CrossRef]
14. Kaiser, G.M.; Heuer, M.M.; Frühauf, N.R.; Kühne, C.A.; Broelsch, C.E. General Handling and Anesthesia for Experimental Surgery in Pigs. *J. Surg. Res.* **2006**, *130*, 73–79. [CrossRef]
15. Smith, A.C.; Ehler, W.J.; Swindle, M.M. Anesthesia and Analgesia in Swine. In *Anesthesia and Analgesia in Laboratory Animals*; Elsevier: Amsterdam, The Netherlands, 1997; pp. 313–336.
16. Hedenqvist, P. Laboratory animal analgesia, anesthesia, and euthanasia. In *Handbook of Laboratory Animal Science*; CRC Press: Boca Raton, FL, USA, 2021; pp. 343–378.
17. Xanthos, T.; Bassiakou, E.; Koudouna, E.; Tsirikos-Karapanos, N.; Lelovas, P.; Papadimitriou, D.; Dontas, I.; Papadimitriou, L. Baseline Hemodynamics in Anesthetized Landrace–Large White Swine: Reference Values for Research in Cardiac Arrest and Cardiopulmonary Resuscitation Models. *J. Am. Assoc. Lab. Anim. Sci.* **2007**, *46*, 21–25. [PubMed]
18. Dyce, K.M.; Sack, W.O.; Wensing, C.J.G. *Textbook of Veterinary Anatomy*; Saunders Company: Philadelphia, PA, USA, 2002; pp. 400–401.
19. Singh, B. *Dyce, Sack, and Wensing's Textbook of Veterinary Anatomy*; Saunders: St. Louis, MI, USA, 2018.
20. Erkert, R.S.; MacAllister, C.G. Use of a eutectic mixture of lidocaine 2.5% and prilocaine 2.5% as a local anesthetic in animals. *J. Am. Vet. Med. Assoc.* **2005**, *226*, 1990–1992. [CrossRef] [PubMed]
21. Swindle, M.M. *Swine in the Laboratory: Surgery, Anesthesia, Imaging, and Experimental Techniques*; CRC Press: Boca Raton, FL, USA, 2007.
22. Hodgkinson, O. Practical Sedation and Anaesthesia in Pigs. *Practice* **2007**, *29*, 34–39. [CrossRef]
23. Swindle, M.M.; Sistino, J.J.; Perioperative Care. *Swine in the Laboratory: Surgery, Anesthesia, Imaging, and Experimental Techniques*; CRC Press: Boca Raton, FL, USA, 2015; p. 39.
24. Clarke, K.W.; Trim, C.M. *Veterinary Anaesthesia E-Book*; Elsevier Health Sciences: Oxford, UK, 2013.
25. Costea, R. *Anestezologie*; Printech: Bucharest, Romania, 2017; pp. 127–131.
26. Sogebi, E.A.; Makinde, O.A.; Cliff, A.I.; Koleoso, S.; Mshelbwala, F.; Olukunle, J.O. Multimodal Approach to Surgical Pain Management in Weaner Pigs: A Clinical Trial. *Alex. J. Vet. Sci.* **2021**, *70*, 151–157. [CrossRef]
27. Bollen, P.J.; Hansen, A.K.; Alstrup, A.K.O. *The Laboratory Swine*; CRC Press: Boca Raton, FL, USA, 2010.
28. Flôres, F.N.; Tavares, S.G.; de Moraes, A.N.; Oleskovicz, N.; Santos, L.C.P.; Minsky, V.; Keshen, E. Azaperone and its association with xylazine or dexmedetomidine in pigs. *Ciência Rural* **2009**, *39*, 1101–1107. [CrossRef]
29. Nussbaumer, I.; Indermühle, N.; Zimmermann, W.; Leist, Y. Piglet Castration by Injection Anaesthesia: Experience with the Azaperone, Butorphanol and Ketamine Combination. *SAT Schweiz. Arch. Für Tierheilkd.* **2011**, *153*, 33–35. [CrossRef] [PubMed]
30. Short, C.E. Preanesthetic medications in ruminants and swine. *Vet. Clin. N. Am. Food Anim. Pract.* **1986**, *2*, 553–566. [CrossRef]
31. Bigby, S.E.; Carter, J.E.; Bauquier, S.; Beths, T. The Use of Alfaxalone for Premedication, Induction and Maintenance of Anaesthesia in Pigs: A Pilot Study. *Vet. Anaesth. Analg.* **2017**, *44*, 905–909. [CrossRef]

32. Kat, I.; Ahern, B.J.; Dhanani, J.; Whitten, G.; Cowling, N.; Goodwin, W. Long Duration Anaesthesia in Pigs with an Infusion of Alfaxalone and Dexmedetomidine. *Vet. Med. Sci.* **2022**, *8*, 2418–2421. [CrossRef]
33. Thurmon, J.C.; Benson, G.J. Anesthesia in Ruminants and Swine. *Curr. Vet. Ther.* **1993**, *3*, 58–76.
34. Nishimura, R.; Kim, H.; Matsunaga, S.; Sakaguchi, M.; Sasaki, N.; Tamura, H.; Takeuchi, A. Antagonism of Medetomidine Sedation by Atipamezole in Pigs. *J. Vet. Med. Sci.* **1992**, *54*, 1237–1240. [CrossRef]
35. Gruen, M.E.; Sherman, B.L.; Papich, M.G. *Drugs Affecting Animal Behavior*; John Willey & Sons: Hoboken, NJ, USA, 2018.
36. Lin, H. Injectable Anesthetics and Field Anesthesia. In *Farm Animal Anesthesia*; John Wiley & Sons: Hoboken, NJ, USA, 2022; pp. 60–100, ISBN 978-1-119-67266-1.
37. Golan, D.E.; Tashjian, A.H.; Armstrong, E.J. *Principles of Pharmacology: The Pathophysiologic Basis of Drug Therapy*; Lippincott Williams & Wilkins: Philadelphia, PA, USA, 2011.
38. Svoboda, M.; Fajt, Z.; Mruvčinská, M.; Vašek, J.; Blahová, J. The Effects of Buccal Administration of Azaperone on the Sedation Level and Biochemical Variables of Weaned Piglets. *Acta Vet. Brno* **2021**, *90*, 47–56. [CrossRef]
39. Svoboda, M.; Blahova, J.; Jarkovsky, J.; Zacharda, A.; Hajkova, S.; Vanhara, J.; Vasek, J. Efficacy of the Intranasal Application of Azaperone for Sedation in Weaned Piglets. *Vet. Med.* **2023**, *68*, 145–151. [CrossRef]
40. Swine, L.M. *Veterinary Anesthesia and Analgesia: The Fifth Edition of Lumb and Jones*; John Wiley & Sons, Inc.: Hoboken, NJ, USA, 2015; pp. 928–940.
41. Moon, P.F.; Smith, L.J. General Anesthetic Techniques in Swine. *Vet. Clin. N. Am. Food Anim. Pract.* **1996**, *12*, 663–691. [CrossRef] [PubMed]
42. McGrath, C.J.; Rempel, W.E.; Addis, P.B.; Crimi, A.J. Acepromazine and Droperidol Inhibition of Halothane-Induced Malignant Hyperthermia (Porcine Stress Syndrome) in Swine. *Am. J. Vet. Res.* **1981**, *42*, 195–198. [PubMed]
43. Lacoste, L.; Bouquet, S.; Ingrand, P.; Caritez, J.C.; Carretier, M.; Debaene, B. Intranasal Midazolam in Piglets: Pharmacodynamics (0.2 vs. 0.4 mg/kg) and Pharmacokinetics (0.4 mg/kg) with Bioavailability Determination. *Lab. Anim.* **2000**, *34*, 29–35. [CrossRef]
44. de Souza Dantas, L.M.; Crowell-Davis, S.L. Benzodiazepines. In *Veterinary Psychopharmacology*; John Wiley & Sons, Inc.: Hoboken, NJ, USA, 2019; pp. 67–102.
45. Lee, J.Y.; Kim, M.C. Anesthesia of Growing Pigs with Tiletamine-Zolazepam and Reversal with Flumazenil. *J. Vet. Med. Sci.* **2012**, *74*, 335–339. [CrossRef]
46. Thurmon, J.C.; Smith, G.W. Swine. In *Lumb and Jones' Veterinary Anesthesia and Analgesia*, 4th ed.; Tranquili, W.J., Thurmon, J.C., Grimm, K.A., Eds.; Blackwell Publishing: Ames, IA, USA, 2007; pp. 747–764.
47. Riebold, T.; Geiser, D.; Goble, D.O. Anesthetic agents and ancillary drugs. In *Large Animal Anesthesia*; Iowa State University Press: Ames, IA, USA, 1995; pp. 11–64.
48. Sakaguchi, M.; Nishimura, R.; Sasaki, N.; Ishiguro, T.; Tamura, H.; Takeuchi, A. Anesthesia Induced in Pigs by Use of a Combination of Medetomidine, Butorphanol, and Ketamine and Its Reversal by Administration of Atipamezole. *Am. J. Vet. Res.* **1996**, *57*, 529–534.
49. Ugarte, C.E.; O'Flaherty, K. The use of a medetomidine, butorphanol and atropine combination to enable blood sampling in young pigs. *N. Z. Vet. J.* **2005**, *53*, 249–252. [CrossRef]
50. Bettschart-Wolfensberger, R.; Stauffer, S.; Hässig, M.; Flaherty, D.; Ringer, S.K. Racemic Ketamine in Comparison to S-Ketamine in Combination with Azaperone and Butorphanol for Castration of Pigs. *Schweiz. Arch. Tierheilkd.* **2013**, *155*, 669–675. [CrossRef]
51. Lester, P.A.; Moore, R.M.; Shuster, K.A.; Myers, D.D. Anesthesia and Analgesia. In *The Laboratory Rabbit, Guinea Pig, Hamster, and Other Rodents*; Elsevier: Amsterdam, The Netherlands, 2012; pp. 33–56.
52. Kumar, A.; Mann, H.J.; Remmel, R.P. Pharmacokinetics of Tiletamine and Zolazepam (Telazol®) in Anesthetized Pigs. *J. Vet. Pharmacol. Ther.* **2006**, *29*, 587–589. [CrossRef]
53. Pavlovsky, V.H.; Corona, D.; Hug, P.J.; Kümmerlen, D.; Graage, R.; Bettschart-Wolfensberger, R. Butorphanol induces anxiety-like behaviour and distress in piglets. *Schweiz. Arch. Für Tierheilkd.* **2021**, *163*, 485–491. [CrossRef] [PubMed]
54. Viscardi, A.V.; Turner, P.V. Efficacy of buprenorphine for management of surgical castration pain in piglets. *BMC Vet. Res.* **2018**, *14*, 318. [CrossRef] [PubMed]
55. Rodriguez, N.A.; Cooper, D.M.; Risdahl, J.M. Antinociceptive activity of and clinical experience with buprenorphine in swine. *J. Am. Assoc. Lab. Anim. Sci.* **2001**, *40*, 17–20.
56. Lujan, S.O.; Habre, W.; Daali, Y.; Pan, Z.; Kronen, P.W. Plasma concentrations of transdermal fentanyl and buprenorphine in pigs (*Sus scrofa domesticus*). *Vet. Anaesth. Analg.* **2017**, *44*, 665–675. [CrossRef] [PubMed]
57. Malavasi, L.M.; Nyman, G.; Augustsson, H.; Jacobson, M.; Jensen-Waern, M. Effects of epidural morphine and transdermal fentanyl analgesia on physiology and behaviour after abdominal surgery in pigs. *Lab. Anim.* **2006**, *40*, 16–27. [CrossRef]
58. Keates, H. Induction of Anaesthesia in Pigs Using a New Alphaxalone Formulation. *Vet. Rec.* **2003**, *153*, 627–628. [CrossRef]
59. Santos, M.; de Lis, B.T.B.; Tendillo, F.J. Effects of Intramuscular Dexmedetomidine in Combination with Ketamine or Alfaxalone in Swine. *Vet. Anaesth. Analg.* **2016**, *43*, 81–85. [CrossRef]
60. Lervik, A.; Toverud, S.F.; Krontveit, R.; Haga, H.A. A Comparison of Respiratory Function in Pigs Anaesthetised by Propofol or Alfaxalone in Combination with Dexmedetomidine and Ketamine. *Acta Vet. Scand.* **2020**, *62*, 14. [CrossRef]
61. Satas, S.; Johannessen, S.I.; Hoem, N.-O.; Haaland, K.; Sorensen, D.R.; Thoresen, M. Lidocaine Pharmacokinetics and Toxicity in Newborn Pigs. *Anesth. Analg.* **1997**, *85*, 306.

62. Romera, A.; Cebollero, M.; Romero-Gómez, B.; Carricondo, F.; Zapatero, S.; García-Aldao, U.; Martín-Albo, L.; Ortega, J.; Vara, E.; Garutti, I.; et al. Effect of intravenous lidocaine on inflammatory and apoptotic response of ischemia-reperfusion injury in pigs undergoing lung resection surgery. *BioMed Res. Int.* **2021**, *2021*, 6630232. [CrossRef]
63. Garutti, I.; Rancan, L.; Simón, C.; Cusati, G.; Sanchez-Pedrosa, G.; Moraga, F.; Olmedilla, L.; Lopez-Gil, M.T.; Vara, E. Intravenous lidocaine decreases tumor necrosis factor alpha expression both locally and systemically in pigs undergoing lung resection surgery. *Anesth. Analg.* **2014**, *119*, 815–828. [CrossRef]
64. Smith, J.S.; Gebert, J.E.; Ebner, L.S.; Bennett, K.O.; Collins, R.J.; Hampton, C.E.; Kleine, S.A.; Mulon, P.-Y.; Smith, C.K.; Seddighi, R. Pharmacokinetics of Intramuscular Maropitant in Pigs (*Sus scrofa domesticus*). *J. Vet. Pharmacol. Ther.* **2023**, *46*, 158–164. [CrossRef]
65. Hay Kraus, B.L. Spotlight on the perioperative use of maropitant citrate. *Vet. Med. Res. Rep.* **2017**, *8*, 41–51. [CrossRef]
66. Bradbury, A.G.; Clutton, R.E. Are neuromuscular blocking agents being misused in laboratory pigs? *Br. J. Anaesth.* **2016**, *116*, 476–485. [CrossRef] [PubMed]
67. Pedersen, K.; Kruhøffer, L.L.; Lykkesfeldt, J.; Kousholt, B.S. Comparison of the neuromuscular effects of two infusion rates of rocuronium in anesthetized pigs. *Acta Vet. Scand.* **2022**, *64*, 38. [CrossRef] [PubMed]
68. Pehböck, D.; Dietrich, H.; Klima, G.; Paal, P.; Lindner, K.H.; Wenzel, V. Anesthesia in Swine. *Anaesthesist* **2015**, *64*, 65–70. [CrossRef] [PubMed]
69. Amornyotin, S. Ketofol: A Combination of Ketamine and Propofol. *J. Anesth. Crit. Care Open Access* **2014**, *1*, 00031.
70. Chum, H.; Pacharinsak, C. Endotracheal Intubation in Swine. *Lab. Anim.* **2012**, *41*, 309–311. [CrossRef] [PubMed]
71. Mirra, A.; Spadavecchia, C.; Micieli, F. Intubation in Swine: What Recumbency to Choose? *Animals* **2022**, *12*, 2430. [CrossRef]
72. Mirra, A.; Arnold, M.; Casoni, D.; Maidanskaia, E.G.; Casalta, L.G.G.; Levionnois, O. Fatal upper airway obstruction in a pig after general anaesthesia. *Vet. Anaesth. Analg.* **2022**, *49*, 145–146. [CrossRef] [PubMed]
73. Janiszewski, A.; Pasławski, R.; Skrzypczak, P.; Paslawska, U.; Szuba, A.; Nicpoń, J. The Use of a Plastic Guide Improves the Safety and Reduces the Duration of Endotracheal Intubation in the Pig. *J. Vet. Med. Sci.* **2014**, *76*, 1317–1320. [CrossRef] [PubMed]
74. Steinbacher, R.; Von Ritgen, S.; Moens, Y.P.S. Laryngeal Perforation during a Standard Intubation Procedure in a Pig. *Lab. Anim.* **2012**, *46*, 261–263. [CrossRef]
75. Morath, U.; Skogmo, H.K.; Ranheim, B.; Levionnois, O.L. The use of bougie-guided insertion of a laryngeal mask airway device in neonatal piglets after unexpected complications. *Vet. Rec. Case Rep.* **2014**, *2*, e000040. [CrossRef]
76. Beths, T. *TIVA/TCI in Veterinary Practice. Total Intravenous Anesthesia and Target Controlled Infusions: A Comprehensive Global Anthology*; Springer: Cham, Switzerland, 2017; pp. 589–618. [CrossRef]
77. Costea, R.; Tanase, A.; Ioniță, L.; Copaescu, C.; Girjoaba, I.; Mocanu, J.; Drugociu, D.S. Inhalatory anaesthesia in pigs for laparascopic surgery. *Lucr. Științifice Med. Vet. Univ. Științe Agric. Și Med. Vet. Ion Ionescu Brad Iași* **2009**, *52*, 503–505.
78. Rosenberg, H.; Pollock, N.; Schiemann, A.; Bulger, T.; Stowell, K. Malignant Hyperthermia: A Review. *Orphanet J. Rare Dis.* **2015**, *10*, 93. [CrossRef]
79. Suckow, M.A.; Stevens, K.A.; Wilson, R.P. *The Laboratory Rabbit, Guinea Pig, Hamster, and Other Rodents*; Academic Press: Cambridge, MA, USA, 2012.
80. Bindu, S.; Mazumder, S.; Bandyopadhyay, U. Non-Steroidal Anti-Inflammatory Drugs (NSAIDs) and Organ Damage: A Current Perspective. *Biochem. Pharmacol.* **2020**, *180*, 114147. [CrossRef]
81. Sutherland, M.A.; Davis, B.L.; Brooks, T.A.; Coetzee, J.F. The Physiological and Behavioral Response of Pigs Castrated with and without Anesthesia or Analgesia. *J. Anim. Sci.* **2012**, *90*, 2211–2221. [CrossRef]
82. Holman, S.D.; Gierbolini-Norat, E.M.; Lukasik, S.L.; Campbell-Malone, R.; Ding, P.; German, R.Z. Duration of Action of Bupivacaine Hydrochloride Used for Palatal Sensory Nerve Block in Infant Pigs. *J. Vet. Dent.* **2014**, *31*, 92–95. [CrossRef]
83. Costea, R.; Degan, A.; Tudor, R. Crystalloids/Colloids Ratio for Fluid Resuscitation during Anesthesia. *Sci. Works. Ser. C Vet. Med.* **2017**, *63*, 65–66.
84. Malavasi, L.M.; Jensen-Waern, M.; Augustsson, H.; Nyman, G. Changes in Minimal Alveolar Concentration of Isoflurane Following Treatment with Medetomidine and Tiletamine/Zolazepam, Epidural Morphine or Systemic Buprenorphine in Pigs. *Lab. Anim.* **2008**, *42*, 62–70. [CrossRef] [PubMed]
85. Allaouchiche, B.; Duflo, F.; Tournadre, J.-P.; Chassard, D. Influence of Sepsis on Sevoflurane (SEV) Minimum Alveolar Concentration (MAC) in a Swine Model. *Eur. J. Anaesthesiol. EJA* **2000**, *17*, 57. [CrossRef]
86. Mirra, A.; Gamez Maidanskaia, E.; Carmo, L.P.; Levionnois, O.; Spadavecchia, C. How is depth of anaesthesia assessed in experimental pigs? A scoping review. *PLoS ONE* **2023**, *18*, e0283511. [CrossRef] [PubMed]
87. Aguilar, A.; Moll, X.; García, F.; Andaluz, A. Neuromuscular block monitoring after the administration of 1 mg/kg intravenous cisatracurium in the anaesthetized pig. *J. Vet. Pharmacol. Ther.* **2019**, *42*, 67–73. [CrossRef] [PubMed]
88. Smith, J.S.; Seddighi, R. Miniature Companion Pig Sedation and Anesthesia. *Vet. Clin. Exot. Anim. Pract.* **2022**, *25*, 297–319. [CrossRef]
89. Lin, H. Comparative Anesthesia and Analgesia of Ruminants and Swine. In *Veterinary Anesthesia and Analgesia: The Fifth Edition of Lumb and Jones*; John Wiley & Sons, Inc.: Hoboken, NJ, USA, 2015; pp. 743–753.
90. Stadlbauer, K.H.; Wagner-Berger, H.G.; Raedler, C.; Voelckel, W.G.; Wenzel, V.; Krismer, A.C.; Klima, G.; Rheinberger, K.; Nussbaumer, W.; Pressmar, D. Vasopressin, but Not Fluid Resuscitation, Enhances Survival in a Liver Trauma Model with Uncontrolled and Otherwise Lethal Hemorrhagic Shock in Pigs. *J. Am. Soc. Anesthesiol.* **2003**, *98*, 699–704. [CrossRef] [PubMed]

91. Swindle, M.M.; Smith, A.C. Best Practices for Performing Experimental Surgery in Swine. *J. Investig. Surg.* **2013**, *26*, 63–71. [CrossRef]
92. Tanwar, P.; Naagar, M.; Malik, G.; Alam, M.S.; Singh, T.; Singh, O.; Maity, M.K. A Review on Malignant Hyperthermia: Epidemiology, Etiology, Risk Factors, Diagnosis, Clinical Management and Treatment Modalities. *World J. Biol. Pharm. Health Sci.* **2023**, *13*, 138–161. [CrossRef]
93. O'Brien, P.J.; Shen, H.; Cory, C.R.; Zhang, X. Use of a DNA-Based Test for the Mutation Associated with Porcine Stress Syndrome (Malignant Hyperthermia) in 10,000 Breeding Swine. *J. Am. Vet. Med. Assoc.* **1993**, *203*, 842–851.
94. El-Hayek, R.; Parness, J.; Valdivia, H.H.; Coronado, R.; Hogan, K. Dantrolene and Azumolene Inhibit [3H] PN200-110 Binding to Porcine Skeletal Muscle Dihydropyridine Receptors. *Biochem. Biophys. Res. Commun.* **1992**, *187*, 894–900. [CrossRef]
95. Do Carmo, P.L.; Zapata-Sudo, G.; Trachez, M.M.; Antunes, F.; Guimarães, S.E.F.; Debom, R.; Rizzi, M.D.R.; Sudo, R.T. Intravenous Administration of Azumolene to Reverse Malignant Hyperthermia in Swine. *J. Vet. Intern. Med.* **2010**, *24*, 1224–1228. [CrossRef] [PubMed]
96. Schütte, J.K.; Becker, S.; Burmester, S.; Starosse, A.; Lenz, D.; Kröner, L.; Wappler, F.; Gerbershagen, M.U. Comparison of the Therapeutic Effectiveness of a Dantrolene Sodium Solution and a Novel Nanocrystalline Suspension of Dantrolene Sodium in Malignant Hyperthermia Normal and Susceptible Pigs. *Eur. J. Anaesthesiol. EJA* **2011**, *28*, 256–264. [CrossRef]
97. Wu, X.; Kochanek, P.M.; Cochran, K.; Nozari, A.; Henchir, J.; Stezoski, S.W.; Wagner, R.; Wisniewski, S.; Tisherman, S.A. Mild Hypothermia Improves Survival after Prolonged, Traumatic Hemorrhagic Shock in Pigs. *J. Trauma Acute Care Surg.* **2005**, *59*, 291–301. [CrossRef] [PubMed]
98. Couto, M.; Cates, C. Laboratory Guidelines for Animal Care. In *Vertebrate Embryogenesis: Embryological, Cellular, and Genetic Methods*; Springer Science+Business Media, LLC: Berlin/Heidelberg, Germany, 2019; pp. 407–430. [CrossRef]
99. Hubrecht, R.C.; Carter, E. The 3Rs and Humane Experimental Technique: Implementing Change. *Animals* **2019**, *9*, 754. [CrossRef] [PubMed]

animals

MDPI

Systematic Review

Pain Assessment in Cattle by Use of Numerical Rating and Visual Analogue Scales—A Systematic Review and Meta-Analysis

Theresa Tschoner [1,*], Kristina R. Mueller [2], Yury Zablotski [1] and Melanie Feist [1]

1 Clinic for Ruminants with Ambulatory and Herd Health Services at the Centre for Clinical Veterinary Medicine, Ludwig-Maximilians-Universität Munich, Sonnenstrasse 16, 85764 Oberschleißheim, Germany; y.zablotski@med.vetmed.uni-muenchen.de (Y.Z.); melanie.feist@lmu.de (M.F.)

2 School of Veterinary Science, Massey University, Private Bag 11 222, Palmerston North 4474, New Zealand; k.mueller@massey.ac.nz

* Correspondence: t.tschoner@lmu.de

Simple Summary: Pain assessment in cattle can be performed using pain scales, e.g., the Numerical Rating (NRS) or Visual Analogue Scale (VAS). Pain scoring via pain scales is subjective to the experience and attitude of the observer. This systematic review and meta-analysis aimed to compare pain evaluation by dairy practitioners of different countries participating in surveys about pain management in cattle. Pain scoring is influenced by different factors, including the participant's age, gender, education, and profession. Female participants gave higher pain scores, as did recently graduated veterinarians. Differences in pain scales, nomenclature of terms, and values used between studies complicate the direct comparison of pain scores. The majority of these articles originated from the European Union. Different legislation and welfare considerations of countries could possibly influence pain scoring. Only a small number of studies could be included in the meta-analysis. Mean values of pain scores given for different procedures and conditions differed significantly, for both Numerical Rating Scales 0–10 and 1–10. The findings of the present study showed that the comparison of pain scores used in different articles is difficult due to the use of different pain scales and nomenclature, and that pain scoring is influenced by different factors, such as age and gender.

Abstract: Subjective pain assessment in cattle is contingent upon the observer's experience and attitude. Studies of pain assessment in cattle by veterinarians and farmers using different pain scales have been published. This systematic review and meta-analysis aimed to describe and compare the pain scores given by veterinarians and producers for different procedures and conditions using either a NRS or VAS. The literature search was conducted with PubMed (MEDLINE) and Agricola, using defined search terms (e.g., peer-reviewed). A total of 842 articles were identified. After screening of duplicates, abstracts, and full texts, a total of 16 articles were included in this systematic review. Different pain scales were used for the included studies (NRS 0–10 for eight studies, NRS 1–10 for six studies, NRS 1–10 and VAS 0–10 for one study, and VAS 0–1 for one study). Most studies (n = 11) originated from the European Union. Mean values for pain scores differed significantly between studies included in the meta-analysis for both NRS 0–10 and 1–10. The findings of this study indicated that comparison of pain scoring used in different studies is difficult due to use of different pain scales and varying nomenclature, and that many variables (such as age and gender) influence pain scoring.

Keywords: acute pain; analgesia; calves; dairy cattle; farmers; NRS; pain management; pain scoring; survey; VAS; veterinarians

1. Introduction

Painful conditions are frequently seen in cattle, caused either by disease or by veterinary or husbandry procedures [1]. Cattle are stoic prey animals; as a result, it is considered that these animals present a higher threshold for pain compared to other species because they show a strong pain-masking behavior [2–4]. Therefore, recognition and behavioral changes and categorization of the degree of pain experienced by cattle is the responsibility of both the producer and the veterinarian to preserve a good welfare status [3,5]. However, analgesic treatment to mitigate the level of acute pain is mainly up to the veterinarian [1], but despite this, it is important to note that this treatment is largely dependent on the full knowledge of the normal behavior of the species. Individual animals may differ in their expression of emotions [6]. However, the assessment and evaluation of pain by use of behavioral parameters, such as ethograms [7,8] or facial grimace scales [5], depends on the observer's experience and attitude [2], and is therefore subjective.

In the last 15 years, many surveys about pain assessment and management in cattle have been published to evaluate attitudes of veterinarians and farmers towards pain and pain management in cattle and assess their use of non-steroidal anti-inflammatory drugs (NSAIDs) in regard to frequency, active components, and occasions [9–11]. Studies of pain assessment for different procedures and/or conditions in cattle were conducted among veterinarians [1,9,12] as well as producers [13,14], using either Numerical Rating (NRS) [9,11] or Visual Analogue (VAS) [15] Scales. The concluding results of these studies showed that there is a wide range of attitudes of veterinarians [9,11,12] as well as practitioners [16] about pain assessment in cattle. However, questionnaires about the assessment of pain during procedures and conditions are described to be a promising method to assess the attitudes of participants towards pain in cattle [17].

A NRS is a scale that can be delivered verbally or graphically, and has two end points ("no pain" and "worst pain") [18]. In bovine medicine, the NRS normally ranges either from 0 (no pain) to 10 (worst pain imaginable) [9,12] or 1 (no pain) to 10 (worst pain imaginable) [1,15]. The VAS is a horizontal line of 100 mm, describing pain limits from "no pain" (0, left side) to worst pain imaginable (10, right side) [15,19]. This scale can also be used in surveys about pain assessment in cattle [15,19,20] and is described to be more informative than the NRS [17].

Scientific evidence shows that in both human and veterinary medicine, there are inherent factors associated with the assessor that can influence the recognition of acute pain, such as the social status, work status, age, gender, degree of empathy, and educational level of the assessor [1,9,12,21,22], as well as inherent factors associated with the animal such as species, age, breed, gender, and even the presence of previous pathologies [23].

However, individual pain scores given for different conditions and procedures throughout the studies show a high variety, from the lowest to the highest score presented to the respondents being selected by individuals [1,9,12].

Numerous studies of pain assessment in cattle have been published [5,24]. However, to this day, there is no systematic review of pain assessment in cattle conducted by veterinarians or producers using different rating scales.

Therefore, the objectives of this systematic review were to (i) describe and compare pain scores and their ranges awarded by veterinarians and producers using either a Numerical Rating or a Visual Analogue Scale, and (ii) compare these scores with a meta-analysis. The aim of this review is to contribute to the current knowledge about pain assessment in cattle, and the possible differences between veterinarians and producers.

2. Materials and Methods

2.1. Search Strategy and Selection Criteria

The systematic review and meta-analysis were performed following the Preferred Reporting Items for Systematic Reviews and Meta-Analysis protocols (PRISMA-P) study protocols [25] as described by Oehm et al. [26] and Tschoner and Feist [27] (Figure 1, Supplemental Table S1). The literature search was conducted using the scientific literature databases PubMed (including MEDLINE) and Agricola on 11 September 2023. A range of years for the analysis of articles was not defined. The search was conducted for all available years. The search terms were separated to include the four components of this review:

1. To identify studies with a study population of veterinarians, farmers, or other people working with cattle: (veterinar* OR farm* OR produc* OR livestock* OR clinic* OR practition* OR caretak*) AND
2. To identify studies performed on cattle: (cattle OR cow OR calves OR calf OR dairy OR beef OR bovine) AND
3. To identify studies where a questionnaire was used: (survey OR question* OR attitud* OR opinion*) AND
4. To identify studies with surveys conducted on pain assessment or management: (pain* OR analges*).

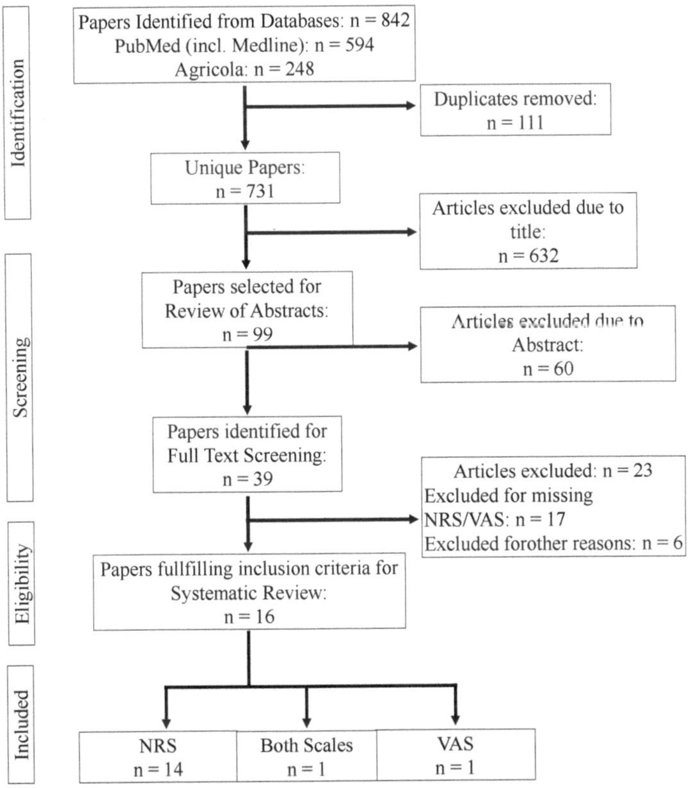

Figure 1. PRISMA (Preferred Reporting Items for Systematic Reviews and Meta-Analysis) flow chart of the literature search and the selection of the studies included in the present systematic review about the comparison of pain assessment using either a Numerical Rating (NRS) or Visual Analogue (VAS) Scale.

Alternative wording was included by using the operator "OR", and all components were combined by the separator "AND". Using an asterisk, the databases were screened for words beginning with these letters.

2.2. Selection of Studies

Initially, studies of all languages and designs describing pain assessment and/or management by veterinarians, farmers, and producers in cattle were included in the study selection. Subsequently, studies that were not written in German or English, or studies that were not accessible in any way, were not included in this review. De-duplication was conducted manually by the first author (TT) using EndNote (Version X9.3.3). The titles were then screened by TT. Reviews and proceedings, as well as titles including other species than cattle, were excluded at this point. The abstracts of the remaining publications were then evaluated by three authors (TT, KM, MF) to assess whether the eligibility criteria where met. The screening criteria for abstracts were the following:

1. The title and abstract were written in either English or German.
2. The study was conducted using a questionnaire or survey.
3. Veterinarians, farmers, producers, or other people handling cattle were involved.
4. The animal population was cattle.
5. Pain assessment was conducted.

All three reviewers were blinded to the decision of the other reviewers until decisions (include, exclude, maybe) had been made. If a study seemed eligible for two of the three reviewers, the full text was retrieved. All full texts were then screened by TT and were included in the present systematic review if the following questions, as described previously [27–29], could be answered with "yes":

(1) Can the full text be obtained?
(2) Is the full text written in English or German?
(3) Is the study population either veterinarians, producers, or farmers?
(4) Is the study design a survey or a questionnaire?
(5) Is the questionnaire or survey about the assessment of painful conditions/procedures?
(6) Is either a Numerical Rating or Visual Analogue Scale used for pain assessment?
(7) Is the questionnaire about cattle?
(8) Is the article peer-reviewed?

To objectively compare pain assessment, only full texts including pain scoring using either a NRS or VAS were included. If the screening author was uncertain whether a study should be included, two other authors (KM and MF) were consulted.

2.3. Extraction of Data

Data extraction was performed by the first author (TT). Data were extracted regarding primary author, year of publication, country, group, and number of participants, return rate and responses included, demographic data of participants, pain scale used, painful condition and procedures assessed in either adult cows or calves, assessment of necessity and/or use of analgesics, and funding information.

2.4. Meta-Analysis and Statistical Analysis

For the meta-analysis, studies with more than three pain scores per condition or procedure were included. Median and mean values as well as SD were collected. If these were not presented in the articles, the first author of the respective article was contacted, with three authors responding and providing the original data. Pain scores were compared with the Kruskal–Wallis test. Studies were divided into articles using NRS 0–10 and NRS 1–10. For the 0 hypothesis, scores of 0 were defined to be no pain. The means of pain scores for professions were compared by the meta-analysis.

3. Results

3.1. Demographic Findings

A PRISMA flow chart presenting an overview of the literature search and study selection is given in Figure 1. A pool of 842 articles was identified by the search terms in the databases; of these, 111 were duplicates. A total of 731 titles were screened, with 632 excluded at this point due to the title. Abstract screening was conducted for 99 references, with a total of 39 references retrieved for full-text screening, including 3 references for which no abstract was available. Studies for which the abstract was not accessible were included in the full-text retrieval and screening. A total of five references were excluded as they were commentaries (n = 2) or books or book chapters with no abstract (n = 3). A total of 16 references met all eight inclusion criteria and were therefore included in the systematic review. All abstracts and full texts screened for this systematic review were written in English. The publication year ranged from 2006 to 2022. A total of three studies were conducted in Finland, two studies each in the United Kingdom, New Zealand, Norway, and Germany, and one study in each of Canada, Denmark, Switzerland, Brazil, and China. Demographic information about the articles is presented in Table 1.

3.2. Material and Methods

The study design of the articles was heterogenous. The study population was veterinarians in seven studies, farmers in four studies, and either veterinary students, veterinarians and veterinary students, or veterinarians and farmers in one study, respectively. In one study each, the study population was either veterinarians, frontline staff, and managers, or veterinarians, farmers, and claw trimmers. Pain scores were given under the assumption that no analgesia was provided for 11 studies. A total of four studies did not define if pain scoring was conducted under the assumption of pain relief, and one study asked for pain scoring depending on the presence or absence of analgesia (Table 1). Surveys were conducted exclusively in paper form for eight studies and online for five studies. A total of four studies used both paper and online surveys. Pain assessment was presented for procedures and conditions in either adult cattle or in calves in two studies each, and for both in eleven studies. For one of the surveys, the age category of animals was not stated. A total of 14 studies were exclusively about pain management in cattle, whereas one study each also included horses, or horses and pigs, in the survey. NRS ranging from 0 to 10 was used in eight studies, NRS ranging from 1 to 10 in six studies, VAS ranging from 0 to 10 in one study, and both NRS (1–10) and VAS (0–10) in one study (Table 1). In one study, use of NRS (0–10) was described in the Material and Methods section, but VAS (0–10) was used in the Results section. In nine studies, sections about availability and use of analgesics, and/or questions regarding pain management, were included in the survey. Detailed information about Material and Methods is presented in Supplemental Table S2.

3.3. Funding Information

Funding information was provided for 87.5% (n = 14) of studies and is presented in Supplemental Table S2.

3.4. Pain Scores for Adult Cattle

Pain Scores awarded for different procedures and conditions in adult cattle are presented in Table 2 (procedures) and Table 3 (conditions). The procedures that were rated to be most painful were caesarean section (median 9 for [1,9,11,12,14]), claw amputation (median 10 for [1,9,12,30] and 9 for [11,14]), extirpation of the eye bulb (median 9 for [11] and 10 for [14]), laparotomy (median 9 for [14]), left displacement of the abomasum (LDA) surgery (median 9 for [9,12]), omentopexy (median 9 for [14]), and treatment of interdigital hyperplasia (median 9 for [30]). The conditions and diseases that were rated most painful were acute toxic (*E. coli*) mastitis (median 9 for [13]), dystocia (mean and SD 9.0 ± 1.83 for [16]), and fracture of tuber coxae (median 9 for [15]). Numerical data were not presented for n = 3 (18.8%) of the included articles [31–33]. Lorena et al. [31] stated that fracture repair was considered the most painful procedure. Pain scores including ranges are given in Supplemental Table S3 (procedures) and Supplemental Table S4 (conditions).

3.5. Pain Scores for Calves

Pain scores awarded for different procedures and conditions in calves are presented in Table 4 (procedures) and Table 5 (conditions). The procedures considered to be most painful were castration (median of 9 both for Burdizzo and surgical [11,14]), disbudding (median 9 for [20] and mean and SD 9.0 ± 1.2 for [22]), laparotomy (median 9 for [11,14]), repair of distal limb fracture (median 10 for [12]), and umbilical hernia surgery (median 9 for [11,14]). The conditions that were rated to be most painful were ileus (median 9 for [11,14]) and distal limb fracture (mean and SD 9.0 ± 1.2 for [22]). Numerical data were not presented for n = 3 (18.8%) articles [31–33]. Pain scores including ranges are given in Supplemental Table S5 (procedures) and Supplemental Table S6 (conditions).

3.6. Differences between Veterinarians, Farmers, and Others

A comparison of pain scoring between veterinarians and farmers was presented in 25% ($n = 4$) of papers. According to Thomsen et al. [13], farmers considered diseases to be more painful than veterinarians. These differences were significant for left displaced abomasum ($p < 0.0001$), digital dermatitis ($p = 0.01$), mastitis ($p < 0.001$), and interdigital necrobacillosis ($p < 0.0001$). Becker et al. found significant differences in pain perception for therapeutic trim of a sole ulcer and treatment for white-line disease between veterinarians, farmers, and claw trimmers [30]. Contrary to that, pain scores given by Bavarian veterinarians and farmers did not differ significantly [14]. Profession of the participants also had no effect on pain scores in a Chinese study [16].

3.7. Influence of Gender on Pain Scoring

Comparison of pain scoring between male and female participants was conducted for 11 studies. A total of three studies [16,19,22] found no differences in pain scoring between genders. In a study from 2006, British female veterinarians awarded significantly higher pain scores for treatment of a sole ulcer ($p < 0.001$), dystocia ($p < 0.001$), fracture of tuber coxae ($p < 0.001$), left displaced abomasum ($p < 0.001$), acute metritis ($p < 0.001$), swollen hock ($p < 0.001$), acute toxic *E. coli* mastitis ($p < 0.001$), mastitis (clots in milk only, $p < 0.001$), and neck calluses ($p < 0.001$) in adult cattle, and surgical castration ($p < 0.001$), disbudding ($p < 0.001$), following dystocia ($p < 0.01$), umbilical abscess ($p < 0.001$), joint ill ($p < 0.001$), and pneumonia ($p < 0.001$) in calves. Male veterinarians ranked claw amputation ($p < 0.01$) and dehorning ($p < 0.01$) as significantly more painful than female veterinarians [9]. Ten years later, Remnant et al. [1] found that female gender of respondents was associated with an increase of 0.36 in pain scoring. According to Laven et al. [12], the Mann–Whitney mean rank score for pain scores was higher for female than male veterinarians from New Zealand, with significant differences for treatment of a white-line abscess, acute metritis, swollen hock, acute toxic mastitis, and white-line disease with sub-sole abscess in adult cattle, and castration (Burdizzo), umbilical abscess, joint ill, and pneumonia in calves ($p < 0.01$, respectively). Pain scores also differed significantly ($p < 0.05$) between Brazil veterinarians, with female veterinarians awarding higher pain scores than male ones for all procedures, expect laparotomy and fractures [31]. Female veterinary students gave median scores that were 0.9 points higher than those of their male colleagues [15]. Bavarian female veterinarians awarded significantly higher pain scores for fetotomy and removal of retained fetal membranes ($p < 0.01$, respectively) [11], whereas Bavarian female farmers scored treatment of interdigital hyperplasia, dehorning, laparoscopic fixation of left displaced abomasum, laparotomy, caesarean section, artificial insemination ($p = 0.01$, respectively), and fetotomy ($p < 0.01$) in adult cattle, and laparotomy in calves ($p = 0.01$) significantly higher [14]. Female veterinarians from New Zealand scored supernumerary teat removal ($p = 0.009$) and disbudding ($p = 0.003$) significantly higher [33] than their male counterparts. However, Kielland et al. [19] found no differences in pain scoring between male and female Norwegian farmers.

3.8. Influence of Age on Pain Scoring

According to Huxley and Whay [9], pain scoring differed significantly between British veterinarians who had qualified in different decades, with higher pain scores awarded for dystocia, fracture of tuber coxae, left displaced abomasum, acute metritis, swollen hock, toxic *E. coli* mastitis, mastitis, and neck calluses in adult cattle, and umbilical hernia surgery, umbilical abscess, joint ill, and pneumonia in calves, by veterinarians who had qualified more recently; veterinarians who had been qualified longer awarded significantly higher pain scores for claw amputation and dehorning in adult cattle, and disbudding in calves. Ten years later, British veterinarians graduating before 1990 awarded pain scores that were 0.48 points lower compared with veterinarians graduating since 2010. There was no significant difference in graduation between these years for either group [1].

In a survey from New Zealand, highest median pain scores for 14 out of 24 procedures and conditions were scored by respondents who graduated from 2000 onwards. However, decade of graduation was only associated with a significant ($p < 0.01$) difference in pain scoring for 4 of these 24 conditions and procedures [12]. However, according to Kielland et al. [19], there was no influence of age on median pain scoring in Norwegian farmers.

3.9. *Influence of Education and Experience on Pain Scoring*

Huxley and Whay [9] found that British veterinarians with postgraduate training or qualification assigned significantly higher ($p \leq 0.01$) pain scores for LDA surgery, and higher pain scores for claw amputation, caesarean section, digital dermatitis, and fracture of a distal limb ($p \leq 0.05$). A larger amount of time spent working with cattle than other species resulted in significantly lower pain scores for cattle diagnosed with LDA ($p \leq 0.01$), and significantly higher pain scores for DD ($p \leq 0.001$). Ten years later, significant differences in pain scoring were observed between year and school of graduation, background of participants prior to university ($p < 0.01$), and holding clinical postgraduate qualifications ($p < 0.05$) [1]. In Bavaria, veterinarians with a graduation date between 1960 and 1970 assigned lower pain scores for 11 out of 33 diseases, and 2 out of 20 procedures, but higher pain scores for claw amputation and dehorning in adult cattle, and surgical castration, tenotomy of contracted tendons, and dehorning in calves [11]. Van Dyke et al. [33] found a significant effect of years since graduation on pain scoring, with male veterinarians awarding lower pain scores with increasing years since graduation, whereas pain scoring was consistent within the group of female veterinarians over the years. Other studies found no influence of level of education [19,22] or experience as a veterinarian [22] on pain scoring.

3.10. *Results of the Meta-Analysis*

A total of eleven articles were included in the meta-analysis: six for NRS 0–10 and five for NRS 1–10. For NRS 0–10, pain scores of 16 procedures (n = 10 for cattle and n = 6 for calves) and 7 conditions (n = 3 for cattle and n = 4 for calves) were compared. For NRS 1–10, pain scores of five procedures and ten conditions in cattle were compared; a total of three conditions (treatment of interdigital hyperplasia, treatment of sole ulcer, treatment of white-line abscess) were excluded, as no SD was available for the pain scores.

Kruskal–Wallis tests showed pain scores were not significantly different between professions (veterinarians and veterinary students, farmers, and practitioners (including frontline staff), neither for NRS 0–10 ($p = 0.42$) nor for NRS 1–10 ($p = 0.33$). The meta-analysis of professions shows a very high heterogeneity within professions, but not significant differences in means between professions. For NRS 0–10, heterogeneity of mean values of pain scores was significant for all procedures and conditions in both calves and cattle ($p < 0.01$, $p = 0.02$ for digital dermatitis, respectively). Forest plots of the meta-analysis for NRS 0–10 are given in Figure 2 (procedures for cattle), Figure 3 (conditions for cattle), Figure 4 (procedures for calves), and Figure 5 (conditions in calves).

For NRS 1–10, heterogeneity of mean values of pain scores was significant ($p < 0.01$) for acute metritis, acute toxic (*E. coli*) mastitis, neck calluses, and swollen hock (Figure 6). For five procedures (claw amputation, dehorning, treatment of sole ulcer, interdigital hyperplasia, and white-line abscess) and one condition (dystocia), the p-values for heterogeneity could not be calculated due to missing SD (claw amputation, dehorning, treatment of sole ulcer, interdigital hyperplasia, and white-line abscess).

Table 1. Demographic information of 16 articles included in the systematic review. All articles were written in English and published in peer-reviewed journals. Articles were included if pain assessment was conducted by use of a survey, and either a Numerical Rating (NRS) or a Visual Analogue (VAS) Scale.

Year	Author	Country	Ref. [1]	Participants	Return Rate (%)	Responses Included	Gender	Rating Scale	Analgesia [2]
2006	Huxley and Whay	United Kingdom	[9]	Veterinarians	26.8	615/641	72.6% male, 27.4% female	NRS (0–10)	No
2007	Hewson et al.	Canada	[10]	Veterinarians	50.1	585/586	65% male, 35% female	NRS (1–10)	No
2009	Kielland et al.	Norway	[15]	Veterinary Students	57 [3]	171/171 [4]	19.9% male, 80.1% female	VAS (0–10) NRS (1–10)	Not Stated
2009	Laven et al.	New Zealand	[12]	Veterinarians	37	166/166	62.7% male, 37.3% female	NRS (0–10)	No
2010	Kielland et al.	Norway	[19]	Dairy Farmers	70	149/154	87% male, 13% female	VAS (0–10)	Not Stated
2012	Thomsen et al.	Denmark	[13]	Veterinarians Dairy Farmers	28 47	137/493 189/401	Not Stated	NRS (1–10)	No
2013	Becker et al.	Switzerland	[30]	Veterinarians Claw Trimmers Dairy Farmers	Not stated	137 32 77	77.4% male, 22.6% female 100% male 89.6% male, 10.4% female	NRS (1–10)	No
2013	Lorena et al.	Brazil	[31]	Veterinarians	Not Stated	713/800	60% male, 40% female	NRS (1–10)	No
2013	Wikman et al.	Finland	[20]	Dairy Farmers	45	439/451	Not stated [5]	NRS (0–10) [6]	Not Stated
2014	Norring et al.	Finland	[22]	Veterinarians Veterinary Students	about 40% about 40% [7]	189 in total	9% male, 91% female	NRS (0–10)	Not Stated
2015	Hokkanen et al.	Finland	[32]	Veterinary Students	45	438/451	Not Stated	NRS (0–10)	No
2017	Remnant et al.	United Kingdom	[1]	Veterinarians	16 [8]	242/247	56% male, 44% female	NRS (1–10)	No
2020	Tschoner et al.	Germany	[11]	Veterinarians	26.2	274/287	82.1% male, 17.5% female	NRS (0–10)	No
2021	Tschoner et al.	Germany	[14]	Dairy Farmers	15.4	492/577	79.5% male, 18.7% female	NRS (0–10)	No
2021	Van Dyke et al.	New Zealand [9]	[33]	Veterinarians	17.6	104/106	48% male, 52% female	NRS (1–10)	No and Yes [10]
2022	Shi et al.	China	[16]	Veterinarians Frontline Staff	24.1	465/666	90.1% male, 9.9% female	NRS (0–10)	No

[1] Reference; [2] pain assessment was either conducted under the assumption that no analgesia or anesthesia was administered; [3] 54.7 for VAS; 59.3 for NRS; [4] 82 responses for VAS; 89 responses for NRS; [5] 255 men and 175 women, 9 not clarified; [6] NRS explained in Material and Methods, but VAS (0–10) used in Result section; [7] approximately 42% of students from preclinical and clinical stage, respectively; [8] no return rate for online survey, as extent of distribution unknown; [9] survey conducted among veterinarians in New Zealand, affiliation of authors in United Kingdom; [10] presence or absence or local anesthesia and/or postoperative analgesia.

Table 2. Pain scoring for different procedures presented to participants of surveys about pain assessment in adult cattle, under the assumption that no analgesics are used. Use of analgesics was not stated for [15,19,20,22]. Pain scoring was conducted by either veterinarians (V), veterinary students (VS), farmers (F), practitioners (and frontline staff, P), and/or claw trimmers (C) by use of a Numerical Rating (NRS, ranging either from 0 to 10 or 1 to 10) or Visual Analogue Scale (VAS, ranging from 0 to 10), or both. Ref. [20] described using a NRS in the Material and Methods section, but indicated use of a VAS in the Results section, and is therefore included as VAS. Left displacement of the abomasum is abbreviated as LDA. If procedures were not presented in the respective reference, this is indicated as -.

				NRS (0–10)						NRS (1–10)						VAS (0–10)	
	[9]	[12]	[22] [1]	[11]	[14]	[16] [1]	[10] [1]	[15]	[13] [2]		[30]			[1]	[15]	[19]	[20]
Professional Group	V	V	V/VS	V	F	V/F/P	V	VS	V	F	V	F	C	V	VS	F	P
							Procedures on the Head										
Dehorning [3]	8	8	-	8	8	7.1 ± 2.52	7.4	4	-	-	-	-	-	8	3	5.1	-
Extirpation of eye bulb	-	-	-	9	10	-	-	-	-	-	-	-	-	-	-	-	-
							Abdominal Surgeries										
Laparoscopic fixation of LDA	-	-	-	-	8	-	-	-	-	-	-	-	-	-	-	-	-
Laparotomy			-	8	9	-	-	-	-	-	-	-	-	-	-	-	-
LDA surgery	9	9	-	-	-	-	-	-	-	-	-	-	-	8	-	-	-
Omentopexy	-	-	-	-	9	-	-	-	-	-		-	-	-	-	-	-
							Orthopedics										
Claw amputation	10	10	-	9	9	-	-	-	-	-	10	10	10	10	-	-	-
Debriding of a digital dermatitis lesion	6	-	-	-	-	-	-	-	-	-	-	-	-	7	-	-	-
Treatment of interdigital hyperplasia [4]	-	-	-	8	7	-	-	-	-	-	8	8	9	-	-	-	-
Treatment of a sole ulcer [4]	6	-	-	7	7	-	-	-	-	-	8	7	7	7	-	-	-
Treatment of white-line abscess [4]	-	4	-	-	-	-	-	-	-	-	8	7	7	-	-	-	-
							Obstetrics and Gynaecology										
Artificial Insemination	-	-	-	1	0	-	-	-	-			-		-	-	-	-
Caesarean section	9	9	-	9	9	8.6 ± 2.12	8	-	-			-		9	-	-	-
Fetotomy	-	-	-	7	7/8	-	-	-	-			-		-	-	-	-
Rectal examination	-	-	-	1/2	1	-	-	-	-			-		-	-	-	-
Removal of retained fetal membranes	-	-	-	3	5	-	-	4	-			-		-	2	2.4	-
							Other										
Needle prick [5]	-		2.5 ± 1.87	-	-	-	-	-	-			-		-	-	-	-

[1] Mean values including standard deviation where indicated; [2] median as well as mean values presented in article, median values were included in the table; [3] horns > 8 cm long for Huxley and Whay (2006) [9], Laven et al. (2009) [12]; in cattle over 6 months of age for Hewson et al. (2007) [10]; [4] excision for Becker et al. (2013) [30], [5] fully grown cattle, into the shoulder muscle for Norring et al. (2014) [22].

Table 3. Pain scoring for different conditions presented to participants of surveys about pain assessment in adult cattle, under the assumption that no analgesics are used. Use of analgesics was not stated for [15,19,20,22]. Pain scoring was conducted by either veterinarians (V), veterinary students (VS), farmers (F), and/or practitioners (and frontline staff, P) by use of a Numerical Rating (NRS, ranging either from 0 to 10 or 1 to 10) or Visual Analogue Scale (VAS, ranging from 0 to 10), or both. Ref. [20] described using a NRS in the Material and Methods section, but indicated use of a VAS in the Results section and is therefore included as VAS. Retained fetal membrane is abbreviated as RFM. If procedures were not presented in the respective reference, this is indicated as -.

	NRS (0–10)							NRS (1–10)				VAS (0–10)		
	[9]	[12]	[22] [1]	[11]	[14]	[16] [1]	[10] [1]	[15]	[13] [2]		[1]	[15]	[19]	[20]
Professional Group	**V**	**V**	**V/VS**	**V**	**F**	**V/F/P**	**V**	**VS**	**V**	**F**	**V**	**VS**	**F**	**P**
Conditions of the Head														
Corneal ulcer	-	-	-	-	-	-	5.5	-	-	-	-			
Fracture of the horn	-	-	-	6	6	-	-	-	-	-	-			
Loss of nose ring	-	-	-	6	6	-	-	-	-	-	-			
Neck calluses	2	-	-	3	3	-	-	4	3	4	3	4		
Uveitis [3]	6	-	-	5	5	-	-	6	8	4	6	5		
Conditions of the Abdomen														
Abomasal displacement	-	-	7.3 ± 1.9	-	-	7.4 ± 2.18	-	-	-		-	-	-	8
Left displaced abomasum	3	6	-	5	5	-	-	6	5	6	-	4	7	-
Oesophageal obstruction	-	-	-	-	-	5.9 ± 2.36	-	-	-		-	-	-	-
Right displaced abomasum	-	-	-	6	6	-	-	-	-		-	-	-	-
Ruminal acidosis	-	-	-	-	-	5.3 ± 2.62	-	-	-		-	-	-	-
Severe tympany in cattle [4]	-	-	7.9 ± 1.6	-	-	6.1 ± 2.18	-	-	-		-	-	-	9
Traumatic pericarditis	-	-	-	-	-	7.8 ± 2.25	-	-	-		-	-	-	-
Orthopedic Conditions														
Decubitus	-	-	-	4/5	4/5	-	-	-	-		-	-	-	-
Digital Dermatitis	6	-	-	7	7	-	-	-	7	7	6	-	-	-
Footrot	-	5	-	-	-	-	-	-	-		-	-	-	-
Fracture of long bone [5]	-	-	-	8	8	8.4 ± 2.08	-	-	-		-	-	-	-
Fracture of tuber coxae [6]	7	8	-	-	-	-	-	9	8	8	8	8	-	-
Hock with hair loss	3	-	-	-	-	-	-	-	-		3	-	-	-
Hoof disease	-	-	-	-	-	6.9 ± 2.18	-	-	-		-	-	-	-
Injuries on hock [7]	-	-	-	-	-	-	-	4	-		-	4	2.9	-
Interdigital necrobacillosis	-	-	-	-	-	-	-		8	8	-	-	-	-
Laminitis	-	-	-	8	8	-	-	7	-		-	7	5.7	-
Rupture of muscle	-	-	-	8	8	-	-	-	-		-	-	-	-
Septic Arthritis/Polyarthritis	-	-	-	8	8	-	-	-	-		-	-	-	-
Sole ulcer	-	-	-	8	8	-	-	7	-		-	6	7.1	-
Swollen hock	5	6	-	-	-	-	-	5	5	5	5	5	-	-
White-line disease [8]	7	-	-	-	-	-	-	-	-		7	-	-	-

Table 3. *Cont.*

	NRS (0–10)							NRS (1–10)				VAS (0–10)		
	[9]	[12]	[22] [1]	[11]	[14]	[16] [1]	[10] [1]	[15]	[13] [2]		[1]	[15]	[19]	[20]
Professional Group	V	V	V/VS	V	F	V/F/P	V	VS	V	F	V	VS	F	P
Mastitis and Udder Health														
Acute mastitis [9]	-	-	7.3 ± 1.4	-	-	-	-				-	-	7.6	8
Acute toxic (E. Coli) mastitis [10]	7	8	-	7	7	7 ± 2.2	-	7	9	9	7	7	-	-
Intertrigo	-	-	-	4	4	-	-	-	-		-	-	-	-
Mastitis (clots in milk only) [11]	3	3	-	1	1	3.4 ± 2.65	-	4	2	3	4	5	-	-
Moderate mastitis	-	-	-	-	-	5.1 ± 2.24	-	-	-		-	-	-	-
Open teat injury	-	-	-	6	6	-	-	-	-		-	-	-	-
Teat injury [12]	-	-	7.4 ± 1.7	-	-	-	-	-	-		-	-	-	8
Obstetrics and Gynaecology														
Acute metritis [13]	4	-	-	5	5	-	-	7	6	6	5	4	-	-
After removal of RFM	-	-	-	-	-	-	-	4	-		-	4	2.4	-
Calving	-	-	-	-	-	8.5 ± 1.99	-		-		-	-	-	-
Dystocia [14]	7	7	7.3 ± 1.7	8	8	9.0 ± 1.83	5.3	8	-		7	8	-	-
Endometritis	-	-	-	-	-	5.9 ± 2.62		-	-		-	-	-	-
Postpartum paralysis	-	-	-	-	-	5.9 ± 3.02	-	-	-		-	-	-	-
Tissue injuries following birth	-	-	-	5	5	-	-	-	-		-	-	-	-
Uterine torsion	-	-	-	6	6	-	-	-	-		-	-	-	-
Uterine prolapse [15]	-	-	6.9 ± 2.0	5	5	7.9 ± 2.3	-	-	-		-	-	-	8
Vaginal prolapse	-	-				6.3 ± 2.66	-	-	-		-	-	-	-
Metabolic and Nutritional Diseases														
Hypocalcemia [16]	-	-	-	1	1	-	-	5	-		-	5	3.3	-
Ketosis	-	-	-	1	1	-	-	4	-		-	4	-	-
Nutritional deficiency disease	-	-	-	-	-	3.6 ± 2.8	-	-	-		-	-	-	-
Other														
Infectious disease	-	-	-	-	-	4.5 ± 2.91	-	-	-		-	-	-	-
Parasitic disease	-	-	-	-	-	4.1 ± 2.57	-	-	-		-	-	-	-

[1] Mean values including standard deviation where indicated; [2] median as well as mean values presented in reference, median values were included in the table; [3] eye infection for Kielland et al. (2009; 2010) [15,19]; [4] ruminal bloat for Shi et al. (2022) [16]; [5] fracture for Shi et al. (2022) [16]; [6] one-sided for Kielland et al. (2010) [19]; [7] skin lesions on hock for Kielland et al. (2010) [19]; [8] with subsole abscess for Huxley and Whay (2006) [9]; [9] fever 41 °C, lumps in milk, hard udder for Norring et al. (2014) [22]; [10] *Escherichia coli* mastitis for Huxley and Whay (2006) [9], serious mastitis for Kielland et al. (2009; 2010) [15,19]; severe mastitis for Shi et al. (2022) [16]; [11] mastitis for Thomsen et al. (2012) [13], mild mastitis for Shi et al. (2022) [16], chronic mastitis for Tschoner (2020; 2021) [11,14]; [12] teat tramping in cows for Norring et al. (2014) [22], teat broken at the roof for Wikman et al. (2013) [20]; [13] puerperal metritis for Tschoner et al. (2020, 2021) [11,14], metritis for Remnant et al. 2017 [1]; [14] fetal-maternal disproportion requiring traction alone for Huxley and Whay (2006) [9], Laven et al. (2009) [12], Tschoner et al. (2020, 2021) [11,14]; strong pull assistance for Norring et al. (2014) [22]; [15] uterine eversion for Tschoner et al. (2020, 2021) [11,14]; [16] milk fever for Kielland et al. (2009; 2010) [15,19].

Table 4. Pain scoring for different procedures presented to participants of surveys about pain assessment in calves, under the assumption that no analgesics are used. Use of analgesics was not stated for [15,19,20,22]. Pain scoring was conducted either by veterinarians (V), veterinary students (VS), farmers (F), and/or practitioners (and frontline staff, P) by use of a Numerical Rating (NRS, ranging either from 0 to 10 or 1 to 10) or Visual Analogue Scale (VAS, ranging from 0 to 10), or both. Ref. [20] described using a NRS in the Material and Methods section, but indicated use of a VAS in the Results section and is therefore included as VAS. Ranges are included in brackets if indicated in the references.

	NRS 0–10						NRS (1–10)			VAS (0–10)		
	[9]	[12]	[22] [1]	[11]	[14]	[16] [1]	[10] [1]	[15]	[1]	[15]	[19]	[20]
Professional Group	V	V	V/VS	V	F	V/F/P	V	VS	V	VS	F	P
						Castration						
Castration [2] up to 6 months	-	-	-	-	-	-	4.9	-	-	-	-	-
Castration [2] over 6 months	-	-	-	-	-	-	5.9	-	-	-	-	-
Castration (Burdizzo)	7	6	-	9	9	-	-	-	6	-	-	-
Castration (Rubber Ring)	6	5	-	-	-	-	-	-	6	-	-	-
Castration (Surgical)	6	8	-	9	9	7.8 ± 2.32	-	-	7	-	-	-
						Dehorning/Disbudding						
Dehorning [3]	-	-	-	8	8	-	6.8	-	-	-	-	-
Dehorning over 6 months	-	-	-	-	-	-	7.4	-	-	-	-	-
Disbudding	7	8	9 ± 1.2	-	-	7.6 ± 2.32	-	-	7	-	-	9
Disbudding (caustic paste)	-	-	-	-	-	5.6 ± 2.52	-	-	-	-	-	-
Disbudding with analgesics [4]	-	-	2.4 ± 1.8	-	-	-	-	-	-	-	-	-
Abdominal Surgery												
Laparotomy	-	-	-	9	9	-	-	-	-	-	-	-
Umbilical hernia surgery [5]	8	8	-	9	9	6.8 ± 2.32	7.3	-	8	-	-	-
Orthopedic Procedures												
Repair of distal limb fracture	-	10	-	-	-	-	-	-	-	-	-	-
Tenotomy of contracted tendons	-	-	-	8	8	-	-	-	-	-	-	-
Other												
Ear tagging	-	-	-	4	4	-	-	-	-	-	-	-

[1] Mean values including standard deviation where indicated; [2] Hewson et al. (2007) [10] did not distinguish between methods of castration; [3] in calves up to 6 months for Hewson et al. (2007) [10]; [4] pain during burning for Wikman et al. (2013) [20] and Norring et al. (2014) [22]; [5] in calves up to 3 months for Hewson et al. (2007) [10].

Table 5. Pain scoring for different conditions presented to participants of surveys about pain assessment in calves, under the assumption that no analgesics are used. Use of analgesics was not stated for [15,19,20,22]. Pain scoring was conducted either by veterinarians (V), veterinary students (VS), farmers (F), and/or practitioners (and frontline staff, P) by use of a Numerical Rating (NRS, ranging either from 0 to 10 or 1 to 10) or Visual Analogue Scale (VAS, ranging from 0 to 10), or both. Ref. [20] described using a NRS in the Material and Methods section, but indicated use of a VAS in the Results section and is therefore included as VAS. Ranges are included in brackets if indicated in the references.

	NRS 0–10						NRS (1–10)			VAS (0–10)		
Reference	[9]	[12]	[22] [1]	[11]	[14]	[16] [1]	[10] [1]	[15]	[1]	[15]	[19]	[20]
Professional Group	V	V	V/VS	V	F	V/F/P	V	VS	V	VS	F	P
Abdominal Conditions												
Enteritis/Diarrhea [2]	-	-	-	5	5	4.6 ± 2.5	-	5	6	6	4.8	-
Ileus	-	-	-	9	9	-	-	-	-	-	-	-
Ruminal acidosis	-	-	-	4	4	-	-	-	-	-	-	-
Umbilical Conditions												
Navel infection	-	-	6.8 ± 1.6	7	7	-	-	5	-	5	5.2	8
Umbilical abscess	5	5	-	-	-	-	-	-	6	-	-	-
Umbilical hernia [3]	-	-	-	-	-	-	-	-	-	-	-	6
Orthopedic Conditions												
Contracted tendons	-	-	-	4	4	-	-	-	-	-	-	-
Joint ill [4]	7	8	-	8	8	-	-	7	7	6	6.7	-
Distal limb fracture [5]	8	-	9 ± 1.2	8	8	-	-	8	9	8	7.6	-
Other												
(Broncho)Pneumonia	6	8	-	6	6	-	-	6	7	6	6.7	-
Following dystocia [6]	4	3	5.9 ± 1.9	5	5	-	-	4	5	4	3.3	-
Meningitis	-	-	-	8	8	-	-	-	-	-	-	-
Needle prick neck	-	-	2.4 ± 1.6	-	-	-	-	-	-	-	-	-

[1] Mean values including standard deviation where indicated; [2] intestinal infection for Kielland et al., 2009 [15], no age defined for Shi et al., 2022 [16]; [3] the size of an apple for Wikman et al. (2013) [20]; [4] septic arthritis/polyarthritis for Tschoner et al. (2020, 2021) [11,14]; [5] broken bone with open fracture on calf's hind leg for Norring et al. (2014) [22], fracture of long bone for Tschoner et al. (2020, 2021) [11,14]; [6] fetal-maternal disproportion requiring traction alone for Huxley and Whay (2006) [9], Tschoner et al. (2020, 2021) [11,14], strong pull assistance for Norring et al. (2014) [22].

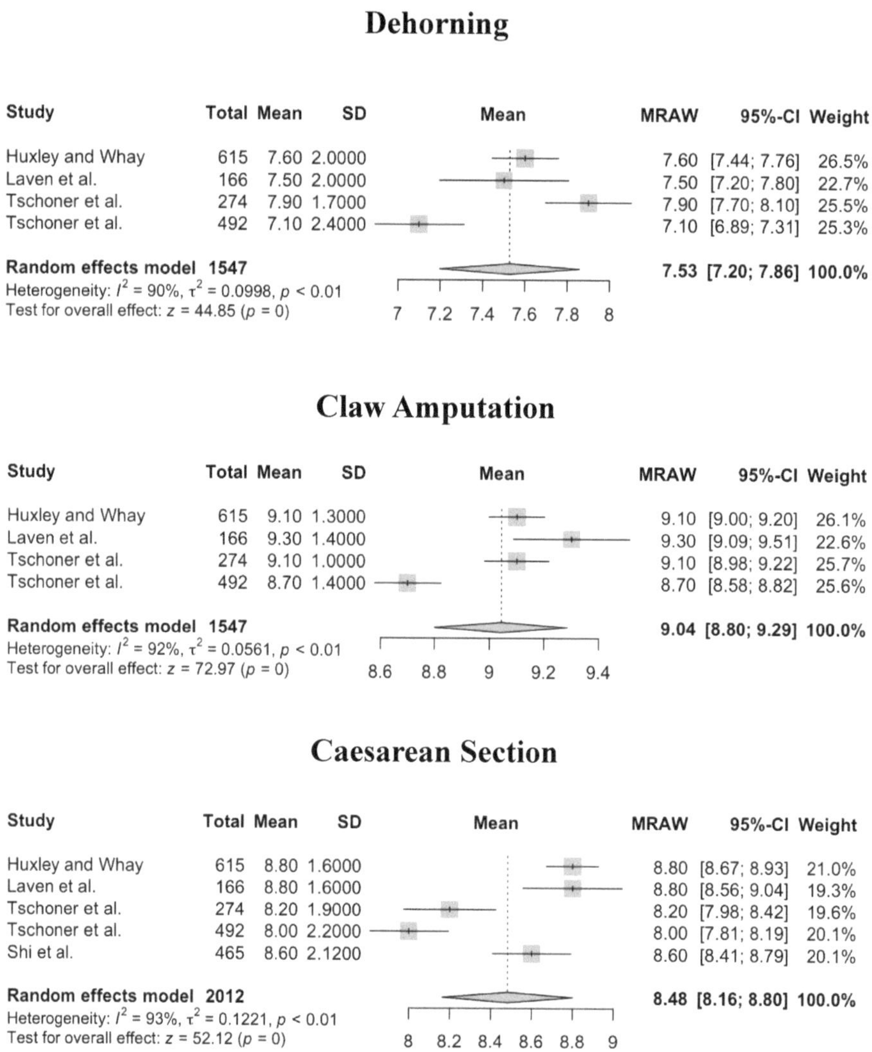

Figure 2. Forest plots for mean and SD of pain scores assigned using a Numerical Rating Scale ranging from 0 to 10 for procedures in cattle. Heterogeneity of mean values of pain scores was significant ($p > 0.01$) for all procedures. Test for overall effect shows a significant pain score measured for cows across studies, where no pain (pain score = 0) is the null hypothesis [9,11,12,14,16].

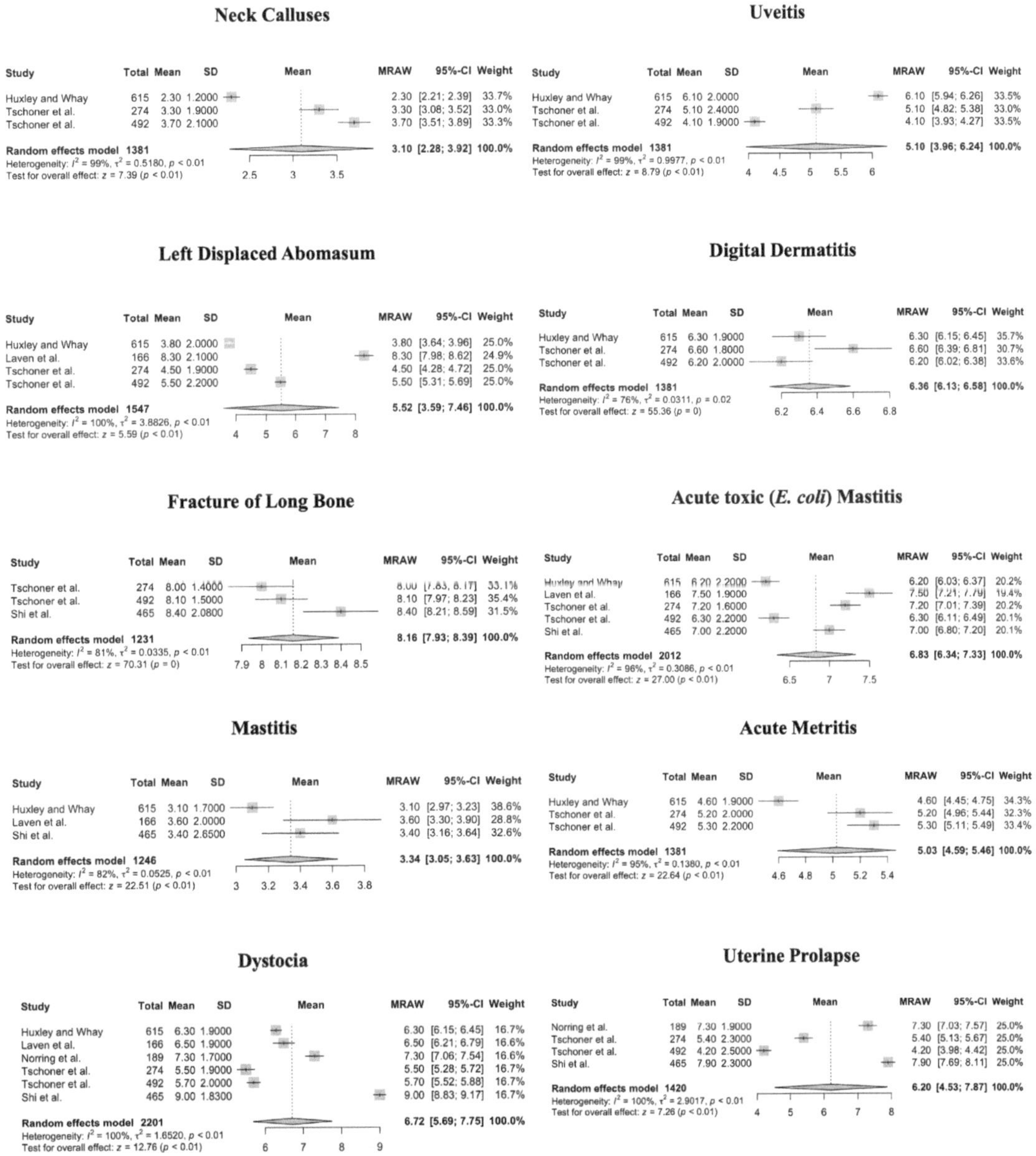

Figure 3. Forest plots for mean and SD of pain scores assigned using a Numerical Rating Scale ranging from 0 to 10 for conditions in cattle. Heterogeneity of mean values of pain scores was significant for all procedures ($p > 0.01$ respectively, $p = 0.02$ for digital dermatitis). Test for overall effect shows a significant pain score measured for cows across studies, where no pain (pain score = 0) is the null hypothesis [9,11,12,14,16,22].

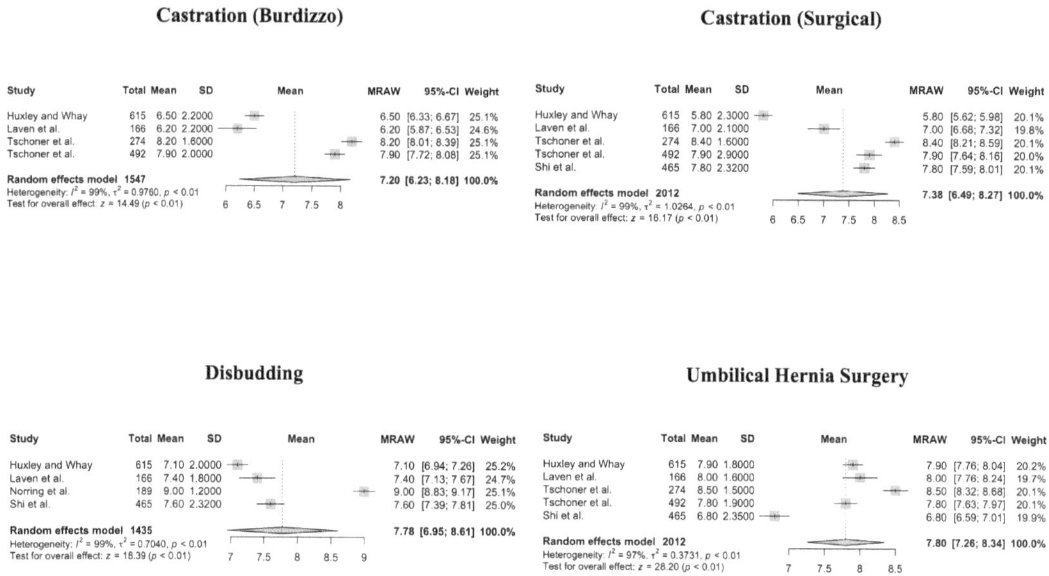

Figure 4. Forest plots for mean and SD of pain scores assigned using a Numerical Rating Scale ranging from 0 to 10 for procedures in calves. Heterogeneity of mean values was significant ($p < 0.01$) for all procedures. Test for overall effect shows a significant pain score measured for cows across studies, where no pain (pain score = 0) is the null hypothesis [9,11,12,14,16,22].

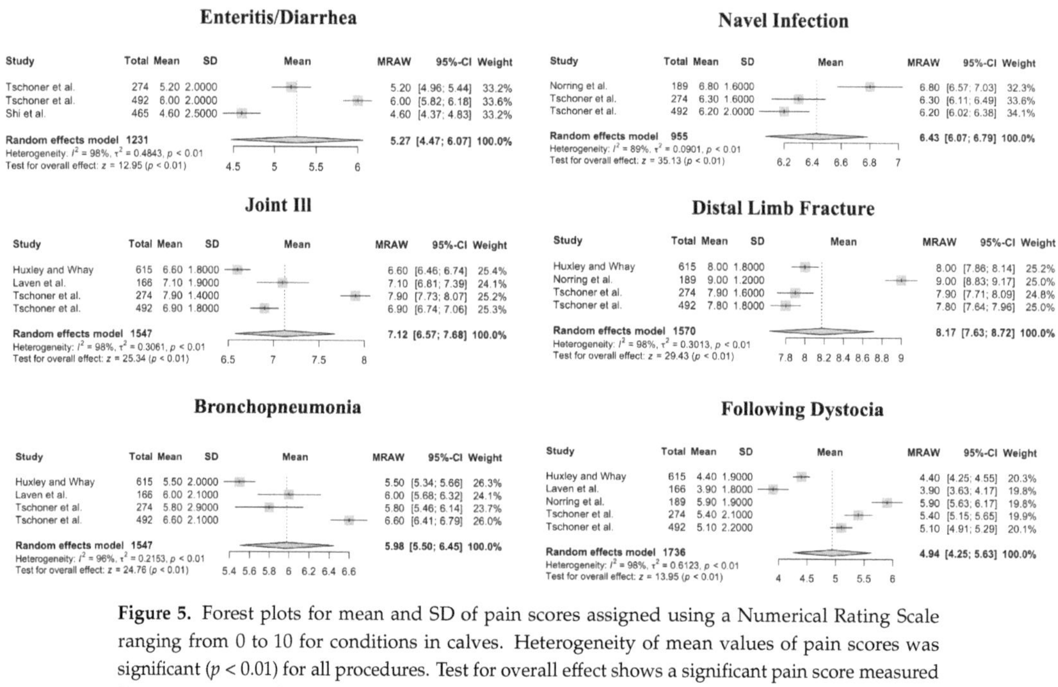

Figure 5. Forest plots for mean and SD of pain scores assigned using a Numerical Rating Scale ranging from 0 to 10 for conditions in calves. Heterogeneity of mean values of pain scores was significant ($p < 0.01$) for all procedures. Test for overall effect shows a significant pain score measured for cows across studies, where no pain (pain score = 0) is the null hypothesis [9,11,12,14,16,22].

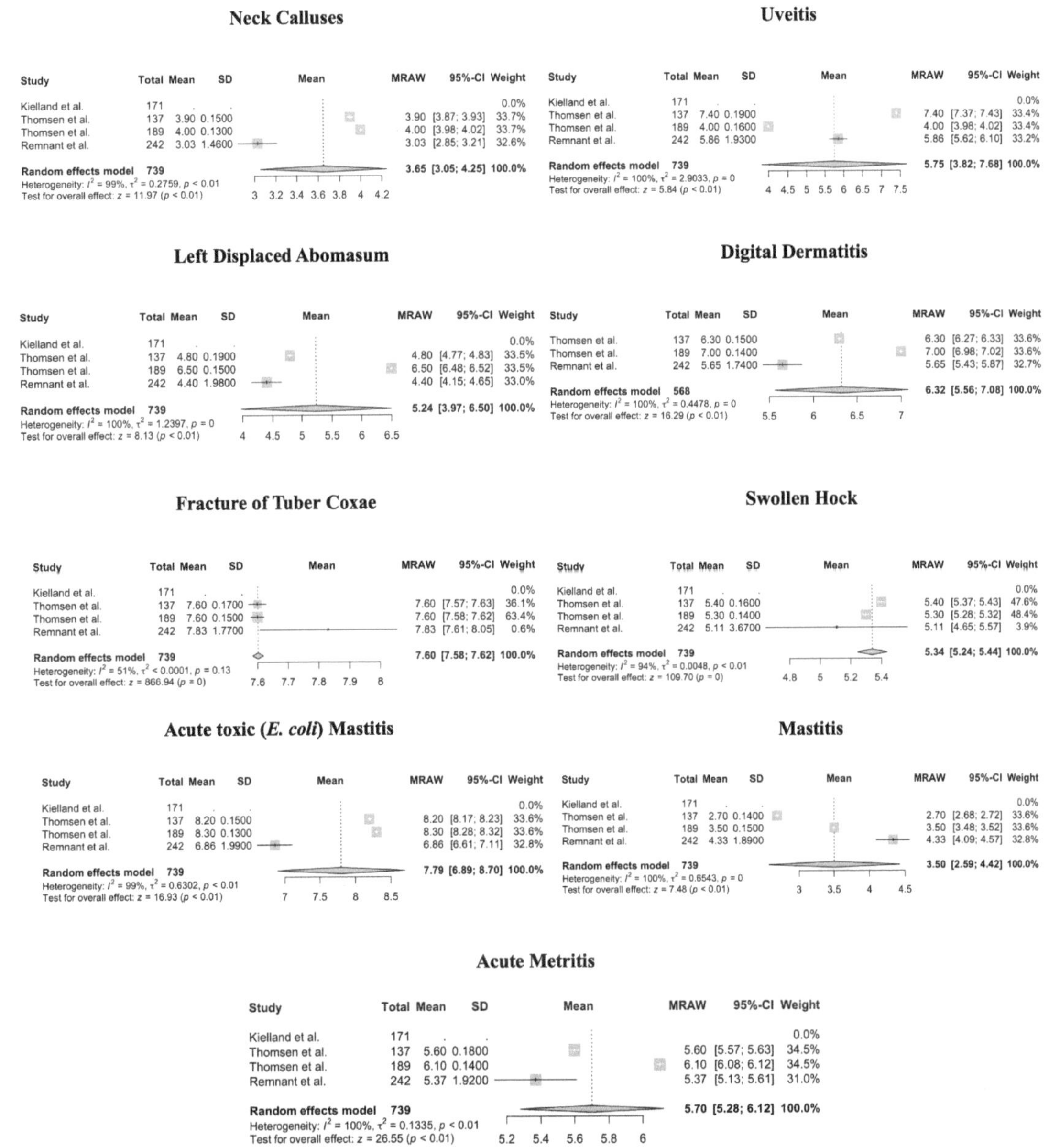

Figure 6. Forest plots for mean and SD of pain scores assigned using a Numerical Rating Scale ranging from 1 to 10 for conditions in cattle. Heterogeneity of mean values of pain scores was significant ($p < 0.01$) for acute metritis, acute toxic (*E. coli)* mastitis, neck calluses, and swollen hock. Test for overall effect shows a significant pain score measured for cows across studies, where no pain (pain score = 0) is the null hypothesis [1,13,15].

4. Discussion

4.1. Findings of the Systematic Review

The objective of the present systematic review was to describe and compare pain scores assigned by veterinarians, veterinary students, and farmers to different procedures and conditions in cattle. We wanted to assess the existing body of research, indicating areas where knowledge could be increased. Pain assessment is influenced by sex and age, and there are significant differences in pain scores given for some painful procedures and conditions between countries.

Even with a large number (n = 842) of articles extracted from the databases, there was a manageable number (n = 99) of references left after removal of duplicates and screening of titles, resulting in a small number (n = 16) of references included in this systematic review after the full-text screening. This provides evidence that research about pain assessment in cattle using NRS or VAS is still rare. Out of the 39 articles retrieved for full-text screening, 13 were only excluded due to not working with a NRS or VAS but would have otherwise fulfilled all other criteria. This is a limitation of the present study, as pain assessment was conducted in these articles, even if not by using NRS or VAS, but agreement with either "Yes/No" [34] or other predetermined statements [35,36], or pain scales ranging from five (e.g., "not important" to "extremely important" [37], "not painful" to "severe pain", including "cannot assess" [38]) to six ("no pain" to "worst pain imaginable" [39,40]) categories.

Another major limitation of comparing the studies included in this systematic review is the difference in the pain scales that were used, making direct comparison of pain scores between studies impossible. A total of eight studies worked with a NRS ranging from 0 to 10; in seven studies, NRS ranged from 1 to 10; and two studies used either both NRS and VAS, or only the VAS. Kielland et al. (2009) [15] compared median pain scores assigned by veterinary students using either a NRS (1–10) or a VAS (0–10), finding that students assigned a score that was 0.9 higher via NRS, which correlated with the different ranges of the two scales. In veterinary medicine, the VAS is described to be more informative than the NRS [17]. Nevertheless, VAS was only used in 2 out of 16 studies. According to a systematic review, correlation between VAS and NRS is good in human medicine, with some discrepancies depending on the situation [41]. Most literature in human medicine only compares different pain scales, such as NRS, VAS, and others [41,42], and not different scales of NRS. For comparability between studies, use of the same pain scale would be advisable.

Another problem of comparing the median or mean pain scores of the included articles is the different professions of participants. For the majority of studies, the questionnaire was sent to veterinarians, but other studies compared pain scoring between veterinarians and farmers [13], or veterinarians, farmers, and claw trimmers [30]. However, as median or mean pain scoring was provided for each profession in these articles, comparability between professions and with other articles was given. One other study combined groups of professions, e.g., veterinarians and veterinary students [22], which could have influenced the pain scoring, as veterinary students have less experience with managing animals in pain and with assessing painful conditions and procedures. Another study combined different professions (frontline staff, managers, veterinarians working with cattle) as dairy practitioners [16]. Research shows that pain perception differs between farmers and veterinarians [13], as well as between veterinarians, farmers, and claw trimmers [30]. Even if profession of Chinese dairy practitioners had no influence on pain scoring [16], and pain scores did not differ significantly between Bavarian veterinarians and farmers, there were differences in the perception of the painfulness of conditions and procedures [11,14]. Therefore, the combination of different professional groups for pain assessment could have an influence on study results, and professional groups should be evaluated individually.

The distribution between participating sexes was different throughout the studies, with a higher proportion of male participants compared with female participants for 10 out of 16 studies. The proportion of female participants was only higher for three studies [15,22,33]. In the two articles distributing the survey among veterinary students, as many as 81% [15] and 91% [22] respondents were female, which could be explained by the fact that nearly 80% of veterinary students are women [43]. According to Irvine et al. [43], about half of practicing veterinarians are female, but women are outnumbered by men in food animal practice, contrary to small animal and equine practice [44], which could explain the uneven distribution of genders throughout the studies. Gender distribution could also be influenced by profession; in surveys distributed to other professions, male participants accounted for 85% [19], 89.6% [30], and 79.5% [14] for Danish, Swiss, and Bavarian farmers, respectively; 100% for Swiss claw trimmers [30]; and 90.1% for Chinese practitioners, which included veterinarians as well as frontline staff [16]. As female participants ranked cattle pain higher [1,9,11,12,31], this imbalance in the gender of the participants could have an influence on the pain scores.

Age and year of graduation seem to have an influence on pain assessment, with higher pain scores awarded by more recent graduates [1,9,12]. The publication range of the articles was from 2006 to 2022. Recognition of pain in cattle has been lagging behind that in companion animals and horses, with pain scoring systems for cattle only published in recent years [5,45], which could explain the higher awareness for pain in cattle by recently graduated veterinarians. Only one study compared pain scores awarded by UK veterinarians to a study published ten years before [1], and found that pain perception in cattle veterinarians increased since the study conducted in 2006 [9], with higher pain scores given to over 40% of the listed procedures and conditions [1]. However, no other studies compared data over a period of time, so no other statement can be made about the development of pain assessment for this systematic review.

Another factor that should be considered is the nationality of participants. In the European Union, there is no species-specific legislation for dairy cattle welfare, except for calves [46–48]. Regulatory regimes of countries are not always in accordance with perspectives of veterinarians [33]. Most studies were conducted in Europe, except for five studies originating from Canada [10], New Zealand [12,33], Brazil [31], and China [16]. According to van Dyke et al. [33], demographic effects influence the perceptions of pain management in NZ veterinarians. Therefore, nationality, as well as origin of participants, and different opinions and attitudes towards pain in cattle, could likely have influenced the pain scoring. For example, on a NRS from 0 to 10, umbilical hernia surgery was scored with a median of 8 for UK and NZ veterinarians [9,12], and 9 for Bavarian veterinarians [11], and a mean of 6.8 for Chinese practitioners [16]; on a NRS from 1 to 10, umbilical hernia surgery was scored with a mean of 7.3 for Canadian veterinarians [10].

Nomenclature, as well as procedures and conditions presented to participants of the surveys, was heterogenous. Additionally, definitions of procedures and conditions differed throughout the studies. For example, mastitis was given as either clots in milk only for [9,12,15,19], which was defined as chronic mastitis by [11,14]. Other authors asked for pain assessment for grade 1 mastitis [1], mastitis [13], serious mastitis [15,19], or mild, moderate, and severe mastitis [16]. Another condition for assessment was acute toxic *Escherichia coli* mastitis [9], acute toxic mastitis [1,12], *Escherichia coli* mastitis [13], acute mastitis (*Escherichia coli*) [11,14], or acute mastitis with 41 °C of fever, lumps in milk, and a hard udder [22]. These different definitions of either conditions or procedures make comparison of pain scales throughout studies complicated. Translation of terms from the original language of the survey into English, as well as presenting different terms to farmers and veterinarians, as was done in one study (e.g., laparotomy for veterinarians, and omentopexy of displaced abomasum and laparotomy for farmers) [11], is another factor influencing the uniformity of nomenclature. Especially when including farmers or claw trimmers in the survey, authors might have chosen to use lay terms for conditions and procedures to make sure those can be understood by the participants.

Return rates differed widely between studies, from as high as 70% for Norwegian farmers [19] to 15.4% in Bavarian farmers [14]. It is reasonable to think that people interested in pain management in cattle, as well as empathic people, are more likely to participate in a survey about pain. The level of empathy of a human being towards an animal might be influenced by the species of the animal the observer evaluates. The concern for the welfare of an animal could result from the evolution of the human trait, which can be strongly influenced by culture [49]. Including only participants who responded to the survey is a possible bias. As empathic veterinarians were found to give higher pain scores [22], empathy and attitudes towards pain could have an influence on the findings in the articles included in this systematic review. None of the authors stated if they were working with a reward system for participating in the studies, which is another factor that could have influenced the results.

Only one study included assessment of behavioral and postural parameters used for pain recognition in cattle, showing that veterinarians and farmers differed significantly in the parameters they use for pain assessment for 19 out of 28 parameters presented in the survey [14]. Given the wide variety of pain scores assessed for different procedures and conditions, questions about tools and methods for pain assessment, or evaluation of parameters used to recognize if cattle are in pain, it would have been interesting to determine in studies if there is a lack in education concerning the recognition of pain. Improvement in pain recognition could result in a higher awareness of cattle being in pain, and in higher pain scores given to cattle.

4.2. Findings of the Meta-Analysis

Meta-analysis was performed for six articles for NRS 0–10 and five articles for NRS 1–10. This small number of articles included in the meta-analysis is due to the fact that we only included pain scores of procedures and conditions included in at least three articles. For the meta-analysis, mean values and SD of pain scores were used; however, in most studies, median values were presented [9,12], as data were not normally distributed, which was not optimal. If mean values and SD were not given in articles, authors were contacted to collect missing data, but only three authors answered to the mails. Data describing pain scores for NRS 1–10 were difficult to interpret, due to the missing SD for the majority of included studies. Therefore, these results should not be relied on, which is a limitation of the meta-analysis. Even if the study design was somewhat similar throughout the studies, heterogeneity of the articles was large, with significant differences between the mean pain scores for all procedures and conditions included for NRS 0–10. Even though individual studies found differences in pain assessment according to profession [13,30], our meta-analysis found no significant differences between mean pain scores, which is in accordance with other studies [14,16]. As articles were heterogenous and SD was not provided for some articles, and professions differed between articles, the results of the present meta-analysis represent insufficiently strong conclusions. That is why we compared the means of pains scores (without SDs) of professions via Kruskal–Wallis, to complement the meta-analysis. However, the low number of studies for a particular condition and the wide discrepancies in pain scores between professions and articles, even for the same diseases, hint at the lack of knowledge in this area and the need to collect more data and conduct additional research to close the knowledge gap. Moreover, the pain scales themselves should be unified to one scoring system instead of the two ranging either from 0 to 10 or 1 to 10. Such unification of pain scores and clarifying the pain heterogeneity is of a huge practical importance for veterinarians and researchers to be able to compare the pain scores given for different procedures and conditions. Therefore, the recommendation should be for researchers (a third person assessing pain in animals in theory without actually looking at a patient) as well as clinical practice (people assessing pain of a patient in their care) to use the same scales for better comparability.

The benefits of conducting studies about pain assessment are collecting data about evaluation of pain and pain management, and learning about areas where more education is needed. Another benefit could be that the individual participant is working through a pain assessment questionnaire, thinking about painful events in a cow, and realizing the number of painful procedures and conditions in a calf's or cow's life, possibly thus exercising empathy towards the animal. As empathic veterinarians score pain in cattle higher [22], increasing empathy towards adult cows and calves could result in improved welfare and pain management for cattle.

4.3. Use of Analgesics

The presentation and description of the use of analgesics was so heterogenous in the papers that an assessment and comparison of analgesic use was not possible. There were only ten papers describing analgesic use overall. Huxley and Whay [9] presented the percentage of analgesic classes used for different procedures and conditions, Hewson et al. [10] described the mean percentage of animals (beef and dairy) receiving any kind of analgesia, including the top two active components, and Becker et al. [30] only reported the percentage of respondents stating if a local anesthesia was reasonable for procedures involving the claw. Lorena et al. [31] presented use of different analgesic classes for cattle and horses combined, and Norring et al. [22] asked how many veterinarians and clinical students would use a combination of sedation, local anesthesia, and analgesia for disbudding in calves. Remnant et al. [1] included a figure overlaying respondents using NSAIDs in 50% of cases over the stated pain scores. Tschoner et al. [11] divided the use of different classes of analgesics into categories (regularly, frequently, occasionally, never) and presented the percentage of respondents agreeing to each category, and asked about agreement if local anesthesia and NSAIDs were necessary during and after painful procedures [14]. Van Dyke et al. [33] only included pain management protocols for four procedures, and Shi et al. [16] presented a figure showing the proportion of which analgesics would be used by respondents (multi-response answer). Therefore, the description of the use of analgesics is hard. It is also not possible to compare the use of analgesics between animal species, as this depends on which analgesics are labeled for use in food-producing animals. Questions about analgesics in cattle usually refer to NSAIDs and local anesthetics. Reviews of pain management in cattle have been published [50,51]—thus, pain management will not be discussed here.

4.4. Methodology and Limitations

The present systematic review was conducted following the PRISMA guidelines [25], as described previously [26,27]. This was done to reduce the possible risk of bias for the study selection process and analysis. Registration via PROSPERO was not possible, as it can only be used for systematic reviews in human medicine and research [27]. To further reduce the risk of bias, the titles and abstracts were screened independently by three authors, and the full-text screening was undertaken using previously specified guidelines. Eligibility for a meta-analysis was discussed with a statistician, as described previously [26].

4.5. Risk of Bias

To reduce the risk of missing any articles, the authors used two search engines. Titles and abstracts were included in the keyword search, which should also result in not missing any relevant articles [27]. All articles included in the systematic search were published in English, and no article had to be excluded due to the language not being English or German. Therefore, a bias because of the language barrier can be excluded. A total of three abstracts could not be accessed; as these articles were then included in full-text retrieval, and all articles selected for full-text screening could be accessed, risk of bias due to limited access can also be excluded. The origin of articles was evenly distributed, with 16 articles originating from 10 countries. Therefore, country of origin should not have influenced the present study. Funding information was provided for thirteen of the sixteen articles, with two articles not stating any funding [15,31], and one receiving no external funding [33]. Pharmaceutical companies were involved in the funding of seven of sixteen articles; however, as the articles focused on pain assessment using pain scales, this is unlikely to have had any influence on the results of the study.

5. Conclusions

This systematic review should aid researchers to identify gaps in the current knowledge for conceptualization of objectives and the study design for future research.

Studies of pain assessment using NRS or VAS are rare in bovine medicine, and use of pain scales is heterogenous, making comparison of pain scores difficult. There are many variables possibly influencing pain assessment, such as gender, age, education, or profession. Studies mainly originate from the European Union, and research about pain assessment in other countries should be conducted, or published if they are performed. Researchers should focus on using one pain scale throughout studies for better comparability, since there seem to be no clear benefits of using less common pain scales over the commonly used 0 to 10 NRS scale. Additionally, the nomenclature of terms should be consistent, and pain scoring conducted by different professions should be assessed individually. We recommend researchers assess behavioral and postural parameters used by veterinarians and farmers to assess pain in cattle, to evaluate if pain assessment can be improved by training and education, thus improving dairy cattle welfare. Future studies could compare bovine pain scales, changes in perceptions of pain levels after a period of clinical training of respondents, and coherent assessment of use of analgesics in cattle. Assessment of pain should not be performed under the assumption that no pain medication was given, as this is not feasible anymore, especially for surgical procedures.

Supplementary Materials: The following supporting information can be downloaded at: https://www.mdpi.com/article/10.3390/ani14020351/s1, Table S1: PRISMA-P checklist for the systematic review "Substance P concentrations in adult cattle and calves during different painful procedures and conditions—a systematic review", Table S2: Findings of Material and Methods section for 16 references about the assessment of pain in cattle, including funding information. Table S3: Pain scoring for different procedures presented to participants of surveys about pain assessment in adult cattle. Table S4: Pain scoring for different conditions presented to participants of surveys about pain assessment in adult cattle; Table S5: Pain scoring for different procedures presented to participants of surveys about pain assessment in calves. Table S6: Pain scoring for different conditions presented to participants of surveys about pain assessment in calves.

Author Contributions: Conceptualization, T.T., K.R.M. and M.F.; methodology, T.T., K.R.M. and M.F; formal analysis, T.T. and Y.Z.; data curation, T.T. and Y.Z.; writing—original draft preparation, T.T.; writing—review and editing, T.T., K.R.M. and M.F.; visualization, T.T. and Y.Z. All authors have read and agreed to the published version of the manuscript.

Funding: Theresa Tschoner is funded by the Deutsche Forschungsgesellschaft (DFG), grant number 505835300.

Institutional Review Board Statement: Not applicable.

Informed Consent Statement: Not applicable.

Data Availability Statement: The data lies with the corresponding author.

Conflicts of Interest: The authors declare no conflicts of interest.

References

1. Remnant, J.G.; Tremlett, A.; Huxley, J.N.; Hudson, C.D. Clinical attitudes to pain and use of analgesia in cattle—Where are we 10-years on? *Vet. Rec.* **2017**, *181*, 400. [CrossRef] [PubMed]
2. Hudson, C.; Whay, H.; Huxley, J. Recognition and managment of pain in cattle. *Practice* **2008**, *30*, 126–134. [CrossRef]
3. Cockcroft, P.D. Pain Managment in Cattle Practice. In *Bovine Medicine*; Cockcroft, P.D., Ed.; Wiley Blackwell: West Sussex, UK, 2015; Volume 3, pp. 238–245.
4. Reader, J. Clinicians' attitudes to pain and the use of anagesia in cattle. *Vet. Rec.* **2017**, *181*, 397–399. [CrossRef] [PubMed]
5. Gleerup, K.B.; Andersen, P.H.; Munksgaard, L.; Forkman, B. Pain evaluation in dairy cattle. *Appl. Anim. Behav. Sci.* **2015**, *171*, 25–32. [CrossRef]
6. Coria-Avila, G.A.; Pfaus, J.G.; Orihuela, A.; Domínguez-Oliva, A.; José-Pérez, N.; Hernández, L.A.; Mota-Rojas, D. The neurobiology of behavior and its applicability for animal welfare: A review. *Animals* **2022**, *12*, 928. [CrossRef] [PubMed]

7. Fraser, A.F.; Broom, D.M. Describing, recording and measuring behaviour. In *Farm Animal Behaviour and Welfare*, 3rd ed.; Fraser, A.F., Broom, D.M., Eds.; CAB International: Wallingford, UK, 1990; pp. 7–16.
8. Johnson, C.B.; Gibson, T.J.; Flint, P.; Wilson, P.W.; Mellor, D.J. New techniques for pain recognition: What are the applications, where are the limits? In Proceedings of the Australian Animal Welfare Strategy International Conference, Gold Coast, QLD, Australia, 31 August–3 September 2008.
9. Huxley, J.; Whay, H. Current attitudes of cattle practitioners to pain and the use of analgesics in cattle. *Vet. Rec.* **2006**, *159*, 662–668. [CrossRef] [PubMed]
10. Hewson, C.J.; Dohoo, I.R.; Lemke, K.A.; Barkema, H.W. Canadian veterinarian's use of analgesics in cattle, pigs and horses in 2004 and 2005. *Can. Vet. J.* **2007**, *48*, 155–164.
11. Tschoner, T.; Peinhofer, V.C.; Sauter-Louis, C.; Feist, M. Attitudes of Bavarian bovine veterinarians towards pain and pain management in cattle. *Vet. Rec.* **2020**, *187*, e90. [CrossRef]
12. Laven, R.A.; Huxley, J.N.; Whay, H.R.; Stafford, K.J. Results of a survey of attitudes of dairy veterinarians in New Zealand regarding painful procedures and conditions in cattle. *N. Z. Vet. J.* **2009**, *57*, 215–220. [CrossRef]
13. Thomsen, P.T.; Anneberg, I.; Herskin, M.S. Differences in attitudes of farmers and veterinarians towards pain in dairy cows. *Vet. J.* **2012**, *194*, 94–97. [CrossRef]
14. Tschoner, T.; Sauter-Louis, C.; Peinhofer, V.; Feist, M. Exploring the attitudes of Bavarian farmers towards pain in cattle and how they differ from the attitudes of bovine veterinarians. *Vet. Rec.* **2021**, *189*, e515. [CrossRef] [PubMed]
15. Kielland, C.; Skjerve, E.; Zanella, A.J. Attitudes of veterinary students to pain in cattle. *Vet. Rec.* **2009**, *165*, 254–258. [CrossRef] [PubMed]
16. Shi, R.; Shu, H.; Yu, R.; Wang, Y.; Zhang, Z.; Zhang, J.; Gu, X. Current Attitudes of Chinese Dairy Practitioners to Pain and Its Management in Intensively Raised Dairy Cattle. *Animals* **2022**, *12*, 3140. [CrossRef] [PubMed]
17. Kielland, C.; Ruud, L.; Zanella, A.; Østerås, O. Prevalence and risk factors for skin lesions on legs of dairy cattle housed in freestalls in Norway. *J. Dairy Sci.* **2009**, *92*, 5487–5496. [CrossRef] [PubMed]
18. Williamson, A.; Hoggart, B. Pain: A review of three commonly used pain rating scales. *J. Clin. Nurs.* **2005**, *14*, 798–804. [CrossRef] [PubMed]
19. Kielland, C.; Skjerve, E.; Østerås, O.; Zanella, A.J. Dairy farmer attitudes and empathy toward animals are associated with animal welfare indicators. *J. Dairy Sci.* **2010**, *93*, 2998–3006. [CrossRef] [PubMed]
20. Wikman, I.; Hokkanen, A.-H.; Pastell, M.; Kauppinen, T.; Valros, A.; Hänninen, L. Dairy producer attitudes to pain in cattle in relation to disbudding calves. *J. Dairy Sci.* **2013**, *96*, 6894–6903. [CrossRef]
21. Haefeli, M.; Elfering, A. Pain assessment. *Eur. Spine J.* **2006**, *15*, S17–S24. [CrossRef]
22. Norring, M.; Wikman, I.; Hokkanen, A.-H.; Kujala, M.V.; Hänninen, L. Empathic veterinarians score cattle pain higher. *Vet. J.* **2014**, *200*, 186–190. [CrossRef]
23. Hugonnard, M.; Leblond, A.; Keroack, S.; Cadore, J.L.; Troncy, E. Attitudes and concerns of French veterinarians towards pain and analgesia in dogs and cats. *Vet. Anaesth. Analg.* **2004**, *31*, 154–163. [CrossRef]
24. Tschoner, T. Methods for Pain Assessment in Calves and Their Use for the Evaluation of Pain during Different Procedures—A Review. *Animals* **2021**, *11*, 1235. [CrossRef] [PubMed]
25. Shamseer, L.; Moher, D.; Clarke, M.; Ghersi, D.; Liberati, A.; Petticrew, M.; Shekelle, P.; Stewart, L.A. Preferred reporting items for systematic review and meta-analysis protocols (PRISMA-P) 2015: Elaboration and explanation. *BMJ* **2015**, *349*, g7647. [CrossRef] [PubMed]
26. Oehm, A.W.; Knubben-Schweizer, G.; Rieger, A.; Stoll, A.; Hartnack, S. A systematic review and meta-analyses of risk factors associated with lameness in dairy cows. *BMC Vet. Res.* **2019**, *15*, 346. [CrossRef] [PubMed]
27. Tschoner, T.; Feist, M. Substance P concentrations in the blood plasma and serum of adult cattle and calves during different painful procedures and conditions—A systematic review. *BMC Vet. Res.* **2022**, *18*, 232. [CrossRef]
28. Winder, C.B.; Miltenburg, C.L.; Sargeant, J.M.; LeBlanc, S.J.; Haley, D.B.; Lissemore, K.D.; Godkin, M.A.; Duffield, T.F. Effects of local anesthetic or systemic analgesia on pain associated with cautery disbudding in calves: A systematic review and meta-analysis. *J. Dairy Sci.* **2018**, *101*, 5411–5427. [CrossRef]
29. Mason, W.; Cuttance, E.; Müller, K.; Huxley, J.; Laven, R. Graduate Student Literature Review: A systematic review on the associations between nonsteroidal anti-inflammatory drug use at the time of diagnosis and treatment of claw horn lameness in dairy cattle and lameness scores, algometer readings, and lying times. *J. Dairy Sci.* **2022**, *105*, 9021–9037. [PubMed]
30. Becker, J.; Reist, M.; Friedli, K.; Strabel, D.; Wüthrich, M.; Steiner, A. Current attitudes of bovine practitioners, claw-trimmers and farmers in Switzerland to pain and painful interventions in the feet in dairy cattle. *Vet. J.* **2013**, *196*, 467–476. [CrossRef] [PubMed]
31. Lorena, S.E.; Luna, S.P.; Lascelles, B.D.; Corrente, J.E. Attitude of Brazilian veterinarians in the recognition and treatment of pain in horses and cattle. *Vet. Anaesth. Analg.* **2013**, *40*, 410–418. [CrossRef]
32. Hokkanen, A.-H.; Wikman, I.; Korhonen, T.; Pastell, M.; Valros, A.; Vainio, O.; Hänninen, L. Perceptions and practices of Finnish dairy producers on disbudding pain in calves. *J. Dairy Sci.* **2015**, *98*, 823–831. [CrossRef]
33. van Dyke, R.; Connor, M.; Miele, A. An Investigation into the Perceptions of Veterinarians towards Perioperative Pain Management in Calves. *Animals* **2021**, *11*, 1882. [CrossRef]

34. O'Callaghan Lowe, K.A.; Murray, R.D.; Cripps, P.J.; Ward, W.R. Working practices of cattle foot trimmers used for footcare in dairy cattle compared with those of veterinary surgeons for treatment of lameness in large animal practice. *J. Vet. Med. A Physiol. Pathol. Clin. Med.* **2004**, *51*, 429–434. [CrossRef] [PubMed]
35. Johnstone, E.C.S.; Coetzee, J.F.; Pinedo, P.J.; Edwards-Callaway, L. Current attitudes of veterinarians and producers regarding the use of local and systemic analgesia in beef and dairy cattle in the United States. *J. Am. Vet. Med. Assoc.* **2021**, *258*, 197–209. [CrossRef] [PubMed]
36. Canozzi, M.E.A.; Borges, J.A.R.; Barcellos, J.O.J. Attitudes of cattle veterinarians and animal scientists to pain and painful procedures in Brazil. *Prev. Vet. Med.* **2020**, *177*, 104909. [CrossRef] [PubMed]
37. Leach, K.A.; Whay, H.R.; Maggs, C.M.; Barker, Z.E.; Paul, E.S.; Bell, A.K.; Main, D.C. Working towards a reduction in cattle lameness: 2. Understanding dairy farmers' motivations. *Res. Vet. Sci.* **2010**, *89*, 318–323. [CrossRef] [PubMed]
38. Staněk, S.; Šárová, R.; Nejedlá, E.; Šlosárková, S.; Doležal, O. Survey of disbudding practice on Czech dairy farms. *J. Dairy Sci.* **2018**, *101*, 830–839. [CrossRef]
39. Edwards-Callaway, L.N.; Keller, K.P.; Oselinsky, K.; Johnstone, E.; Cramer, C.; Román-Muñiz, N.; Stallones, L.; Coetzee, J.F. A nationwide survey on producer and veterinarian perceptions of the painfulness of procedures and disease states in dairy and beef cattle. *Front. Pain Res.* **2023**, *4*, 1059224. [CrossRef]
40. Mijares, S.; Edwards-Callaway, L.; Roman-Muniz, I.N.; Coetzee, J.F.; Applegate, T.J.; Cramer, M.C. Veterinarians' perspectives of pain, treatment, and diagnostics for bovine respiratory disease in preweaned dairy calves. *Front. Pain Res.* **2023**, *4*, 1076100. [CrossRef]
41. Karcioglu, O.; Topacoglu, H.; Dikme, O.; Dikme, O. A systematic review of the pain scales in adults: Which to use? *Am. J. Emerg. Med.* **2018**, *36*, 707–714. [CrossRef]
42. Ahlers, S.J.; van Gulik, L.; van der Veen, A.M.; van Dongen, H.P.; Bruins, P.; Belitser, S.V.; de Boer, A.; Tibboel, D.; Knibbe, C.A. Comparison of different pain scoring systems in critically ill patients in a general ICU. *Crit. Care* **2008**, *12*, R15. [CrossRef]
43. Irvine, L.; Vermilya, J.R. Gender work in a feminized profession: The case of veterinary medicine. *Gend. Sic.* **2010**, *24*, 56–82. [CrossRef]
44. Gilliam, G.; Fajt, V.; Wagner, S.; White, B.; Apley, M. Perceptions of gender bias among members of the American Association of Bovine Practitioners in bovine practice in the United States in 2018. *Bov. Pract.* **2021**, *55*, 98–103. [CrossRef]
45. de Oliveira, F.A.; Luna, S.P.L.; do Amaral, J.B.; Rodrigues, K.A.; Sant'Anna, A.C.; Daolio, M.; Brondani, J.T. Validation of the UNESP-Botucatu unidimensional composite pain scale for assessing postoperative pain in cattle. *BMC Vet. Res.* **2014**, *10*, 200. [CrossRef] [PubMed]
46. Nalon, E.; Stevenson, P. Protection of dairy cattle in the EU: State of play and directions for policymaking from a legal and animal advocacy perspective. *Animals* **2019**, *9*, 1066. [CrossRef] [PubMed]
47. TierSchNutztV. Verordnung zum Schutz Landwirtschaftlicher Nutztiere und Anderer zur Erzeugung Tierischer Produkte Gehaltener Tiere bei ihrer Haltung (Tierschutz-Nutztierhaltungsverordnung-TierSchNutztV). Available online: https://www.gesetze-im-internet.de/tierschnutztv/BJNR275800001.html (accessed on 2 January 2024).
48. Council Directive 2008/119/EC. Available online: https://eur-lex.europa.eu/eli/dir/2008/119/oj (accessed on 2 January 2024).
49. Bradshaw, J.W.; Paul, E.S. Could empathy for animals have been an adaptation in the evolution of Homo sapiens? *Anim. Welf.* **2010**, *19*, 107–112. [CrossRef]
50. Mota-Rojas, D.; Velarde, A.; Marcet-Rius, M.; Orihuela, A.; Bragaglio, A.; Hernández-Ávalos, I.; Casas-Alvarado, A.; Domínguez-Oliva, A.; Whittaker, A.L. Analgesia during parturition in domestic animals: Perspectives and controversies on its use. *Animals* **2022**, *12*, 2686. [CrossRef]
51. Edmondson, M.A. Local, Regional, and Spinal Anesthesia in Ruminants. *Vet. Clin. N. Am. Food Anim. Pract.* **2016**, *32*, 535–552. [CrossRef]

animals

MDPI

Article

Tramadol and M1 Bioavailability Induced by Metamizole Co-Administration in Donkeys (*Equus asinus*)

Gabriel Araújo-Silva [1], Luã B. de Macêdo [2], Andressa N. Mouta [2], Maria Gláucia C. de Oliveira [2], Kathryn N. Arcoverde [2], Lilian G. S. Solon [3], José T. Perez-Urizar [4] and Valéria V. de Paula [2,*]

1 Organic Chemistry and Biochemistry Laboratory, Universidade do Estado do Amapá, UEAP, Avenida Presidente Vargas, 650, Macapá 68900-070, Brazil; gabriel.silva@ueap.edu.br
2 Department of Animal Science, Universidade Federal Rural do Semi-Árido, 572, Rua Francisco Mota, Presidente Costa e Silva, Mossoró 59625-900, Brazil; luanb.macedo27@gmail.com (L.B.d.M.); andressanmouta@hotmail.com (A.N.M.); glauciacarlos@hotmail.com (M.G.C.d.O.); kathrynnobrega@gmail.com (K.N.A.)
3 Post-Graduate Program in Pharmaceutical Sciences, PPGCF, Universidade Federal do Amapá, UNIFAP, Rod Juscelino Kubitschek, Km2, Macapá 68903-419, Brazil; liliansolon@yahoo.com.br
4 Facultad de Ciencias Químicas, Universidad Autónoma de San Luis Potosí, 6, Avenida Doutor Manuel Nava, Zona Universitaria, San Luis 78210, Mexico; jpurizar@uaslp.mx
* Correspondence: valeria@ufersa.edu.br; Tel.: +55-8499942-0775

Simple Summary: The combination of an opioid compound with a nonsteroidal anti-inflammatory drug can lead to additive or enhanced effects while minimizing adverse reactions. Recently, we reported the pharmacokinetic profiles of metamizole and tramadol in donkeys at single doses and without association. However, no studies have reported on the pharmacokinetic profile of the combination of tramadol and metamizole. The objective of this research was to assess pharmacokinetic profile of metamizole co-administered with tramadol at single dose. Behavioral changes in the animals were observed, and at specific intervals, blood samples were taken for subsequent analysis. Analyses were performed using ultra-high-performance liquid chromatography-tandem mass spectrometry. The findings indicate that tramadol and its metabolite presented modified profiles, implying that metamizole and tramadol interact and affect each other's metabolic processes at the dosages used in this study. Clinical researches are necessary to determine the optimal that effectively addresses the pain relief requirements of the species.

Abstract: Our objective was to assess the pharmacokinetic characteristics of metamizole when administered together with tramadol in a single intravenous dose to donkeys. Ten male animals received 10 mg·kg^{-1} of dipyrone associated with 2 mg·kg^{-1} of tramadol (T_2M_{10}) and 25 mg·kg^{-1} of dipyrone with 2 mg·kg^{-1} of tramadol (T_2M_{25}). Venous blood samples were taken from groups to determine the pharmacokinetics after drug administration, using initial brief intervals that were followed by extended periods until 48 h. Restlessness and ataxia were observed in two animals in the T_2M_{25} group. Analysis revealed prolonged detectability of tramadol, 4-methylamine antipyrine, 4-aminoantipyrine (up to 24 h), and O-desmethyltramadol (up to 12 h) after administration. Although metamizole and its metabolites showed no significant pharmacokinetic changes, tramadol and O-desmethyltramadol exhibited altered profiles, likely because of competition for the active sites of CYP450 enzymes. Importantly, the co-administration of metamizole increased the bioavailability of tramadol and O-desmethyltramadol in a dose-dependent manner, highlighting their potential interactions and emphasizing the need for further dose optimization in donkey analgesic therapies. In conclusion, metamizole co-administered with tramadol interferes with metabolism and this interference can change the frequency of drug administration and its analgesic efficacy.

Keywords: pharmacokinetic profile; drug interaction; analgesics; metabolites

1. Introduction

Donkeys are animals primarily used for manual labor in developing countries, where they transport materials and people in challenging conditions such as extreme heat and humidity, noise, and urban pollution. They are often exposed to the risk of accidents owing to heavy traffic [1,2]. These factors make donkeys more susceptible to trauma and other pathologies that can result in pain and increased stress levels.

Metamizole (MET), commonly known as dipyrone, is often categorized as a pyrazolone derivative, functioning as a non-opioid analgesic, and is of particular importance to veterinarians in the field of equine care because it presents good analgesic effects and excellent antipyretic properties, and is recommended for managing colic syndromes, alleviating muscle pain and fever, and recovery after surgical interventions [3,4]. However, it was withdrawn from markets in the United States, Japan, Iran, and the United Kingdom because of its rare side effect in humans, a hematological condition called agranulocytosis [5]. Metamizole is rapidly hydrolyzed to its metabolite 4-methylaminoantipyrine (MAA) and its by-product, 4-aminoantipyrine (AA). The pharmacological effects are attributed to MAA, which is formed in much larger amounts than other smaller metabolites are [6].

Tramadol is an opioid analgesic that is widely prescribed by veterinarians [7]. It is routinely used to treat acute and chronic pain in animals [8,9]. Pharmacokinetic research on tramadol and its primary metabolite, O-desmethyltramadol (M1), has highlighted interspecies variations in drug metabolism, underscoring the importance of conducting pharmacokinetic studies to establish appropriate dosage schedules for various species [10–12].

The concurrent administration of an opioid and a pyrazolone derivative NSAID can yield synergistic benefits, enhancing the therapeutic effects while mitigating the adverse reactions associated with each medication [13]. In a previous study, 25 combinations of different doses of the combination of metamizole (56.2–562.3 $mg \cdot kg^{-1}$) and tramadol (3.2–56.2 $mg \cdot kg^{-1}$) were evaluated through a single administration, with each combination assessed for additive or potentiated anti-nociceptive effect in mice when compared to the effects of treatment using singular drugs [14].

Research on the pharmacology of numerous drugs in donkeys remains limited, highlighting a significant gap in the existing literature. The drugs used to treat these animals have been frequently developed and recommended for horses [15]. Donkeys differ from horses in behavioral, physiological, and pharmacological aspects [16]. Therefore, knowledge of the pharmacokinetic properties of drugs is fundamental to guarantee effective and safe therapeutic administration [7].

This study aimed to evaluate the pharmacokinetic profile of a single intravenous dose of metamizole co-administered with tramadol in donkeys. We hypothesized that metamizole and tramadol could compete for the same enzymes, causing changes in the concentrations of metabolites of both metamizole and tramadol.

2. Materials and Methods

2.1. Animals and Experimental Design

This study was approved by the Institutional Animal Use Ethics Committee of Universidade Federal Rural do Semi-Árido (Approval number 23091.006896/2019-47). For this study, ten adult male northeastern Brazilian donkeys aged between 2 to 14 years (mean age of 6.4 ± 3.1 years), with weights ranging from 110 to 145 kg (average weight 126 ± 11.8 kg), were selected. These donkeys were sourced from the Apodi Animal Protection Association (APA). Eligibility for participation in the study necessitated that the donkeys were deemed to be in apparent good health as determined by comprehensive physical and laboratory assessments. Physical examination was based on cardiorespiratory and abdominal auscultation, capillary refill time, mucous membrane color, and fecal characteristics. Laboratory examinations included hemograms and urea, creatinine, alanine aminotransferase, aspartate aminotransferase, and total protein plasma concentrations.

Four weeks before the study began, the animals were treated with an oral dose of ivermectin, 1 g per 100 kg of body weight (Piraverme® Lavizoo, Registro, Brazil) for deworming and vaccinated against rabies using (Rai-Vet Líquida®, Vaxxinova, São Paulo, Brazil). They were accommodated in groups of four within an outdoor enclosure measuring 17 m by 13 m, equipped with shaded areas. The diet for the donkeys consisted of 7.5 kg of Napier grass (*Pennisetum purpureum*) per 100 kg body weight and 1.1 kg of a concentrate mix (comprising ground corn, soybean, wheat bran, common salt, and calcitic limestone) provided twice daily, alongside unlimited access to water. A four-week acclimatization period was allotted for the donkeys to adjust to their new surroundings and human handling.

The day before starting the treatment, the animals were moved to individual stalls, where they were fasted for 10 h and water was withheld for six hours, respectively. Following this preparatory phase, the animals underwent two distinct treatments: in the T_2M_{10} treatment, the ten donkeys received 10 mg·kg^{-1} of metamizole (D-500®, Zoetis, São Paulo, Brazil, 500 mg/mL) associated with 2 mg·kg^{-1} tramadol (Tramadon®, Cristália—Produtos Químicos Farmacêuticos Ltda, São Paulo, Brazil, 50 mg/mL), whereas in the T_2M_{25} treatment, the same ten animals received 25 mg·kg^{-1} of metamizole (D-500®, Zoetis, São Paulo, Brazil, 500 mg/mL) associated with 2 mg·kg^{-1} tramadol (Tramadon®, Cristália Produtos Químicos Farmacêuticos Ltda, São Paulo, Brazil, 50 mg/mL), in each case delivered separately, one in each jugular vein, in a double administration of 10 mL (metamizole) and 20 mL (tramadol) made with 0.9% NaCl solution. For the intravenous (IV) administration of metamizole and tramadol, as well as for the collection of blood samples, the administration of the drugs was consistently carried out by the same individual. For the purposes of infusion, thorough antisepsis was conducted similar to surgical preparations, including skin cleaning, followed by the placement of a 16G caliber catheter attached to a 3-way tap, which was then securely inserted into both the animals' jugular veins. The administration of the drugs was achieved using two infusion pumps (Syringe Pump ST670®, Samtronic, São Paulo, Brazil) over a span of two minutes. Following administration, the animals were monitored for signs of adverse effects such as ataxia, restlessness, salivation, sweating, and muscle spasms. The occurrence and duration of these effects were evaluated. One hour after the drug administration, water was made available to the animals, and food was offered at regular intervals thereafter.

Blood samples, each amounting to 10 mL, were drawn at specific intervals: immediately prior to the administration of the drug (0 or baseline); at 5, 10, 20, 30, 40, and 50 min; and then at 1.0, 1.15, 1.3, 1.45, 2.0, 2.5, 3.0, 4.0, 6.0, 8.0, 12.0, 24.0, and 48.0 h following the administration. These samples were promptly placed into tubes containing ethylenediaminetetraacetic acid (EDTA). Within 30 min of collection, plasma was extracted by centrifuging the samples at 1715× *g* for 10 min at ambient temperature. The separated plasma was then frozen and stored at −80 °C in cryogenic vials until further analysis.

2.2. Sample Extraction Procedure

The aliquots of the plasma samples (250 µL) were supplemented with 10 µL of 0.1 mg/mL metoprolol solution (internal standard) and 800 µL of acetonitrile, followed by vortex homogenization for 60 s, and then the samples were centrifuged for 5 min at 14,200 rpm. The supernatant (900 µL) was transferred to vials for injection in the chromatographic equipment.

2.3. Instrumentation and LC and MS Conditions

For the chromatographic analysis, the ultra-performance liquid chromatography system, which was coupled with mass spectrometry (UHPLC-MS/MS) system was used, consisting of Nexera 2 UHPLC coupled to an LCMS-8040 mass spectrometry detector (Shimadzu®, Kyoto, Japan) and BEH C18 column (1.7 µm, 2.1 × 75 mm) (Shimadzu®, Kyoto, Japan). The mobile phase consisted of acetonitrile and 0.1% formic acid solution (75:25, *v*/*v*) at 0.3 mL/min. The running time was 2.0 min; the volume of sample injection was 5.0 µL. The column temperature was adjusted to 40 °C and the automatic sampler refrigerator was set to 5 °C. For tramadol, M1, MAA, and AA, the mass spectrometer was adjusted in the multiple reaction monitoring (MRM) mode, utilizing positive electrospray ionization (ESI). The collision energy and cone voltage were 12 V and 19 V, respectively. The flow rates of the cone gas and desolvation were calibrated to 150 L/min and 600 L/min, respectively, using Argon as the collision gas at a flow rate of 0.15 mL/min. A mass spectrometer was configured to monitor the transitions of the main ion and fragment ion ranges, the mass and mass-to-charge ratio (m/z) of the main ions for tramadol, M1, metoprolol, MAA and AA was 264.0, 250.0, 268.1, 218.20, 204.20, respectively, whereas the corresponding transitions for the fragment ions were 264.0 > 58.0, 250.0 > 58.0, 268.1 > 131.1, 218.20 > 159.10, 204.20 > 76.90. With a residence time of 0.3 s. MRM data were acquired and analyzed using the Labsolution software 6.9 (Shimadzu®, Kyoto, Japan).

2.4. Validation

The analytical method was validated according to the criteria established by the International Conference on Harmonization [17] and the Brazilian National Health Surveillance Agency (ANVISA) in Resolution of the Collegiate Board (RDC) number 166/2017 [18]. The following factors were evaluated: linearity (MAA and AA: 800–40,000 ng/mL; Tramadol and M1: 5–5000 ng/mL; metoprolol: 1000 ng/mL), repeatability, reproducibility, and selectivity in solution and plasma; stability by short- and long-term methods; freezing and thawing cycles with lower and upper limit control samples; and controls at low, medium, and high concentrations (MAA and AA: 2400; 12,000 and 30,400 ng/mL; Tramadol and M1: 15; 1000 and 3750 ng/mL). Standard drug solutions were added to drug-free plasma to create a calibration curve. In addition, quality control (QC) samples were prepared, which were utilized to assess absolute recovery, accuracy, and both intra- and interday precision. The selectivity of the method was evaluated by establishing the lower limit of quantitation (LLOQ) using drug-free plasma. The limit of detection (LOD) and lower limit of quantification were evaluated based on the signal-to-noise ratio (SNR) of three replicates of blank specimens fortified with decreasing quantities of each compound. LOD and LLOQ (metamizole: 840 ng/mL and tramadol: 5.25 ng/mL). Stability (long term in biological matrix at −70 °C; bench temperature at room temperature (20 °C); three freeze–thaw cycles and samples processed in the automatic sampler) was also evaluated.

2.5. Pharmacokinetic Analysis and Statistics

Pharmacokinetic profiles were obtained using a non-compartmental analysis with WinNonlin 6.2.1 software (Pharsight, Mountain View, CA, USA, 2011). The observed variables were: the maximum plasma concentration (Cmax), extrapolated concentration without time 0 (C_0), time to reach Cmax (Tmax), area under the plasma concentration curve from time zero until the moment of the last measurable concentration ($AUC_{0 \to t}$), and the extrapolation of the AUC to infinity (AUC0 $\to \infty$), volume of distribution (Vz), clearance (Cl), elimination half-life ($T_{1/2}$); mean residual time until the last measurement (MRT0 $\to$ t), and mean residual time from zero to infinity (MRT0 $\to \infty$).

Statistical analyses were performed using BioEstat® version 5.0,(Instituto Mamirauá, Belém, Brazil). The normality of all parameters was analyzed using the Shapiro–Wilk test. Except for Tmax which was evaluated using the Mann–Whitney test and expressed as a median, all other parameters were compared using the T test and expressed as mean and standard deviation. Differences were considered statistically significant at $p < 0.05$.

3. Results

After optimizing the chromatographic parameters and defining the analytical variables, the method was validated to ensure data accuracy. All calibration curves showed correlation coefficient (R^2) values greater than 0.99, demonstrating the linearity between the equipment responses and curve concentrations. This method showed good selectivity, reproducibility, and repeatability, with a relative standard deviation of less than 5%. Moreover, the samples were stable under the conditions used for analysis.

The mean plasma concentrations of tramadol, M1, MAA, and AA were plotted on a comparative chart of elapsed time (Figures 1–4).

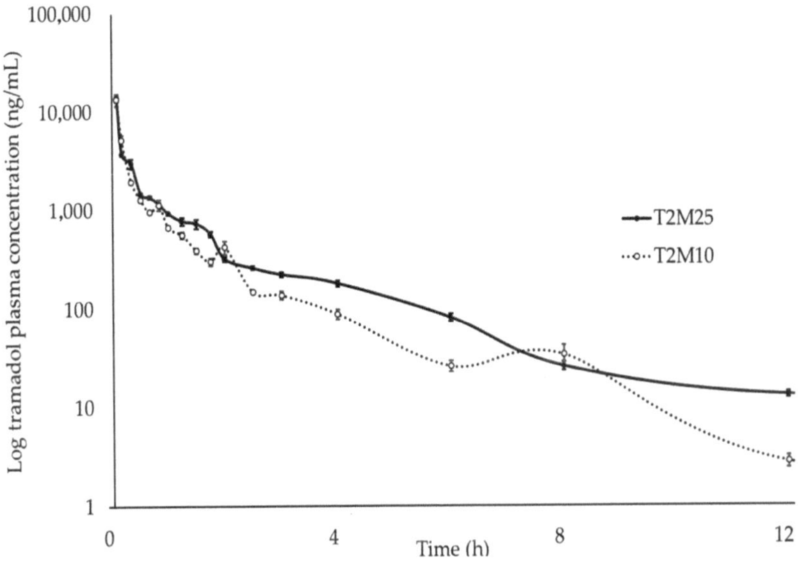

Figure 1. Log_{10} plasma concentration vs. time for tramadol following intravenous administration of metamizole (25 mg·kg^{-1} and 10 mg·kg^{-1}) and tramadol (2 mg·kg^{-1}) in ten healthy northeastern Brazilian donkeys. Notes: Black circles (•) represent the mean and standard deviation results of the group treated with 2 mg·kg^{-1} tramadol and 25 mg·kg^{-1} metamizole; white circles (○) indicate the mean and standard deviation of the group treated with 2 mg·kg^{-1} tramadol and 10 mg·kg^{-1} metamizole.

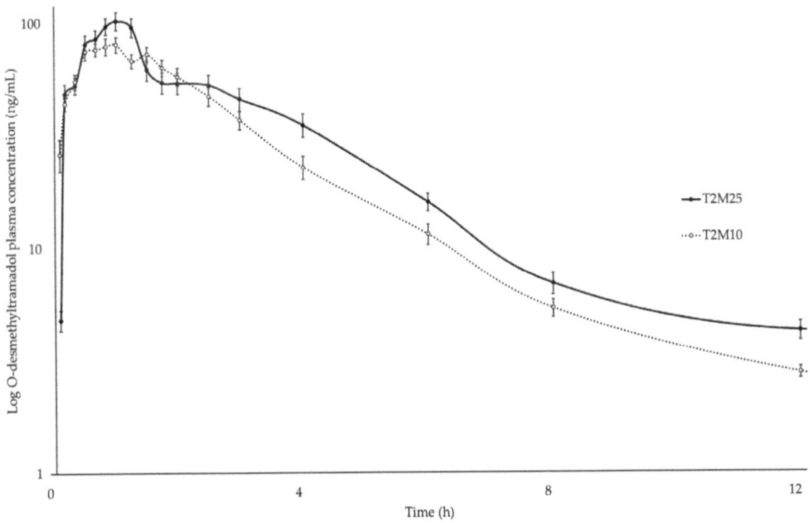

Figure 2. Log_{10} plasma concentration vs. time for O-desmethyltramadol (M1) following intravenous administration of metamizole (25 mg·kg^{-1} and 10 mg·kg^{-1}) and tramadol (2 mg·kg^{-1}) in ten healthy northeastern Brazilian donkeys. Notes: Black circles (•) represent the mean and standard deviation results of the group treated with 2 mg·kg^{-1} tramadol and 25 mg·kg^{-1} metamizole; white circles (○) indicate the mean and standard deviation of the group treated with 2 mg·kg^{-1} tramadol and 10 mg·kg^{-1} metamizole.

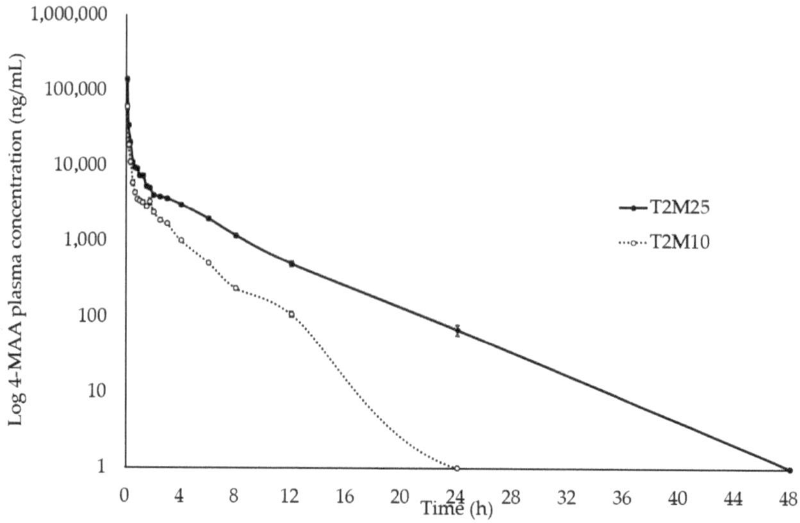

Figure 3. Log_{10} plasma concentration vs. time for N-methyl-4-aminoanthypyrine (MAA) following intravenous administration of metamizole (25 mg·kg^{-1} and 10 mg·kg^{-1}) and tramadol (2 mg·kg^{-1}) in ten healthy northeastern Brazilian donkeys. Notes: Black circles (•) represent the mean and standard deviation results of the group treated with 2 mg·kg^{-1} tramadol and 25 mg·kg^{-1} metamizole; white circles (○) indicate the mean and standard deviation of the group treated with 2 mg·kg^{-1} tramadol and 10 mg·kg^{-1} metamizole.

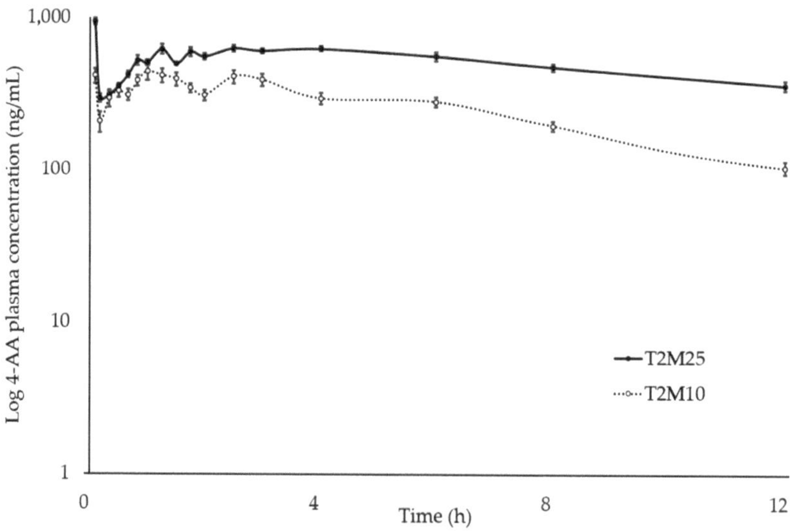

Figure 4. Log_{10} plasma concentration vs. time for 4-aminoanthypyrine (AA) following intravenous administration of metamizole (25 mg·kg^{-1} and 10 mg·kg^{-1}) and tramadol (2 mg·kg^{-1}) in ten healthy northeastern Brazilian donkeys. Notes: Black circles (•) represent the mean and standard deviation results of the group treated with 2 mg·kg^{-1} tramadol and 25 mg·kg^{-1} metamizole; white circles (○) indicate the mean and standard deviation of the group treated with 2 mg·kg^{-1} tramadol and 10 mg·kg^{-1} metamizole.

After intravenous administration, tramadol, MAA, and AA were detected during 24 h of analysis, and M1 during 12 h.

The pharmacokinetic parameters obtained for tramadol, M1, MAA and AA after intravenous administration of metamizole at doses of 10 mg·kg^{-1} or 25 mg·kg^{-1} in association with tramadol (2 mg·kg^{-1}) are shown in Tables 1–4, respectively. Additionally, for analysis, the obtained data were compared with results from previous studies conducted by the same research group.

Table 1. Pharmacokinetic parameters of tramadol (2 mg·kg^{-1}) in association with metamizole (10 mg·kg^{-1} or 25 mg·kg^{-1}) intravenous (IV) administration in ten donkeys.

Tramadol	(Tramadol 2 mg·kg^{-1} IV) Mouta et al., 2021 [19]	T_2M_{10} (Tramadol 2 mg·kg^{-1} and Metamizole 10 mg·kg^{-1} IV)	T_2M_{25} (Tramadol 2 mg·kg^{-1} and Metamizole 25 mg·kg^{-1} IV)
C_0 (ng/mL)	6150 ± 1717 [b]	9995 ± 2095 [a]	13,776 ± 3254 [a]
AUC0 → ∞ (h.ng/mL)	2663 ± 1828 [b]	3882 ± 764 [a]	5008 ± 808 [b]
$T_{1/2}$ (h)	0.97 ± 0.17 [a]	8.12 ± 1.55 [b]	13.62 ± 1.26 [a]
Vz (L/h/kg)	1.32 ± 0.63 [a]	9.15 ± 7.01 [b]	8.75 ± 7.21 [b]
Cl (L/h/kg)	1.01 ± 0.50 [a]	0.76 ± 0.32 [b]	0.42 ± 0.18 [c]
MRT0 → ∞ (h)	1.34 ± 0.36 [a]	2.10 ± 0.48 [b]	8.35 ± 2.44 [c]

Notes: Results are presented as the mean ± SD of n = 10. [a, b] and [c]: Different subscript letters show statistical differences between treatments ($p < 0.05$). C_0: extrapolated concentration without time 0; AUC0→ ∞: area under the curve from time zero to infinity; $T_{1/2}$: elimination half-life; Vz: volume of distribution; Cl: clearance; MRT0 → ∞: mean residence time from zero to infinity.

Table 2. Pharmacokinetic parameters of O-desmethyltramadol (M1) after intravenous (IV) administration of tramadol (2 mg·kg^{-1}) in association with metamizole (10 mg·kg^{-1} or 25 mg·kg^{-1}) in ten donkeys.

M1	(Tramadol 2 mg·kg^{-1} IV) Mouta et al., 2021 [19]	T_2M_{10} (Tramadol 2 mg·kg^{-1} and Metamizole 10 mg·kg^{-1} IV)	T_2M_{25} (Tramadol 2 mg·kg^{-1} and Metamizole 25 mg·kg IV)
Cmax (ng/mL)	90 ± 61 [a]	94 ± 22 [a]	124 ± 30 [b]
Tmax (h)	1.00 ± 0.16 [a]	0.72 ± 0.14 [a]	0.91 ± 0.08 [a]
AUC0 → ∞ (h.ng/mL)	379 ± 238 [a]	303 ± 76 [a]	584 ± 412 [b]
$T_{1/2}$ (h)	8.43 ± 3.57 [a]	6.11 ± 1.35 [a]	13.50 ± 2.62 [b]
Vz (L/h/kg)	NA	8.05 ± 5.12 [a]	8.32 ± 7.73 [a]
Cl (L/h/kg)	NA	9.32 ± 6.51 [a]	5.48 ± 2.54 [a]
MRT0 → ∞ (h)	10.80 ± 4.30 [a]	5.09 ± 0.91 [b]	18.25 ± 4.35 [c]

Notes: Results are presented as the mean ± SD of n = 10. [a, b] and [c]: different subscript letters show statistical differences between treatments ($p < 0.05$). Cmax: maximum concentration; Tmax: time to peak concentration; AUC0 → ∞: area under the curve from time zero to infinity; $T_{1/2}$: elimination half-life; Vz: volume of distribution; Cl: clearance; MRT0 → ∞: mean residence time from zero to infinity; NA: not applicable.

Table 3. Pharmacokinetic parameters of 4-methylaminoantipyrine (4-MAA) after intravenous (IV) administration of tramadol (2 mg·kg^{-1}) in association with metamizole (10 mg·kg^{-1} or 25 mg·kg^{-1}) in ten donkeys.

4-MAA	**(Metamizol 10 mg·kg^{-1} IV) Macêdo et al., 2021 [20]**	**T_2M_{10} (Tramadol 2 mg·kg^{-1} and Metamizol 10 mg·kg^{-1} IV)**	**(Metamizol 25 mg·kg^{-1} IV) Macêdo et al., 2021 [20]**	**T_2M_{25} (Tramadol 2 mg·kg^{-1} and Metamizol 25 mg·kg^{-1} IV)**
C_0 (µg/mL)	31 ± 9.7 ^{a}	109 ± 29 ^{b}	100 ± 34 ^{b}	128 ± 30 ^{b}
AUC0 → ∞ (h.µg/mL)	14.51 ± 1.9 ^{a}	53.70 ± 7.3 ^{b}	44.78 ± 5.5 ^{b}	52.2 ± 4.4 ^{b}
$T_{1/2}$ (h)	2.69 ± 0.34 ^{a}	2.05 ± 0.19 ^{a}	3.62 ± 0.24 ^{b}	4.51 ± 0.94 ^{b}
Vz (L/h/kg)	NA	1.6 ± 0.1 ^{a}	NA	3.4 ± 0.2 ^{b}
Cl (L/h/kg)	NA	5.0 ± 0.5	NA	4.9 ± 0.5
MRT0 → ∞ (h)	2.84 ± 0.3 ^{a}	1.92 ± 0.24 ^{a}	3.92 ± 0.36 ^{b}	3.99 ± 0.74 ^{b}

Notes: Results are presented as the mean ± SD of n = 10. $^{a, b}$: Different subscript letters show statistical differences between treatments ($p < 0.05$). C_0: extrapolated concentration without time 0; AUC0 → ∞: area under the curve from time zero to infinity; $T_{1/2}$: elimination half-life; Vz: volume of distribution; Cl: clearance; MRT0 → ∞: mean residence time from zero to infinity; NA: not applicable.

Table 4. Pharmacokinetic parameters of 4-aminoantipyrine (4-AA) after intravenous (IV) administration of tramadol (2 mg·kg^{-1}) in association with metamizole (10 mg·kg^{-1} or 25 mg·kg^{-1}) in ten donkeys.

4-AA	**(Metamizol 10 mg·kg^{-1} IV) Macêdo et al., 2021 [20]**	**T_2M_{10} (Tramadol 2 mg·kg^{-1} and Metamizol 10 mg·kg^{-1} IV)**	**(Metamizol 25 mg·kg^{-1} IV) Macêdo et al., 2021 [20]**	**T_2M_{25} (Tramadol 2 mg·kg^{-1} and Metamizol 25 mg·kg^{-1} IV)**
Cmax (µg/mL)	1598 ± 0.25 ^{a}	1941 ± 0.46 ^{a}	2855 ± 0.55 ^{b}	1067 ± 0.14 ^{a}
Tmax (h)	0.22 ± 0.06 ^{a}	1.56 ± 0.65 ^{b}	0.15 ± 0.06 ^{a}	0.91 ± 0.36 ^{b}
AUC0→ ∞ (h.µg/mL)	6801 ± 1569 ^{a}	14,175 ± 4367 ^{b}	12,494 ± 1532 ^{b}	11,981 ± 2583 ^{b}
$T_{1/2}$ (h)	6.37 ± 1.30 ^{a}	9.41 ± 2.42 ^{b}	7.11 ± 1.01 ^{a}	10.47 ± 0.98 ^{b}
Vz (L/h/kg)	NA	33.2 ± 3.4 ^{a}	NA	37.1 ± 3.5 ^{a}
Cl (L/h/kg)	NA	3.1 ± 0.3 ^{a}	NA	2.5 ± 0.3 ^{a}
MRT0 → ∞ (h)	10.95 ± 1.61 ^{a}	14.04 ± 3.54 ^{a}	11.20 ± 1.43 ^{a}	15.55 ± 1.41 ^{a}

Notes: Results are presented as the mean ± SD of n = 10. $^{a, b}$: different subscript letters show statistical differences between treatments ($p < 0.05$). Cmax: maximum concentration; Tmax: time to peak concentration; AUC0 → ∞: area under the curve from time zero to infinity; $T_{1/2}$: elimination half-life; Vz: volume of distribution; Cl: clearance; MRT0 → ∞: mean residence time from zero to infinity; NA: not applicable.

Analyzing Tramadol, it was observed that AUC0 → ∞ and Cl were higher for T_2M_{10} than for T_2M_{25}, while $T_{1/2}$; MRT0 → ∞; MRT0 → t were significantly higher in the T_2M_{25} group. Regarding its metabolite, M1, $T_{1/2}$; MRT0 → ∞; MRT0 → t were significantly higher for T_2M_{25}.

Regarding the Vz, MRT0 → ∞, $T_{1/2}$ of MAA and MRT0 → t of AA varied significantly between groups, being higher in animals that received 25 mg·kg^{-1} of metamizole.

Adverse effects were noted after the intravenous administration of metamizole and tramadol. Restlessness and ataxia were observed in two animals in the T_2M_{25} group but not in the T_2M_{10} group.

4. Discussion

In this study, the evaluation of the pharmacokinetic profile of two doses of metamizole in co-administration with 2 mg·kg^{-1} of tramadol in donkeys were investigated.

The combination of metamizole and tramadol has been used to treat moderate-to-severe pain in animals affected by neoplasms or arthritis, or undergoing castration surgery [21–23]. The results showed that this association has the potential to improve antinociceptive effects [14]. However, these studies were conducted on small animals or laboratory animals.

In the realm of veterinary medicine, donkeys frequently receive therapeutics prescribed according to dosages and intervals recommended for horses, primarily due to the scant availability of donkey-specific drug labeling guidelines [24]. This study marks a pioneering effort to investigate the pharmacokinetic profiles of tramadol and metamizole specifically in donkeys. The findings from this research could serve as a foundational scientific basis for subsequent pharmacodynamic investigations and clinical trials aimed at optimizing pain management strategies for this species.

Metamizole is rapidly hydrolyzed into its two metabolites (MAA and AA) [25] and therefore, it was not possible to obtain the minimum concentrations for quantification at predetermined times, making it impossible to construct the pharmacokinetic profile of this prodrug. Therefore, its metabolites (MAA and AA) are used as markers for pharmacokinetic studies of this drug [6]. Tramadol is a drug carried to the liver, where it is metabolized in M1 and can be measured using the method employed.

The metabolites 4-methylaminoantipyrine and O-desmethyltramadol, the active metabolites of metamizole and tramadol, respectively, have more potent analgesic activity than those of their parent drugs [4]. Since both drugs are metabolized by the enzyme CYP3A4, increasing the metamizole dose from 10 to 25 $mg{\cdot}kg^{-1}$ extends their blood presence. This is shown by the marked increase in MRT and $T_{1/2}$ in groups treated with 25 $mg{\cdot}kg^{-1}$ of metamizole and in donkeys administered 2.5 $mg{\cdot}kg^{-1}$ of tramadol intravenously, suggesting the combination could boost their therapeutic effects [4,26–28].

Tramadol clearance was higher in the animals that received 10 $mg{\cdot}kg^{-1}$ of metamizole. Both drugs are metabolized by the same enzymes. These results suggest that increasing the dose of metamizole promotes competition for the CYP3A4 binding site, reducing the metabolism of tramadol and prolonging its elimination.

Tramadol exhibits a half-life of 0.97 h at the dosage used in the present study and 1.48 h at a dose of 4 $mg{\cdot}kg^{-1}$; however, these data were obtained when tramadol was used alone [19]. In another study, similar values were found, and a half-life of 1.55 h was obtained in donkeys, wherein the drug was rapidly metabolized to N-desmethyltramadol, an inactive metabolite which contributed to the drug being less effective in this species than in others [26]. In the present study, concomitant administration of metamizole increased the half-life of tramadol to 8.12 h (10 $mg{\cdot}kg^{-1}$ metamizole) and 13.62 h (25 $mg{\cdot}kg^{-1}$), as well as that for M1 (an active metabolite in animals).

The combination of these drugs possibly increases the clinical efficacy of tramadol in donkeys. However, clinical trials are required to determine whether this increased half-life is useful for pain management. An increase in the metamizole dose promoted an increase in the half-lives of tramadol and O-desmethyltramadol. These three substances are metabolized in the liver under the action of cytochrome P450 enzyme variation 3A4, causing it to become overloaded and thus reducing the speed of metabolization of these drugs. However, this half-life extension cannot definitively be deemed beneficial from a clinical point of view. The continuous administration of tramadol and metamizole in rats subjected to the hot plate test initially promoted an improvement in the response; however, with subsequent doses, its efficiency reduced by up to 40%, probably due to the increased opioid tolerance in the animal [14].

Regarding the pharmacokinetic parameters of MAA in donkeys, MRT was higher in animals that received 25 $mg{\cdot}kg^{-1}$ of metamizole. The authors of [28] used metamizole in horses and observed an MRT of 3.70 h. This was similar to our data at the same dose and suggests that, regardless of whether it was used alone, the duration of the effect of metamizole did not vary. In addition, the half-life of MAA presented increased with a dose of 25 $mg{\cdot}kg^{-1}$. Giorgi et al. [29] reported a half-life of 3.34 h, which differs from our study, which found 4.51 h. This variation may be associated with competition for the active site of the CYP3A4 enzyme, suggesting an enzymatic interaction in which tramadol and MAA compete for the same metabolic pathway [30].

Stewart et al. (2011) [31], in their study on the pharmacokinetics and adverse effects of tramadol when administered intravenously to horses, noted that 7 out of the 12 animals experienced muscle fasciculations within three minutes of receiving a 5 $mg{\cdot}kg^{-1}$ dose. Meanwhile, another research [19] focusing on the intravenous application of tramadol in donkeys found adverse reactions in just one animal at a 2 $mg{\cdot}kg^{-1}$ dosage and in 7 animals when the dosage was increased to 4 $mg{\cdot}kg^{-1}$.

However, no adverse reactions were observed in donkeys treated with tramadol and in mice treated with tramadol and metamizole [13,26]. The reactions found in the two donkeys that received metamizole and tramadol (T_2M_{25}) were probably due to the speed or volume of administration. The Cmax of tramadol and M1 did not differ between the groups treated with 10 $mg{\cdot}kg^{-1}$ or 25 $mg{\cdot}kg^{-1}$ of metamizole. Moreover, the intravenous administration of drugs in large volume and at a fast rate demonstrably causes adverse effects, mainly, of neurological order [31,32], as reported in our study.

The analgesic efficacy and side effects induced by both tramadol and metamizole vary among different species, influenced not only by intrinsic hepatic metabolism but also by genetic polymorphisms within the CYP450 subfamilies [33]. For instance, while the impact of tramadol can be significantly altered by the efficiency and quantity of a specific CYP450 enzyme in an individual, metamizole, known for its analgesic and antipyretic properties, may also have its efficacy and side effect profile modified by similar genetic and metabolic factors. These variations in CYP450 phenotypes affect the metabolism, accumulation, or elimination rates of both substances, directly influencing the success or failure of analgesic outcomes and the potential for adverse effects [14]. According to Ruel and Steagall (2019) [34], some medical centers are now integrating computerized clinical decision support systems that include pharmacogenomics tools to customize treatment with tramadol and metamizole. This strategy is based on individual pharmacogenomic profiles (e.g., extensive, intermediate, or poor metabolizers) to predict the safety and efficacy of the combined or individual therapy of these drugs.

The combination of drugs can lead to complex interactions that affect their metabolism, potentially resulting in increased plasma concentrations and bioavailability, which can lead to larger drug distribution volumes and adverse effects. This phenomenon occurs when drugs compete for the same metabolic pathways, especially those involving cytochrome P450 enzymes, leading to slower metabolism and a prolonged presence in the plasma of the drug and its metabolites. Such interactions highlight the importance of understanding the pharmacokinetic profiles of drugs when used in combination [35].

As this study aimed to evaluate the pharmacokinetic profile of analgesic drugs widely used in companion species, it is only possible to determine whether there is an interaction within the scope of the metabolism of these drugs, and we could not infer which of the two associations is more beneficial in the treatment of pain. Further clinical studies are needed to resolve these issues in northeastern Brazilian donkeys.

5. Conclusions

Pharmacokinetic research on donkeys is limited, making this pioneering study on the drug interactions between tramadol and metamizole in donkeys significant. It indicates that these drugs may impact each other's metabolism. Further clinical research is essential to determine the optimal dosages for effective analgesia in donkeys, reducing the need to rely on dosages extrapolated from horse studies.

Author Contributions: Conceptualization, V.V.d.P.; data curation, L.B.d.M.; funding acquisition, G.A.-S. and V.V.d.P.; investigation, A.N.M. and M.G.C.d.O.; methodology, G.A.-S. and J.T.P.-U.; supervision, V.V.d.P.; validation, L.G.S.S.; visualization, L.B.d.M. and K.N.A.; writing—original draft, G.A.-S. and L.B.d.M.; writing—review and editing, J.T.P.-U. All authors have read and agreed to the published version of the manuscript.

Funding: The present work was conducted with financial support from CAPES, Coordination for the Improvement of Higher Education Personnel, Brazil, within the scope of the General Program of International Cooperation (AUX 395/2018 PGCI). JBS Found for the Amazon Rainforest.

Institutional Review Board Statement: The study was approved by the Ethics Committee on Animal Use of the Universidade Federal Rural do Semi-Árido (CEUA-UFERSA protocol number 23091.006896/2019-47).

Informed Consent Statement: Written informed consent has been obtained from the owner of the animals involved in this study.

Data Availability Statement: The data presented in this study are available on request from the corresponding author.

Acknowledgments: The authors express their gratitude to Jeferson Dombroski and the Laboratory Technician Paulo Sérgio. We extend our gratitude to Fernandes das Chagas from the Laboratory of Plant Ecophysiology at CPVSA for granting us access to the mass spectrometer.

Conflicts of Interest: The authors declare no potential conflicts of interest with respect to the research, authorship, or publication of this article.

References

1. Regan, F.H.; Hockenhull, J.; Pritchard, J.C.; Waterman-Pearson, A.E.; Whay, H.R. Behavioural repertoire of working donkeys and consistency of behaviour over time, as a preliminary step towards identifying pain-related behaviours. *PLoS ONE* **2014**, *9*, e101877. [CrossRef]
2. Dai, F.; Dalla Costa, E.; Murray, L.M.A.; Canali, E.; Minero, M. Welfare conditions of donkeys in Europe: Initial outcomes from on-farm assessment. *Animals* **2016**, *6*, 5. [CrossRef]
3. Cook, V.L.; Blikslager, A.T. The use of nonsteroidal anti-inflammatory drugs in critically ill horses. *J. Vet. Emerg. Crit. Care* **2015**, *25*, 76–88. [CrossRef]
4. Lutz, M. Metamizole (dipyrone) and the liver: A review of the literature. *J. Clin. Pharmacol.* **2019**, *59*, 1433–1442. [CrossRef] [PubMed]
5. Hedenmalm, K.; Spigset, O. Agranulocytosis and other blood dyscrasias associated with dipyrone (metamizole). *Eur. J. Clin. Pharmacol.* **2002**, *58*, 265–274. [CrossRef] [PubMed]
6. Kim, T.W.; Łebkowska-Wieruszewska, B.; Sitovs, A.; Poapolathep, A.; Owen, H.; Lisowski, A.; Abilova, Z.; Giorgi, M. Pharmacokinetic profiles of Metamizole (dipyrone) active metabolites in goats and its residues in milk. *J. Vet. Pharmacol. Ther.* **2018**, *41*, 699–705. [CrossRef] [PubMed]
7. Abbiati, R.A.; Cagnardi, P.; Ravasio, G.; Villa, R.; Manca, D. A physiologically based model for tramadol pharmacokinetics in horses. *J. Theor. Biol.* **2017**, *429*, 46–51. [CrossRef] [PubMed]
8. Udegbunam, R.I.; Onuba, A.C.; Okorie-Kanu, C.; Udegbunam, S.O.; Anyanwu, M.U.; Ifeanyi, O.L. Effects of two doses of tramadol on pain and some biochemical parameters in rabbits post-gastrotomy. *Comp. Clin. Pathol.* **2015**, *24*, 783–790. [CrossRef]
9. Bortolami, E.; Della Rocca, G.; Di Salvo, A.; Giorgi, M.; Kim, T.W.; Isola, M.; De Benedictis, G.M. Pharmacokinetics and antinociceptive effects of tramadol and its metabolite O-desmethyltramadol following intravenous administration in sheep. *Vet. J.* **2015**, *205*, 404–409. [CrossRef]
10. Cagnardi, P.; Villa, R.; Zonca, A.; Gallo, M.; Beccaglia, M.; Luvoni, G.C.; Vettorato, E.; Carli, S.; Fonda, D.; Ravasio, G. Pharmacokinetics, intraoperative effect and postoperative analgesia of tramadol in cats. *Res. Vet. Sci.* **2011**, *90*, 503–509. [CrossRef]
11. Sheikholeslami, B.; Gholami, M.; Lavasani, H.; Rouini, M. Evaluation of the route dependency of the pharmacokinetics and neuro-pharmacokinetics of tramadol and its main metabolites in rats. *Eur. J. Pharm. Sci.* **2016**, *92*, 55–63. [CrossRef]
12. Evenson, E.; Mans, C. Antinociceptive efficacy and safety of subcutaneous tramadol in chinchillas (*Chinchilla lanigera*). *J. Exot. Pet. Med.* **2019**, *28*, 98–104. [CrossRef]
13. Moreno-Rocha, L.A.; López-Muñoz, F.J.; Medina-López, J.R.; Domínguez-Ramírez, A.M. Effect of tramadol on metamizol pharmacokinetics and pharmacodynamics after single and repeated administrations in arthritic rats. *Saudi Pharm. J.* **2016**, *24*, 674–684. [CrossRef]
14. Moreno-Rocha, L.A.; Domínguez-Ramírez, A.M.; Cortés-Arroyo, A.R.; Bravo, G.; López-Muñoz, F.J. Antinociceptive effects of tramadol in co-administration with metamizol after single and repeated administrations in rats. *Pharmacol. Biochem. Behav.* **2012**, *103*, 1–5. [CrossRef]
15. Aupanun, S.; Laus, F.; Poapolathep, A.; Owen, H.; Vullo, C.; Faillace, V.; Giorgi, M. Pharmacokinetic assessment of the marker active metabolites 4-methyl-amino-antipyrine and 4-acetyl-amino-antipyrine after intravenous and intramuscular injection of Metamizole (dipyrone) in healthy donkeys. *J. Equine Vet. Sci.* **2016**, *47*, 55–61. [CrossRef]
16. Matthews, N.; van Loon, J.P.A.M. Anaesthesia and analgesia of the donkey and the mule. *Equine Vet. Educ.* **2013**, *25*, 47–51. [CrossRef]

17. International Conference on Harmonisation of Technical Requirements for Registration of Pharmaceuticals for Human Use. *Harmonised Tripartite Guideline for Good Clinical Practice ICH E6(R2)*; Integrated Addedum: Geneva, Switzerland, 2016; Available online: https://www.ema.europa.eu/en/documents/scientific-guideline/ich-guideline-good-clinical-practice-e6r2-step-5_en.pdf (accessed on 1 November 2020).
18. ANVISA; BRASIL. Agência Nacional de Vigilância Sanitária. Resolução da Diretoria Colegiada-RDC N° 166. 2017. Available online: https://www.gov.br/anvisa/pt-br (accessed on 1 November 2020).
19. Mouta, A.N.; de Oliveira Lima, I.; de Oliveira, M.G.C.; Alves, L.P.; de Macêdo, L.B.; Araujo-Silva, G.; Pérez-Urizar, J.; de Paula, V.V. Pharmacokinetic properties of tramadol and M1 metabolite in Northeast Brazilian donkeys (Equus asinus). *J. Vet. Pharmacol. Ther.* **2021**, *44*, 318–325. [CrossRef] [PubMed]
20. De Macêdo, L.B.; Mouta, A.N.; Araújo-Silva, G.; Perez-Urizar, J.T.; de Paula, V.V. Pharmacokinetic properties of metamizole active metabolites in Northeastern Brazilian donkeys (*Equus asinus*). *J. Vet. Pharmacol. Ther.* **2021**, *44*, 842–849. [CrossRef] [PubMed]
21. Imagawa, V.H.; Fantoni, D.T.; Tatarunas, A.C.; Mastrocinque, S.; Almeida, T.F.; Ferreira, F.; Posso, I.P. The use of different doses of metamizolfor post-operative analgesia in dogs. *Vet. Anaesth. Analg.* **2011**, *38*, 385–393. [CrossRef] [PubMed]
22. Flôr, P.B.; Yazbek, K.V.; Ida, K.K.; Fantoni, D.T. Tramadol plus Metamizole combined or not with anti-inflammatory drugs is clinically effective for moderate to severe chronic pain treatment in cancer patients. *Vet. Anaesth. Analg.* **2013**, *40*, 316–327. [CrossRef] [PubMed]
23. Teixeira, R.C.; Monteiro, E.R.; Campagnol, D.; Coelho, K.; Bressan, T.F.; Monteiro, B.S. Effects of tramadol alone, in combination with meloxicam or dipyrone, on postoperative pain and the analgesic requirement in dogs undergoing unilateral mastectomy with or without ovariohysterectomy. *Vet. Anaesth. Analg.* **2013**, *40*, 641–649. [CrossRef] [PubMed]
24. Grosenbaugh, D.A.; Reinemeyer, C.R.; Figueiredo, M.D. Pharmacology and therapeutics in donkeys. *Equine Vet. Educ.* **2011**, *23*, 523–530. [CrossRef]
25. Morresey, P.R.; White, G.W.; Poole, H.M.; Hu, T.; Yin, M.; Sundman, E.A. Randomized blinded controlled trial of dipyrone as a treatment for pyrexia in horses. *Am. J. Vet. Res.* **2019**, *80*, 294–299. [CrossRef] [PubMed]
26. Giorgi, M.; Del Carlo, S.; Sgorbini, M.; Saccomanni, G. Pharmacokinetics of tramadol and its metabolites M1, M2, and M5 in donkeys after intravenous and oral immediate release single-dose administration. *J. Equine Vet. Sci.* **2009**, *29*, 569–574. [CrossRef]
27. Saussele, T.; Burk, O.; Blievernicht, J.K.; Klein, K.; Nussler, A.; Nussler, N.; Hengstler, J.G.; Eichelbaum, M.; Schwab, M.; Zanger, U.M. Selective induction of human hepatic cytochromes P450 2B6 and 3A4 by metamizole. *Clin. Pharmacol. Ther.* **2007**, *82*, 265–274. [CrossRef] [PubMed]
28. Barbosa, J.; Faria, J.; Queirós, O.; Moreira, R.; Carvalho, F.; Dinis-Oliveira, R.J. Comparative metabolism of tramadol and tapentadol: A toxicological perspective. *Drug Metab. Rev.* **2016**, *48*, 577–592. [CrossRef]
29. Giorgi, M.; Aupanun, S.; Lee, H.K.; Poapolathep, A.; Rychshanova, R.; Vullo, C.; Laus, F. Pharmacokinetic profiles of the active Metamizole metabolites in healthy horses. *J. Vet. Pharmacol. Ther.* **2016**, *40*, 165–171. [CrossRef]
30. Zhou, S.F.; Zhou, Z.W.; Yang, L.P.; Cai, J.P. Substrates, inducers, inhibitors and structure-activity relationships of human Cytochrome P450 2C9 and implications in drug development. *Curr. Med. Chem.* **2009**, *16*, 3480–3675. [CrossRef]
31. Stewart, A.J.; Boothe, D.M.; Cruz-Espindola, C.; Mitchum, E.J.; Springfield, J. Pharmacokinetics of tramadol and metabolites O-desmethyltramadol and N-desmethyltramadol in adult horses. *Am. J. Vet. Res.* **2011**, *72*, 967–974. [CrossRef]
32. McMillan, C.J.; Livingston, A.; Clark, C.R.; Dowling, P.M.; Taylor, S.M.; Duke, T.; Terlinden, R. Pharmacokinetics of intravenous tramadol in dogs. *Can. J. Vet. Res.* **2008**, *72*, 325–331.
33. Dhanjal, J.K.; Wilson, D.V.; Robinson, E.; Tobin, T.T.; Dirikolu, L. Intravenous tramadol: Effects, nociceptive properties, and pharmacokinetics in horses. *Vet. Anaesth. Analg.* **2009**, *36*, 581–590. [CrossRef] [PubMed]
34. Ruel, H.L.; Steagall, P.V. Adjuvant analgesics in acute pain management. *Vet. Clin. Small Anim. Pract.* **2019**, *49*, 1127–1141. [CrossRef] [PubMed]
35. Bibi, Z. Role of cytochrome P450 in drug interactions. *Nutr. Metab.* **2008**, *5*, 27. [CrossRef] [PubMed]

www.ingramcontent.com/pod-product-compliance
Lightning Source LLC
Chambersburg PA
CBHW041344210926
51850CB00013B/2987
* 9 7 8 3 7 2 5 8 4 2 9 1 9 *

MDPI AG
Grosspeteranlage 5
4052 Basel
Switzerland
Tel.: +41 61 683 77 34

Animals Editorial Office
E-mail: animals@mdpi.com
www.mdpi.com/journal/animals